Warfighting and Disruptive Technologies

Disguising innovation

Occasionally, militaries during times of peace achieve major warfighting innovations. Terry Pierce calls these 'disruptive innovations'. The more common innovation phenomena, however, have been that of integrating new technologies to help perform existing missions better and to not change them radically. The author calls these 'sustaining innovations'. The central theme of this book is that senior leaders who successfully have managed disruptive innovations disguised them as sustaining to ensure their innovations survived. The recent innovation history suggests two interesting questions. First, how can senior military leaders achieve a disruptive innovation when they are heavily engaged around the world and they are managing sustaining innovations? Second, what have been the external sources of disruptive (and sustaining) innovations?

This book will be essential reading for professionals and students interested in national security, military innovation and organizational theory.

Captain Terry C. Pierce is serving with the US Navy as Chief of Staff for Amphibious Forces 7th Fleet. He holds Doctorate and Master's degrees from the John F. Kennedy School of Government, Harvard University. He also holds a Master's degree from the Naval Postgraduate School in National Security Affairs, Strategic Planning. This book is based on his doctoral thesis and his career military experience.

Cass Series: Strategy and History
Series Editors: Colin Gray and Williamson Murray

This new series will focus on the theory and practice of strategy. Following Clausewitz, strategy has been understood to mean the use made of force, and the threat of the use of force, for the ends of policy. This series is as interested in ideas as in historical cases of grand strategy and military strategy in action. All historical periods, near and past, and even future, are of interest. In addition to original monographs, the series will from time to time publish edited reprints of neglected classics as well as collections of essays.

1 **Military Logistics and Strategic Performance**
Thomas M. Kane

2 **Strategy for Chaos**
Revolutions in military affairs and the evidence of history
Colin Gray

3 **The Myth of Inevitable US Defeat in Vietnam**
C. Dale Walton

4 **Astropolitik**
Classical geopolitics in the space age
Everett C. Dolman

5 **Anglo-American Strategic Relations and the Far East, 1933–1939**
Imperial crossroads
Greg Kennedy

6 **Power and Policy in the Space and Information Age**
Pure strategy
Everett C. Dolman

7 **The Red Army, 1918–1941**
From Vanguard of world revolution to US ally
Earl F. Ziemke

8 **Britain and Ballistic Missile Defence, 1942–2002**
Jeremy Stocker

9 **The Nature of War in the Information Age**
Clausewitzian future
David J. Lonsdale

10 **Strategy as Social Science**
Thomas Schelling and the nuclear age
Robert Ayson

Warfighting and Disruptive Technologies

Disguising innovation

Captain Terry C. Pierce
United States Navy

FRANK CASS
LONDON • NEW YORK

First published in 2004 in Great Britain
by Frank Cass
2 Park Square, Milton Park, Abingdon, Oxon, OX14 4RN

Simultaneously published in the USA and Canada
by Frank Cass
270 Madison Ave, New York NY 10016

Frank Cass is an imprint of the Taylor & Francis Group

Transferred to Digital Printing 2005

Copyright © 2004 Terry C. Pierce

Typeset in Times New Roman by
Newgen Imaging Systems (P) Ltd, Chennai, India

British Library Cataloguing in Publication Data
A catalogue record for this book is available
from the British Library

Library of Congress Cataloging-in-Publication Data
A catalog record for this book has been requested

ISBN 0–714–65547–3
ISBN 0-415-70189-9

Contents

Foreword   vii
Series editor's preface   viii
Preface   x

Maih points on topic

1   Introduction   1

2   Explaining disruptive innovations   19

3   US Marine Corps innovation: the development
    of amphibious warfare   51

4   Post-World War II Marine Corps disruptive innovations:
    (I) helicopter warfare   70

5   Post-World War II Marine Corps disruptive innovations:
    (II) MAGTF warfare – combined arms operations   79

6   US Marine Corps inchoate disruptive innovation:
    maneuver warfare   85

7   US Marine Corps sustaining innovations and summary
    of disruptive Marine Corps cases   104

8   US Navy sustaining innovation: continuous aim gunfire   116

9   US Navy disruptive innovation: carrier warfare   121

10 Disruptive innovation: Japanese carrier warfare   132

11 US Navy disruptive innovation: CWC – naval combined
    arms warfare   145

12   US Navy sustaining innovation: carrier battle
     group concept   152

13   US Navy disruptive innovation – aborted:
     Project 60 – defensive sea control warfare   155

14   US Navy disruptive innovation: maritime action
     groups – surface land attack warfare   164

15   US Navy disruptive innovation: tactical collaborative
     network and a summary of disruptive Navy cases   176

16   Conclusion   192

     Notes   202
     Bibliography   237
     Index   249

Foreword

This work by Captain Terry Pierce advances our understanding of military innovation in several important ways. First, it makes good use of the concept of disruptive innovation. A concept which emerged from the business school literature. Previous work, including my own, had difficulty in defining just what an 'innovation' was, as opposed to incremental improvements in existing tasks. The works by Major Suzanne Nielsen, USA, on military reform, on the one hand, and now by Captain Pierce on disruptive innovation, on the other hand, give us a much better empirical and conceptual understanding of the full range of changes in military behavior, and the issues particular to the different kinds of change. Captain Pierce, specifically, shows us which kinds of change are more likely to encounter organizational opposition. Second, Captain Pierce also helps us understand the organizational strategies that have been successful in bringing about disruptive innovation. These are the kinds of innovation Machiavelli had in mind, I believe, when he wrote that new orders have enemies in those who see their interests damaged by change. Concealing the long-term implications of disruptive innovations has been an important element in bringing them about. It will be interesting to see how publishing this finding will affect the dynamic between military innovators and advocates of incremental change. Third, Captain Pierce successfully brings together the study of technological and doctrinal change within military innovations. Beyond that, Captain Pierce has added several new historical cases of military innovation, previously unexamined to my knowledge, for us to study along with the more familiar cases.

This book emerges from a dissertation that Captain Pierce wrote under my supervision. Its quality will speak for itself. Let me only conclude that in the course of writing his dissertation, Captain Pierce faced and overcame greater obstacles than are usually encountered in getting a Ph.D.

Stephen Peter Rosen
Harvard Professor of Political Science

Series editor's preface

Over the past decade and a half military historians and political scientists have become increasingly interested in the problems that military institutions have confronted in the twentieth century in innovating in an uncertain and ambiguous world. As a result, the simplistic picture presented in many post-World War II generals' memoirs as well as in the works of military pundits like J.F.C. Fuller and B.H. Liddell Hart have given way to an understanding of how difficult it was in the past to glimpse the possibilities that technology and innovating tactics offered.

The old belief that military organizations usually study the last war and that is why they do badly in the next war has been overturned by the recognition that the only military study to study World War I honestly and ruthlessly was by the German Army. Thus, its doctrine and preparation for war reflected what actually had been occurring on the battlefields of World War I rather than what was occurring in Britain and France where generals and pundits manipulated the evidence to fit their own paradigms.

Military innovation must rest on a willingness to study the past dispassionately, to test assumptions about the present ruthlessly and honestly, and to create an atmosphere within the organization that accepts and adapts to change. None of this is easy. In fact, innovation for military organizations is more difficult than in the business world, because the latter receives little feedback on how well they are doing until the next conflict. Yet, failure to innovate intelligently and effectively can lead to disastrous results as the defeat of the French Army in 1940 underlines in spades.

Captain Terry Pierce, whose naval career has thus far reflected a deep interest in innovation and change in the modern era, has here in this major work presented a new and fresh take on the complexities of innovation in the distant past of the 1920s and 1930s as well as the recent past. I find Captain Pierce's arguments about and discussion of disruptive technological change compelling. Equally compelling is his discussion of how in a number of cases innovators in a number of officer corps have managed to conceal the implications of change from their superiors and their colleagues.

Human beings have consistently throughout the ages found it difficult to accept change. Yet increasingly rapid change has been the mark of the twentieth century, and now the twenty-first century. Like it or not, military institutions have had to deal with that world. In this work Captain Pierce has provided us with new and valuable insights into not only how to understand successful change in the past, but how to craft organizational culture and thought processes to the business of military innovation in a rapidly changing world. He has provided us with invaluable insights and understanding of how innovation has occurred in the past. Now it is up to us to use those insights in thinking about future innovation. This book will be of considerable interest not just to academics, but also to those military officers who must deal with change over the course of the coming decade.

Williamson Murray

why read or recomend to others as a veluable book.

Preface

The ability to create disruptive ways of warfighting is a critical function in any military organization. By contrast, those who rely exclusively on using novel technologies to sustain old ways of fighting are at a significant disadvantage. Understanding the factors that cause disruptive innovation should be a vital concern for senior military leaders and most policy makers. By elaborating and testing a number of propositions about championing disruptive innovation, this study aims at improving that understanding.

This project began with several puzzles. A sharp discrepancy exists between what organizational theory predicts about military innovation and the record of major transformations in the way militaries fight. For example, innovation theory posits that innovation in military doctrine should be rare because huge bureaucracies are designed not to change.[1] Yet, the United States Navy and Marine Corps have achieved eight major warfighting changes in the last century. Similarly, although organization theory predicts that new technology will normally be assimilated to an old doctrine, such as the way the British employed naval aircraft and the French employed tanks after World War I, it does not explain how these same breakthrough technologies brought about dramatic innovations in carrier warfare for Japan and the United States and in armored warfare for Germany.

Disagreements also have emerged over the impact of civilian intervention and maverick officers on the cause of disruptive innovation. One opinion contends that civilian leaders use military mavericks as agents to cause major changes in fighting, while another perspective posits that civilians play a relatively minor role in the initiation and advancement of disruptive innovation, while military mavericks cannot create new ways of fighting. Besides this debate, another contentious notion is that militaries innovate in a disruptive way only after they have suffered defeat.[2] Finally, the innovation literature identifies an unexplained technology phenomenon whereby disruptive innovations are often adopted first by someone other than the nation inventing the striking new technology that underlies the innovation. Sorting out which views are the most accurate is worthwhile for those who seek to develop military innovation to impose the will of a nation on its enemies when all else has failed.

Additional puzzles emerged when examining the theory and practice of innovation and the considerable observations about the United States naval

[handwritten: Pierce attempts to improve upon this theory]

services. Arguably, Steve Rosen's intraservice competition theory (a theory that views military organizations as complex political communities) is the most useful general theory available, but it treats doctrine and technology innovation separately. According to Rosen, peacetime doctrine innovation is a two-step process. In the intellectual process, innovators respond to broad structural changes in the security environment by identifying new strategic requirements and translating them into new tasks. Accompanying the intellectual innovation is a political process that changes the way the officers and men sponsoring the new tasks live their professional lives.[3] But if that is so, why did Admiral Zumwalt, the champion of Defensive Sea Control, and General Gray, the champion of Maneuver Warfare, who both apparently followed Rosen's two-step process, fail to achieve significant progress toward institutionalizing these disruptive innovations? These failures make one question what other factors contribute to causing disruptive innovation. Although intraservice competition theory accounts for the why and when of Admiral Zumwalt and General Gray's innovation efforts, it fails to consider how they unsuccessfully promoted the disruptive innovation. This raises the question of whether senior leaders, who, unlike Zumwalt and Gray, successfully achieved disruptive innovations, managed them differently than they did sustaining ones? If so, what are the differences? Is it possible they could follow Rosen's model and still fail because they were good at leading evolutionary changes but poor at championing revolutionary ones? Is it possible that the same decisions for promoting successful sustaining innovations cause military organizations to fail in achieving disruptive innovations?[4] Furthermore, if intraservice competition theory is a theory primarily about new ways of warfighting, then how do we explain the way senior leaders manage technology advances so that they bring about dramatic changes in military operations? Finally, is it reasonable to incorporate these factors with the intellectual and political processes to achieve a more robust model that senior leaders can use to effect disruptive innovations?

Solution: disruptive innovation theory

The solution presented in this book is a reformulation of intraservice competition theory into a new one called 'disruptive innovation theory'. Rosen's two-step intellectual and political processes are important elements of disruptive innovation. Additionally, however, successful disruptive innovators form small groups to develop and nurture innovations. They also protect their inchoate innovation from traditionalists by disguising it as a sustaining innovation. By incorporating these factors that are necessary (but not sufficient by themselves) to cause a revolutionary transformation, disruptive innovation theory provides a better explanation for how leaders champion new ways of war than does intraservice competition theory.

Management of uncertainty[5]

The pages that follow argue that managing innovation involves the management of uncertainty. Given the impossibility of predicting precisely how the enemy will employ old and new technologies and how he will fight the next war, the best innovative strategy is flexible and useful in several possible contingencies. Uncertainty can be managed in this fashion when senior military leaders champion several different types of innovation simultaneously. The impact of these assorted innovations on the organization, however, depends on which management style they select.[6] A central theme of this book is that successful product champions manage disruptive innovations differently from sustaining innovations. Although external factors provide the impetus for innovation, managing innovation is primarily an internal matter. Thus, this work presents a disruptive theory as a useful addition to external innovation theories.

Understanding the impact of technological innovation on organizations provides new insights into the role management plays. Since Archimedes introduced pulleys, war has seemed to be 'permeated by technology to the point that *every* single element is either governed by or at least linked to it'.[7] Militaries recognize that failure to achieve advances in technology may result in defeat in war; hence, they are searching constantly for technological developments that could revolutionize military operations. Military leaders thus have often concluded (incorrectly) that war is an extension of technology.[8] They proffer that victory requires the acquisition and maintenance of technological superiority.[9] These technocratic views are so pervasive that they no longer are presented as theory but as fact. They are wrong.

While it is true that military organizations use technology as instruments of warfighting, scholars such as Steve Rosen, Williamson Murray, and Colin Gray have illustrated that technology is an extension of war. They recognized that technological change, especially a single technological breakthrough, generally does not result in a new way of war.[10] The mistaken belief that technological breakthroughs can win wars fails to recognize the importance of doctrine and organization in translating technology into advantage.[11] As Martin van Creveld argues, 'It was not the intrinsic superiority of the longbow that won the battle of Crecy, but rather the way in which it interacted with the equipment employed by the French on that day at that place.'[12]

This book is written from the camp of the neo-Clausewitzians, who recognize that the fundamental nature of the conduct of war has not changed. All else being equal, the human element is the most important component in war and in innovation development. This work concentrates on strategic management of military innovation as related to doctrine and technological change. In doing so it critiques competing external theories as well as current internal technology-driven and doctrine-driven views of how senior military leaders manage innovation. It is not a discourse on the workings of naval technology; nor does it chronicle all naval hardware origins and how they might have influenced naval warfare.[13] Instead, it focuses on how technologies – both new and old – fit into the overall framework of changes in the way wars are conducted. It also examines the consequences of senior military leaders relying solely on pursuing sustaining innovations, and it offers an approach that permits the military to capitalize on the benefits of pursuing disruptive innovation. Disruptive innovation theory is described in terms of the development of German armor warfare. Using disruptive and sustaining constructs, the theory correlates what Williamson Murray and Colin Gray call the 'revolutionary' and 'evolutionary' phenomenon of innovation.[14]

did we do a reading on this?

Afterwards, the book focuses on an in-depth analysis of several disruptive and sustaining cases in the US naval service, in order to establish what researchers refer to as internal validity of the disruptive framework. In order to establish external validity of the disruptive framework, the book also performs an in-depth case study analysis of a very different navy culture that achieved a major disruptive innovation. Japanese carrier warfare was chosen because it was a case that Rosen wanted to study, but could not because there did not exist a comprehensive and reliable doctrine and technological history written in English of Japanese Naval Air development.[15] Since Steve Rosen published *Winning the Next War*, two seminal works that address many of the intellectual and some of the political processes of disruptive innovation have been published by Mark Pettie, *Sunburst: The Rise of Japanese Naval Air Power, 1909–1941* and in collaboration with David C. Evans *Kaigun: Strategy, Tactics, and Technology in the Imperial Japanese Navy 1887–1941*. Combined with Chuck Prangle's *At Dawn We Slept*, which gives an excellent account of the innovation politics just prior to the attack on Pearl Harbor, and the excellent biographies of key Japanese naval leaders, it is possible to provide the first complete intellectual and political innovation account of Japanese naval aviation.

Portions of the book are based upon my award-winning Arleigh Burke article, 'Sunk Cost Sink Innovation', which appeared in *Proceedings*, May 2002, and I am grateful for Fred Rainbow's (editor-in-chief) for permission to rework this material.

There are a number of people whose assistance was critical to the completion of this project. They can be divided into four general groups: my thesis advisors; selected members of the faculty and staff of the John F. Kennedy School of Government at Harvard; senior navy officers; and various scholars at Navy and Marine Corps research and historical centers.

My greatest debt is to Stephen Rosen, Political Science Professor at Harvard University, and Stephen Walt, Political Science Professor at the John F. Kennedy School of Government at Harvard, and Henry Chesbrough, Technology Innovation and Business Professor at Harvard School of Business, who have served as my dissertation advisors. These three men were instrumental in my graduate education and the development of this project; they each shaped my thinking in their classes, and took the time to work with me when this project was in the nascent stage and many of the ideas were unclear. In addition to these mentors, Professors Graham Allison, Owen Cote, Rebecca Henderson, and Williamson Murray helped throughout my doctoral research to keep my thinking sharp.

In the Navy I would like to thank following senior officers for supporting my studies at John F. Kennedy, Harvard University and for the subsequent support in writing this book: Admiral Walter Doran, Vice Admiral Scott Fry, Vice Admiral Robert Willard, Vice Admiral Tim LaFleur, Vice Admiral Jim Stavridis, Rear Admiral Joseph Sestak, Rear Admiral Stufflebeem, Rear Admiral Rick Ruehe, and Rear Admiral Gary Jones. In the Marine Corps I would like to thank General Al Gray, General Chuck Krulak, Lieutenant General Wallace Gregson, and Major General Tim Ghromely.

I found the Navy and Marine Corps historians at the Navy Yard in Washington, DC and Quantico at the research center to be very generous with their time, and would like to thank Kathline Lloyd, Regina Akers, Ken Johnson, and Bernard Cavalcante at the Naval Historical Center for their help. At the Marine Corps Historical Center I would like to thank Lieutenant Colonel Jon Hoffman and at the Marine Corps Research Center at Quantico I would like to thank Kerry Strong and Jim Ginther. Also, I would like to thank Peter Swartz of the Center for Naval Analysis, Bradd Hayes of the Naval War College, and Gia Harrigan of the Naval Undersea Warfare Center for their support and suggestions for the thesis.

I am indebted to many supporters who took the risk of admitting and supporting a middle-aged student's way into and through the John F. Kennedy Doctoral program. I would like to thank Sue Williamson, Edith Stokey, Louisa Van Baleen, Graham Allison, and Dean Joseph Nye. I owe a special dept of gratitude to Beth Kier and Jon Mercer, now professors at the University of Washington, who provided me with both encouragement and a great deal of advice during my studies.

My doctoral journey taught me the importance of loyal allies and friends. I was fortunate to have many. Five merit my deepest thanks: Rear Admiral 'Rabbit' Christenson, Mackie Christenson, Tim Lewis, Molly Lewis, and Commander Bill Parker, classmate and friend.

Finally, I would like to thank my parents Fred and Rochelle Pierce and my in-laws Bill and Kay Sisk for their encouragement and support. A special thanks goes to my two sons, Andrew and Nathan, who patiently supported their father. Lastly, I would like to thank my wife Lynne, who has been a consistently reliable friend and supporter. Without her encouragement and tolerance, I could not have completed the project.

Notes

1 Barry Posen, *The Sources of Military Doctrine: France, Britain, and Germany between the World Wars* (New York: Cornell University Press, 1984) 54, and Steven Rosen, *Winning the Next War: Innovation and the Modern Military* (New York: Cornell University Press, 1991), 2.
2 Posen, *The Sources of Military Doctrine*, 57.
3 Rosen, *Winning the Next War, Innovation and the Modern Military* (New York: Cornell University Press, 1991), 75.
4 See Clayton M. Christensen, *The Innovator's Dilemma* (Cambridge, MA: Harvard Business School, 1997). The concept of disruptive technologies, first developed by Harvard Business School Professors Richard S. Rosenbloom, Joseph L. Bower, and Clayton M. Christensen, is a new product or innovation that sneaks into an established market because industry leaders failed to recognize the threat it poses. Christensen makes the argument that the good practices that propel sustaining innovations will impede disruptive innovations.
5 For a discussion of Managing Uncertainty see Rosen, *Winning The Next War*, 243–50.
6 See Steve Rosen, 'New Ways of War: Understanding Military Innovation', *International Security*, vol 13, no.1 (Summer 1988).

7 Martin Van Creveld, *Technology and War* (New York: The Free Press, 1989), 311.

8 Ibid., 319.

9 As an illustration of this point, The Joint Chiefs of Staff in their *Joint Vision 2010* list the following specific technological premises to maintain superiority: improve computers, sensors, land vehicles, ships, rockets, aircraft, robotics, and new types of weaponry such as advanced biological agents, energy beams, and space stations. See Michael O'Hanlon, *Technological Change and the Future of Warfare*, Washington, DC: Brookings Institution, 2000, 2–3. Also see Van Creveld, *Technology and War*, 312.

10 Admittedly, the impact of some weapons such as the machine gun can lead to a tactical innovation. In war, tactical innovation is a change in the application of combat power – individual weapons – to a target to defeat the enemy in engagements and battles. Arguably, wonder weapons such as the atomic bomb and advance biological agents transcend tactics and operate in the realm of the operational level of war or winning campaigns. I argue that innovations at the campaign level of war are major military innovations. I also argue, however, that these operational level wonder weapons can only be deployed in an integrated system. As such they are not stand alone weapons.

11 See James William Gibson, *The Perfect War: Technowar in Vietnam* (New York: The Atlantic Monthly Press, 1986), who argues that 'the United States could not have defeated Vietnam even by further escalation because of fundamental flaws in its concepts of war as a kind of high-technology, capital-intensive production process...' p. vii. Also see Van Creveld, *Technology and War*, 319.

12 Van Creveld, *Technology and War*, 319.

13 For a recent study of naval technology see William M. McBride, *Technological Change and the United States Navy, 1865–1945* (Baltimore, MD: Johns Hopkins University Press, 2000). Also see Karl Lautenschlager, 'Technology and the Evolution of Naval Warfare', *International Security* (Fall 1983), 3–51.

14 Colin S. Gray, *Strategy For Chaos: Revolutions in Military Affairs and the Evidence of History* (London: Frank Cass Publishers, 2002), 282; Williamson Murray, 'Innovation: Past and Future', in Williamson Murray and Allan Millett, *Military Innovation in the Interwar Period* (New York: Cambridge University Press, 1996), 306–10.

15 Rosen, *Winning The Next War*, 5.

1   Introduction

This work begins with a puzzle. How can militaries achieve major warfighting innovations when their senior leaders are engaged heavily around the world and focused primarily on improving current capabilities? A related question is whether the approach for managing sustaining innovations is useful for championing new ways of war?

Answering these questions is vital to the security of the state. The ability to establish durable, relevant militaries depends in large part on how senior military leaders integrate breakthrough technologies and transform them into new ways of war. This study focuses on successful, peacetime cases of 'major innovation'. These new ways of war will be called 'disruptive innovations'. A disruptive innovation is defined as an improved performance along a war fighting trajectory that traditionally has not been valued. It involves a 'a change in one of the primary combat arms of a service in the way it fights or alternatively, the creation of a new combat arm'.[1] An example is *Blitzkrieg* tactical warfare, which is disruptive because it required an adversary to respond in an equally novel way or suffer defeat. The disruptive innovation also requires new metrics to measure its performance, because it both bypasses and, eventually, surpasses traditional warfare methods.[2]

By 'surpass', this does not mean to suggest that the new innovation will necessarily outperform the old one using the current measures of performance. Instead, the new innovation wins on some new, heretofore unimportant dimension. As an illustration, the French Maginot Line outperformed the *Blitzkrieg* in holding established positions and territory; where it fell down was in its lack of mobility – a performance dimension that was not highly valued in the decisions that led to the investments that created the Maginot Line. In this case, *Blitzkrieg* warfare was disruptive because it merely bypassed the French wall of high technology.

In contrast to disruptive transformation, a 'sustaining innovation' results in improved performance along a trajectory that traditionally has been valued.[3] Most innovations in the military are sustaining in nature. An example is continuous aim gunfire, which dramatically improved naval gunfire but did not radically change the mission.

Disruptive innovation is a new way to understand self-initiated change in bureaucracies that challenges existing military innovation theory. It is the dependent variable and lies at the heart of this book; but because it is a new term borrowed from the business literature and adapted for use in security studies, its precise meaning may be in doubt.[4] For purposes of this study, disruptive innovation can best and most easily be described using a basketball analogy – the three-point shot.[5]

Reintroduced in 1967 by the old American Basketball Association (ABA), the three-point shot was immediately derided by the established National Basketball Association (NBA) as a gimmick that added no value to the game.[6] As with all truly important, breakthrough transformations – or disruptive innovations – mainstream customers (NBA fans) and the organizational culture (the NBA) initially rejected the shot because they saw little use for it. The NBA opted to sustain the game the way that mainstream fans historically had valued it – by recruiting taller players who were capable of scoring two-point shots from near the basket.

The result of the sustaining innovation (i.e., recruiting ever-taller players) was to turn the NBA into a slowdown league dominated by big men. Naturally, these men had not been recruited for their abilities to shoot long-range shots. Thus, by ignoring the three-point-shot innovation, the NBA prevented itself from creating new markets and attracting new fans who were interested in a higher scoring game played by mid-sized players who excelled at both long-range and above the rim – the type of players the NBA ignored as untalented rogues who could not cut it in the big league.[7]

The NBA's neglect cleared the way for the ABA to create an entirely new game of basketball. In this game, no leads were safe as smaller, quicker players changed the tempo of the contest. As with all disruptive innovations, the ABA initially underperformed the NBA, but because it had exciting features such as the three-point shot that a few fringe and new fans valued, the ABA survived for nine full seasons.

In June 1976, the two rival professional leagues merged, with the four strongest ABA teams – and the three-point shot – joining the NBA. That disruptive innovation (the three-point shot) not only immediately and forever changed the way basketball would be played in the NBA, but also transformed college and high school play as well.

External and internal causes of innovation

A study of modern warfare suggests that whoever is first to combine new technologies with disruptive doctrine can gain a decisive advantage. Conversely, a military that is slow to adapt new ways of fighting to technological advance opens itself to catastrophic defeat. The historical record shows that disruptive innovation is an exceedingly difficult task. In fact, organizational theory teaches that noninnovation or stagnation is more or less the norm in the military and innovations will be rare.[8] Innovation scholars blame uniformed leaders, claiming they resist change because they are overworked and lack the time or the desire to transform.[9] James Q. Wilson, however, proposes a different reason why senior military leaders are biased toward stagnation. He notes in his illuminating study of military bureaucracy that senior military leaders 'are supposed to resist' innovation because they are responsible for enforcing the standard operating procedures (SOP) that provide the organization's stability.'[10] Considering these negative propositions, security scholars concentrate on the external causes of military innovation (the why and when). External engines of change explanations include: Barry Posen's civilian intervention; Owen Cote's interservice competition; Steven Rosen's intraservice rivalry perspective; and Elizabeth Kier's cultural perspective. Although understanding the external factors (the why and when) that ignite and fuel the engine of innovation are important, failure to understand the internal factors (the how) that controls the engine can be fatal.

First was the advent of the steel ship in the Navy. Although it was investing aggressively in new technologies, the Navy resisted innovations that it viewed as disruptive.[11] Eventually, of course, the obvious advantages of this technological advance overcame internal inertia and caused a paradigm shift – a complete turnaround in the way the Navy fought as well as in the way it structured and organized its units.

Second was President John F. Kennedy's attempt to convince Army leadership to develop a counterinsurgency capability, which was defeated by classic internal inertia.[12] Senior Army officers simply refused to comply because they believed superior, conventionally-trained infantry could win under any circumstances.[13]

Third was carrier warfare, which is an example of a military inventing a technology but failing to exploit it fully. The British Navy invented the aircraft carrier in 1914 and conducted the first carrier air raid in history. Despite its head start (possessing nearly a dozen carriers of one sort or another at a time when no other naval power had even one), the United Kingdom fell short when it came to developing carrier warfare – something the Japanese and Americans would do with great success during World War II.[14]

In each of the three examples, failure resulted from internal, not external, mechanisms. The ignition was working and there was plenty of fuel, but a disruptive innovation did not occur because the throttle was stuck in idle. When the external causes are sufficiently present, the success of innovation depends on which hypotheses senior military leaders accept when determining how they will manage change. The aim of this study is to help determine which hypotheses are correct and whether new ones are needed to explain how they should champion disruptive innovation.

The argument

When confronted by military stagnation, scholars generally explain it by focusing on the lack of necessary external stimuli to spur and sustain innovation. In accepting this explanation, one assumes that military innovation is best understood by examining external causes alone. In many successful cases of innovation, this assumption is correct; exploring external drivers is one useful perspective for understanding innovation. This assumption fails, however, to explain those times when external drivers are present yet stagnation persists. External drivers are necessary but not sufficient for disruptive innovation.

This study argues that internal drivers are equally as important as external drivers for innovating. First, it suggests that 'good management' of sustaining innovations is the root cause of stagnation.[15] In contrast to the traditional doctrine-driven and technology-driven theorists who explain innovation by the presence of certain external factors, these factors may be present and stagnation still may result because of the way product champions manage the innovation process. Second, the evidence shows that management that excels in promoting sustaining improvements along a trajectory valued by the organization almost always fails to support disruptive innovations. Indeed, only when confronted by a disruptive innovation – one that introduces a new performance trajectory – do senior leaders recognize the threat.[16] Third, military leaders fail not because they lack foresight or management savvy, but because they champion disruptive innovations as though they were managing sustaining improvements.[17] Taken together, these results help explain why the naval services' capacity to bring about a disruptive innovation is unfavorable and likely to remain so as long as senior naval leaders continue to manage technology-driven changes as sustaining innovations.

Traditional innovation engine models

Ignition and fuel: the why and when

Modern social scientists have generated a number of works examining the origins of disruptive innovations. Although the literature on disruptive innovations is enormous, much of it falls within four main theoretical schools concerning military innovation: civil-military conflict, intraservice conflict, interservice conflict, and organizational culture. Each of these models is derived from two larger theoretical perspectives – balance of power theory and organizational theory – that vie to explain state behavior using different structural sources.[18] Balance of power theorists argue that state actions are the result of rational thought, while organizational theorists argue that such actions are best understood as derived from standard patterns of behavior rather than from deliberate choices.[19] The first three frameworks argue that disruptive military innovation results from conflict in decisive relationships, and that sustaining innovation results when conflicts are suppressed.[20]

Civil-military competition

The first school, represented by Barry Posen, attempts to understand innovation as it relates to major changes in the international balance of power and/or the political framework within which wars occur.[21] Drawing on this structural realist perspective, Posen identifies external threat and civilian intervention as the greatest determinants leading to innovation. Arguing that balance of power theory has greater explanatory power than organizational theory, he claims that a state's ability to innovate is a function of its security environment. Since states behave rationally, they react to insecurity by improving either the external balance – by acquiring allies – or the internal balance – by strengthening their militaries.

When security threats are low, Posen argues, civilian leaders are content with incremental improvement. When threats to security are high, however, so are the incentives to achieve a disruptive innovation, and civilian leaders may directly intervene to impose and audit disruptive innovation. Posen offers the development of the *Blitzkrieg* in Germany during the late 1930s as an example: 'In my judgment, to the extent that the German Wehrmacht achieved a doctrinal innovation that can be called *Blitzkrieg*, Hitler's intervention was decisive. In the absence of his intervention, it seems likely that normal organizational dynamics would have been determinative…[and] events in the Low Countries might have turned out very differently indeed.'[22] Posen suggests that civilian intervention produces military innovation either directly or indirectly through officers he calls military 'mavericks'. Mavericks provide civilians with the military expertise they lack as well as with an insider who can steer the organization down the desired innovation trajectory.

In sum, Posen's model predicts that insecurity motivates civilian leaders to intervene directly or indirectly using military mavericks as proxies to force the military to change dramatically.[23] When security threats are low, civilian leaders are content with incremental improvements.

Intraservice competition

The second broad school of thought, as applied by Steve Rosen, attempts to understand innovation by examining competition between branches of the same service. Drawing on organizational theory, Rosen believes military organizations are capable of innovating on their own. He sees the impetus for reform as coming from within,[24] and posits not only that civilian intervention is not required but also that it generally fails.[25] Rosen agrees that military organizations are stimulated by changes in the security environment, but he believes that innovation results when branches of the same service vie to become their service's dominant guarantor of security. When their capabilities overlap, competition arises, and senior military leaders both encourage and moderate these internecine squabbles. Innovation results when an emerging warfighting concept gains support among senior military leaders and then is endorsed by civilian leaders. Rosen also asserts that innovation requires 'product champions' – senior officers who advocate innovative approaches to warfare and open promotion paths for other reformers.[26]

Having analyzed cases from the US Navy and the Marine Corps, Rosen argues that 'mainstream' senior military officers consciously adopt a two-part strategy to foster innovation.[27] The first part is to challenge old methods for waging war and propose new concepts to replace them. The second part focuses on managing the political struggle inherent in any attempt to implement new concepts. Successful implementation, Rosen posits, requires the creation of stable career paths to flag rank for younger officers who opt to experiment with the new concepts and develop innovative tactics and techniques.

Rosen's final proposition is that civilian intervention can be effective in promoting innovation if it supports senior military leaders in their pursuit of new warfighting methods.[28] He differentiates this type of intervention from the civilian intervention model advocated by Posen by stressing that the new concepts come from within the military. Rosen also disagrees that military mavericks are effective advocates of transformation. Britain's development of integrated air defense in the years immediately preceding World War II illustrates how Rosen and Posen disagree on this point. In Posen's view, British civilian executives visualized an innovative system of air defense and employed a 'maverick' officer, Air Marshal Hugh Dowding, to achieve their aim. Rosen, on the other hand, emphasizes internal interwar Royal Air Force (RAF) activities that formed the framework for the innovation, which civilian leaders later supported.[29]

Rosen also disputes Posen's depiction of Air Marshal Dowding as a maverick innovator in the traditional sense of a Billy Mitchell or a Hyman Rickover. 'There is a great deal of doubt that Dowding was a "maverick" surviving in the RAF only because he had civilian protection,' writes Rosen. 'Though not well liked, Dowding had been chosen in 1930 by his RAF superiors, not civilians, to be the air member for supply and research in the Air Council, from which position he had successfully championed modern fighter aircraft development.'[30] Rosen points out that Dowding was not alone in supporting air defense. The two chiefs of the Air Staff who succeeded the RAF's founding father also advocated the innovation.[31]

In sum, Rosen argues that competition *between* branches of a service stimulates innovation. Rosen predicts that when intraservice competition is high, innovation results because each branch tries to dominate the other.[32]

Interservice competition

The third broad school of thought, advocated by Owen Cote, attempts to understand innovation by examining competition between services. Vincent Davis was the first to recognize the importance of interservice competition.[33] Subsequent studies, first by Bradd Hayes and then by Owen Cote, supported the Davis proposition. Drawing on organizational theory, both Hayes and Cote assert that strong interservice competition generates an environment conducive to innovation as senior naval leaders attempt to secure their piece of the security pie.

Cote accurately notes that neither Posen nor Rosen assign causal significance to differing patterns of interservice competition when explaining military innovation.[34] In his study, Cote argues that innovation and stagnation can be best explained by interservice competition.[35] He draws support by analyzing the development of the Navy's Polaris and Trident II submarine-launched ballistic missile weapon systems and the concomitant changes in US nuclear doctrine.

Essentially, Cote takes Rosen's argument and deduces that competition between services could have an even greater effect on innovation. He argues that the structural dynamics of intra- and interservice competition are different because power distribution is more evenly split between services than it is within.[36] This leads to different internal dynamics and outcomes in the innovation process. According to Cote, internal innovative groups normally are low in the hierarchy and their struggle to gain resources and autonomy is difficult and takes time.[37] Interservice competition, however, faces no such internal inertia. Those who might otherwise attempt to obstruct change generally champion internal groups promoting doctrinal innovations with interservice ramifications – in other words, they find it easier to support efforts aimed at goring another service's ox.[38] Two ramifications are particularly supported. The first is an innovation that exports the cost of innovating to another service, and the second is when the innovation promises to increase the service's portion of the budget pie.

As an illustration, Cote cites the interwar case of carrier aviation. Innovation was crippled in Britain because the RAF controlled all aviation – both land and sea based. In contrast, US aviation was split among the services, and competition contributed to innovation. As Cote notes, US 'interservice competition accelerated intraservice processes of innovation which were already underway'.[39]

In sum, Cote highlights conflict between the services.[40] This third school of thought views innovation as the product of intense competition between services.[41]

Organizational culture

The fourth school of thought, represented by Elizabeth Kier, examines military innovation through a cultural lens. Kier rejects external security threats as the prime driver for innovation and suggests that a state's choices in military doctrine and innovation are shaped by cultural factors.[42] She contends, for example, that interwar British and French warfare developments were driven primarily by the organizational culture of their respective armies.[43] Both countries developed defensive doctrine prior to World War II, and Kier insists that this can be explained only by analyzing the military cultures that shaped the thinking of leaders tasked to prepare to fight the next war.[44] She argues that their organizational cultures inhibited them from thinking about offensive operations and, thus, made them vulnerable to defeat by the more offensive-minded Germans.

Kier's work is important because it is an attempt by a political scientist to bridge the gap between them and historians such as Williamson Murray who argue that a military's culture is one of the most important components in successful military innovation.[45] As Williamson Murray posits, 'The strategic and political environment can indeed create a climate conducive to innovation. The elements in such change, however, occur within organizations themselves. It is the interplay between past experiences, individual leaders and innovators, and the culture climate within military organizations that determines how successfully innovation proceeds.'[46]

In sum, Kier relies on organizational culture to explain military innovation. The fourth school moves away from conflict relationships and uses a cultural perspective to explain military innovation. In doing so, Kier suggests that a state's choices in military doctrine and innovation are shaped by cultural factors.[47]

Summary of four models

Though they differ in particulars, these schools of military innovation have an important similarity. They generally succeed in explaining why and when military innovation has occurred in the Navy and Marine Corps, but to the limited extent they explain how innovations are implemented and managed, they are theoretically indeterminate. Unfortunately, little work by political scientists has been done on the issue of *how* senior military leaders manage the impact of technological innovation on their organizations.[48]

Even though scholars recognize that military innovation is the product of multiple causal factors operating on several levels of analysis, they do not account for the role senior leaders play in managing technological advances and championing military innovations. Emerging civilian innovation research, however, suggests that the *how* may be just as important as the *when* and *why*. This book posits that disruptive innovation theory, which focuses on both external and internal causes of disruptive innovation, is a better alternative than the theories that focus only on external causes. The next section analyses the two most significant internal innovation theories.

Internal causes of military innovation

Managing doctrine-driven innovation

In examining the structure of peacetime military innovation, Rosen focused on the factors within military organizations that help explain variation in behaviors and preferences.[49] Rosen concluded that senior military leaders hold the key to peacetime innovation because they have the best opportunity to change others' minds.

In looking at military organizations as complex political communities, Rosen observed that each service has its own culture and distinct way of thinking about the way war should be waged.[50] Sometimes, there are even differences within a service. The Navy, for example, is composed of three distinct branches – air, surface, and subsurface – as is the Marine Corps, which is composed of the infantry, aviation, and logistics branches. Given this, Rosen contends that each branch also has a distinct view of how it and other branches should fight.[51]

Rosen argues that each branch has a political as well as a warfighting character. The political component of each branch worries about who should command and who should be promoted. Rather than a band of brothers working together in harmony, Rosen depicts competitive brothers arguing over control of the family business.[52] Since there is no permanent warfighting norm (the Navy, e.g., has seen battleship, carrier, and submarine admirals dominate the service), internecine competition is likely to continue.[53]

Consequently, doctrine, which defines how the branches perform their roles and missions during wartime, is fluid as each branch competes for warfighting supremacy within its service. Rosen sees this competition as good for innovation, since this ideological struggle pits senior military leaders against each other as they offer novel concepts that advance their warfighting specialties. Rosen states that the winning ideology will offer 'a new theory of victory, an explanation of what the war will look like and how it will be won'.[54] This ideological struggle culminates when product champions translate the emerging theory of victory into concrete tasks that become the standard in peace and war.[55] Once an innovation is in place, the means by which senior military leaders ensure it endures is by controlling promotion policies.

Rosen's peacetime innovation model includes two processes. The initiating process involves senior 'mainstream' military leaders formulating a new vision of fighting in reaction to structural changes in the security environment. The political implementing process involves senior military leaders directing the ideological struggle to redefine organizational values required to support new concepts. This normally involves opening promotion paths for supportive junior officers.[56]

Rosen's findings conflict with those of Barry Posen. Posen contended that military organizations innovate only in response to civilian intervention.[57] Rosen agrees that civilians can play an important role, but it is a supporting rather than a principal role. Rosen believed there was no evidence showing that civilians spurred innovation either directly or through military mavericks.

Managing technology driven innovation

Rosen concluded correctly that the successful implementation of technological advances depends on how senior military leaders manage the process. 'Technological innovation', he writes, 'is strongly characterized by the need to develop strategies for managing uncertainty.'[58] In other words, Rosen found no single best approach to technology innovation.[59] In reaching this conclusion, Rosen challenged several widely-held beliefs. One such theory is that arms races spur technological innovation and that military intelligence plays an important role as adversaries react to each other's technological advances.[60] Rosen found, however, that US Army and Air Force innovators did *not* have access to good intelligence about Soviet military technology developments.[61] Thus, he concluded that technological innovation is not an action/reaction response to an arms race.

Rosen also tested the theory that technological innovation is based on projections of the cost and utility of alternative technologies. Again, he could find no evidence to support this theory. Finally, Rosen explored the role of military demand (technology pull) and scientific invention (technology push) as drivers for technological innovation. Rosen could not find conclusive evidence that supported either of these propositions as a principal driver.

Rosen notes, '...a US Department of Defense study, Project Hindsight, reviewed the history of twenty major US weapons programs. It identified 710 discrete events in the history of the development of these weapons and found that "a clear understanding of a DOD need motivated 95 percent of all events (73% of all science events, and 97% of all technology events)".'[62] To counter the Project Hindsight report, Rosen quotes Martin van Creveld who states, 'During the twentieth century...none of the important devices that have transformed war – from the airplane through the tank, the jet engine, radar, the helicopter, the atom bomb and so on all the way down to the electronic computer – owed its origins to a doctrinal requirement laid down by people in uniform.'[63] Of note, Jared Diamond, in his Pulitzer-Prize-winning book, *Guns, Germs, and Steel*,[64] echoes van Creveld's position when he states, 'The starting point for our discussion is the common view expressed in the saying "Necessity is the mother of invention." That is, inventions supposedly arise when a society has an unfulfilled need: some technology is widely recognized to be unsatisfactory or limiting.' Differing factors such as money or fame motivated inventors to 'come up with a solution superior to the existing, unsatisfactory technology. Society adopts the solution if it is compatible with the society's values and other technologies. Quite a few inventions do conform to this commonsense view of necessity as invention's mother'. Examples given are the 1942 US Manhattan Project, Eli Whitney's 1794 invention of the cotton gin, and James Watt's 1769 invention of the steam engine to solve the problem of pumping water out of coal mines. Diamond postulates 'these familiar examples deceive us into assuming that other major inventions were also responses to perceived needs. In fact, many or most inventions were developed by people driven by curiosity or by a love of tinkering, in the absence of any initial demand for the product they had in mind'. Diamond contends that the critical element in the successful adaptation of the technology depends upon the inventor finding an application for it. After a lengthy process of consumers using the invention do 'consumers come to feel that they "needed" it. Still other devices, invented to serve one purpose, eventually found most of their use of other, unanticipated purposes'. He concludes that 'It may come as a surprise to learn that these inventions in search of a use include most of the major technical breakthroughs of modern times, ranging from the airplane and automobile, through the internal combustion engine and electric light bulb, to the phonograph and transistor. Thus, invention is often the mother of necessity, rather than vice versa.'[65]

Rosen concluded that technological innovation is only understood when viewed as a series of strategies for managing uncertainty. Two such strategies are: 'let the scientists choose' and 'low cost bets'. In the first strategy, Rosen discovered that neither the civilian scientific community nor the military R&D community demonstrated an advantage in selecting which technologies to develop.[66] In the second strategy, the evidence suggests, especially in the area of US Navy and Air Force guided missiles, that adopting a flexible strategy that reduces uncertainty by sponsoring several projects and then choosing the most viable one might be the best approach. In this case, prototypes of the weapon system are developed and then tested for performance and deployed into field exercises to explore which doctrines they best support.[67] After successful performance testing and field experimentation, the final phase of this strategy is full production.

Peacetime innovation management: technology driven

Vincent Davis provides a different model of how innovation finds its way into the military than either Rosen or Posen. Davis developed a technology-driven model based on case studies of Navy efforts to develop a nuclear weapon delivery capability by carrier-based aircraft, nuclear propulsion, and fleet ballistic missiles. He also examined several pre-World War II cases.[68] In contrast to Rosen, Davis defines innovation as the implementation of new technologies that are used to help perform existing missions better.[69] Disruptive theory categorizes Davis' innovations as sustaining. Whereas Rosen asserted innovation comes from within the military through top-down influence of senior military officers, Davis believes it comes from within the military through the assiduous efforts of middle-grade officers. Their focus is quite different as well. Rosen looks primarily at doctrinal innovation whereas Davis focuses primarily on technological innovation – the building and integrating of technology into warfighting doctrine. Another key difference between them is that Rosen insists that a 'new theory of victory' must precede innovation and Davis does not.[70]

Technology-driven hypothesis

The belief that militaries innovate by first building technologies and then integrating these technologies into doctrine lies at the heart of Davis' innovation model. According to the hypotheses of technology-driven innovation, reform advocates push forward their technology without a new clear vision of victory. They are also zealots, who continue to promote their products even at the expense of their own careers.[71]

The innovation advocate in the Navy is usually an officer holding the middle rank of lieutenant commander, commander, or captain.[72] He is more likely to be a line officer than a civilian Navy scientist. Senior military leaders, defined as flag officers, are usually one-star rear admirals and they serve as the spokesmen for groups of younger officers.

The technology-driven hypothesis stands in sharp contrast to the hypotheses considered in the last section. As a result, most scholars who favor the doctrine-driven theory of innovation discount the importance of technology in military innovation. In Deborah Avant's study, *Political Institutions and Military Change*, she states, 'New technologies may simply be used to support existing practices better.' [73] The dependent variable examined in this study, however, is not technological innovation but a change in the orientation of a service or service brand in response to a new or changed adversary in both peacetime and wartime.

There remain a number of scholars who believe that technology innovations can influence doctrine, such as Martin van Creveld, who insists, 'War is completely permeated by technology and governed by it.'[74] The high priests of technology innovation point to the two Gulf Wars, dominated by United States led high-tech gadgets, novel weapons, and exotic systems, to support these claims.[75]

Davis reached his conclusions about the impact of technology on innovation by examining how various advocates managed to push their concepts into the fleet. In every case, advocates managed to establish both horizontal alliances (among peers) and vertical alliances (among seniors) to achieve their objectives. The story of Rear Admiral Harold Bowen, for example, is a noteworthy case in which an important technology innovation was successfully introduced. Bowen waged a bitter struggle against the influential General Board of the Navy, which included most of the bureau chiefs and key leaders of the shipbuilding industry, in his 1930s campaign to install high-pressure boilers and high-temperature steam systems in new destroyers. In his capacity as Chief of the Bureau of Engineering, Bowen served as the senior spokesman for a group of advocates consisting mostly of younger officers. He also managed to create an alliance with a few important industry executives. Ultimately, he won his case and the Navy transitioned to the new propulsion technology, which proved crucial during World War II. Bowen's career, however, suffered as a result of the hostility he generated and he was denied important fleet commands during the war.[76]

Captain Washington Irving Chambers offers another example.[77] As leader of a group of middle-grade reformers, Chambers used political techniques similar to those utilized by Bowen to successfully introduce aviation into the Navy. In 1910, Chambers and his small group became pioneer enthusiasts for naval aviation. Chamber's breakthrough came when he gained the support of the Navy's most influential senior officer – Admiral George Dewey. Although set back when the Secretary of the Navy refused to create a permanent naval aviation structure, Chambers did succeed in gaining congressional and civilian support. The latter came in the person of Glenn Curtiss, one of the most famous aviators of the day. Tireless in his efforts, Chambers finally saw the Navy's Office of Aeronautics established during the third year of World War I. The personal price of success was again high. Chambers was passed over for flag rank and forced to retire. Fortunately, an equally able advocate – Lieutenant Commander John Towers – replaced him as the immediate assistant to the head of the Office of Aeronautics. Towers secured the support of Rear Admiral Leigh Palmer, who was serving in the powerful position of Chief of the Bureau of Navigation. Palmer allowed Towers to handle personnel matters relating to the wartime expansion of naval aviation. The relationship between Palmer's and Towers' commands formed the genesis of a semi-separate aviation corps within the Navy – a status formalized in 1921 by legislation establishing the Bureau of Aeronautics. Many important technological innovations were introduced during this period, due in large part to the favorable organizational arrangements pushed for by Chambers and Towers.[78]

These examples illustrate how a passionate inside reformer can create effective horizontal and vertical alliances to achieve a technology innovation. Ronald Kurth's study, *The Politics of Technical Innovation*, supports Davis' proposition concerning advocates and notes that technology innovators in the Navy 'have been judged historically to be ambitious, hard-working, middle-men'.[79]

The implications of doctrine-driven and technology-driven innovation

Doctrine-driven and technology-driven theories of transformation offer contrasting hypotheses about military innovation. The contrasts are as much about timing as substance (i.e., they generate a 'chicken and egg' debate over whether software or hardware comes first). Resolving which proposition is more accurate is important because each suggests a different approach for how senior military leaders should manage innovation.

The fact that the proponents of the schools offer competing definitions of innovation makes reconciling the differences more challenging. Continuous aim naval gunfire, as presented in Chapter 8, illustrates the problem. Davis would classify it as a major innovation because it improved naval gunfire by 3,000 per cent. Rosen would not classify it as a major innovation because it did not change the way the Navy fought. Admittedly, the innovation increased the prestige of gunners mates – a mild social change, which is a key Rosen concept for innovation – but it created no new social orders. Thus, Rosen labels continuous aim gunfire a tactical innovation.[80] This study labels it a sustaining innovation because it increased performance along a warfighting trajectory that the Navy had traditionally valued. Of course, subsequent sustaining innovations along this trajectory eventually lead to the cult of the modern battleship, which could lob a 16-inch shell over a distance of 20 miles with an accuracy within 200 yards of its target.

Rosen admits that his definition has the slightly paradoxical effect of excluding some dramatic changes in military technology from the term 'major military innovation'. Some examples that Rosen excludes are the introduction of atomic bombs into the US naval aviation strike force and the introduction of nuclear propulsion systems into the US submarine force. Rosen excludes them because they performed existing missions better but did not change them radically. Davis does include them because they had a profound effect on how the Navy conducted its business.[81]

Neither the Rosen nor Davis approach yielded much in the way of a unifying theory for innovation. Their cumulative impact has been hindered by the lack of consistent definitions, which in turn has yielded a variety of competing causal explanations. As recent innovation scholars suggest, Rosen's restrictive definition that excludes some technological changes as major innovations has created a typology gap.

Rosen admits that the study of innovation would be incomplete without exploring the impact of new military technologies.[82] He examines several 'radical' technologies, such as radar, guided missiles, and proximity fuses, but he fails to identify a causal mechanism that explains how some new technologies culminate in a new warfighting reality. Both general hypotheses agree that new technology alone does not revolutionize naval warfare. Part of the disagreement is over where to set the impact threshold that new capabilities make on the service. As Karl Lautenschlager states, 'Significant changes in the military and political capabilities of naval forces have come when long-existing technologies were eventually refined and integrated. It is the final integration of several technologies that came quickly in some cases. In other cases an essential component was lacking from the ensemble, but by itself would have been useless. Certainly no single technological "breakthrough" has brought immediate change in naval capability.'[83]

Each of the following innovation studies attempts to reconcile the conundrum. In a Naval War College study on *The Politics of Naval Innovation*, Bradd Hayes argues that Rosen's doctrinal definition excludes a technology approach to studying innovation. He makes the case that a major military innovation can result from either a technological or conceptual breakthrough. He therefore accepts Andy Ross' definition of innovation that 'includes not only the actual instruments or artifacts of warfare, but the means by which they are designed, developed, tested, produced, and supplied – as well as the organizational capabilities and processes by which hardware is absorbed and employed'. Hayes claims, 'This definition avoids the tendency of many analysts to focus on hardware rather than on organization and doctrine; it also overcomes the restrictions associated with Rosen's doctrinal definition.'[84]

In a study highly praised by the Director of Net Assessment, Andrew Marshall, titled *Peacetime Military Innovation*, the authors included Rosen's technology category as a major 'peacetime' innovation. They stated, 'While [our] definition may resemble the definition of a major innovation – "...a change that forces one of the primary combat arms of a service to change its concepts of operation and its relation to other combat arms, and to abandon or downgrade traditional missions" – attributed to Stephen Rosen...it does not exclude technological innovation or make it a separate entity.'[85]

Emily Goldman, in her study, 'Mission Possible: Organizational Learning in Peacetime', attempts to expand Rosen's definition in order to capture minor innovations to existing goals, strategies, and/or structures. She calls these innovations 'adaptive strategic adjustments'. Goldman calls major innovations, such as those included by Rosen, 'innovative strategic adjustments'. She supports her position for including adaptive adjustments by arguing, 'While the tendency has been to view innovation as positive and forward-looking, and adaptation as a negative attachment to outmoded ideas or procedures, neither is inherently good nor bad.'[86]

In the next section, the study argues that disruptive innovation theory improves on Rosen's doctrine-driven theory by providing greater explanatory power with equal parsimony. By using disruptive theory advocates can understand not only innovations defined by Rosen's restrictive definition, but they also can explain how Davis' sustaining innovations impact (usually negatively) on Rosen's disruptive innovations. Disruptive theory, as will be discussed, avoids the 'chicken and egg' debate by subsuming both technological and behavioral innovations.

Typology considerations

This study uses a different typology to select its cases than Rosen. When Rosen was designing his typology, his central question was, 'What can be said about innovation in military bureaucracies?' In answering this question, he assumed that 'different kinds of innovation occur for different reasons in the same organization, and different organizations will handle innovation very differently'.[87] Lacking a grand theory of innovation, Rosen categorized results as either major military innovations or technological innovations. He further subdivided major military innovations into peacetime and wartime processes. He defined a major innovation as either a change in how one of the primary combat arms of a service fights or the creation of a new arm. Technological innovation, on the other hand, may or may not change how a service fights. New technologies may simply be used to improve existing practices.

In other words, Rosen sees innovation as either changing operational behavior or building new machines. His operational behavior approach examines the impact of social innovation on the organization. Social innovation is defined as changing the way men and women in organizations behave. In contrast, techno-logical innovation is concerned with building hardware. Rosen posits that tech-nological innovation has an important political component, but no social component.[88] Since Rosen was concerned with major military innovations (i.e., new ways of war), he did not address 'minor' military innovations, which could have involved new technologies.

This study examines the impact of technological innovation on the organization and has both social and political components. This component-linkage approach, which introduces architectural innovation into the social science literature, cate-gorizes innovations according to changes in architectural linkages among com-ponents and changes in the components themselves. The importance of architectural innovation is that it is equivalent to military doctrine innovation and the theories that underlie it help explain why different kinds of innovation occur for different reasons in the same organization, and why different organizations handle innovation differently. The advantage of the architectural typology is that it includes not only Rosen's 'major military innovations' (disruptive innovation) but also Davis' 'minor military innovations' (sustaining innovation) and the impact they have on senior military leaders managing major military innovations.

Architectural typology

The new typology is derived from the studies of Rebecca Henderson and Kim Clark, which distinguish between component technologies and system integra-tion. This taxonomy essentially expands the 'technical-production' into two dimensions. One dimension categorizes component core technologies as being either reinforced or overturned, and a second dimension categorizes the linkages between components as being either changed or unchanged.

The importance of this model is that it explains why seemingly insignificant improvements in technology can result in a warfighting paradigm shift. It does this by disaggregating technology into its components and examining systems architectures that link components together. Disaggregating systems is a useful strategy for understanding and managing the complexity inherent in technical innovation.[89]

This two-dimensional approach for viewing new technologies produced a new framework for differentiating between two types of innovation – radical and architectural. Instead of focusing on the product as a whole system, Henderson and Clark use a component perspective that focuses on the parts that comprise the product. They posit that each product embodies a set of components that interact according to a unique system architecture. Innovations are classified by the degree to which they reinforce or render obsolete an organization's composition along these two dimensions – architectural design and component technology.

The Henderson and Clark model leads to four distinct types of innovation, as shown in Figure 1.1.[90] Each of these is defined in terms of components and system architectures. *Components* are distinct physical building blocks of the product that perform a well-defined function. The *architecture* of a product defines how linked components work together.

Incremental innovation merely refines and extends the dominant design, such as an upgrade to a weapon system. Modular innovation changes a core design without changing internal linkages, as would occur if one shifted from an analog to a digital ship's steering system. Radical innovation occurs when a new dominant design is linked together in a totally new structure. Aircraft and submarines are radical innovations.

Architectural innovation changes the way components are linked together while leaving core design concepts (and thus the basic knowledge underlying the components) untouched.[91] An example of a *subsystems* architectural innovation is continuous aim gunfire. Continuous aim gunfire improved the Navy's hit rate by 3,000 per cent and was achieved by fitting naval guns with elevating gears, which made adjustment easier, and by adding a telescopic sight.

By moving from the subsystem to system level, the architectural model can explain why major warfare innovations occur. An example of a system-level architectural innovation is the German *Blitzkrieg*, which combined existing core technologies (such as tanks, aircraft, and radios) in a novel way – a doctrine shift.

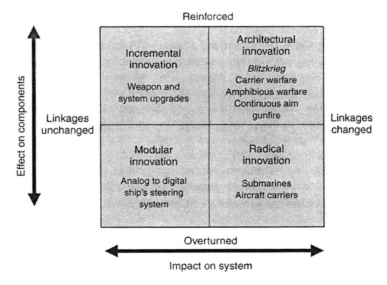

Figure 1.1 A new typology for defining *technology* and *military* innovation.

Source: Henderson & Clark's Typology (1990).

Examining architectural subsystems at a higher level builds on the work of Clayton Christensen, who points out the usefulness of viewing products as a *nested* set of components held together by a unique system architecture. In Christensen's approach, components *themselves* are systems, comprised of subcomponents whose relationships to each other are defined by an architectural design. Thus, the tank, airplane, and radio, which can be viewed as complex architectural systems at one level, act as components in the *Blitzkrieg* at a higher level.[92]

The component technology and system integration typology offers a classification scheme that categorizes all innovation types within one framework. This unified typology, which includes incremental, modular, radical, and architectural innovation in a single model, is a breakthrough in military innovation theory.

The principle focus of this study is on architectural and radical innovations. Since architectural innovations can be either sustaining or disruptive, cases are selected from both categories. (Cases in the study are referred to as either disruptive or sustaining, but technically, they are disruptive 'architectural' innovations and sustaining 'architectural innovations'.)

Case selection

The book focuses on successful Navy and Marine Corps disruptive and sustaining innovations between 1899 and 2003. Seventeen innovation cases were identified by architectural typology, which are discussed in the next chapter. The study compares the results of senior military leaders' management styles in each case with the predicted outcome based on competing hypotheses. The aim is to identify those hypotheses that explain the greatest number of innovations.

To enhance internal validity, the study uses the approach that Davis and Rosen used in determining their cases. It focuses on successful instances of innovation, not on failures, because as Rosen notes, 'the absence of innovation is the rule, the natural state'.[93] It examines primarily cases from the naval services (i.e., the Navy and Marine Corps) and limits analysis to management styles of product champions who actively promoted or opposed a given innovation.

Table 1.1 is a typology measure of the dependent variable – disruptive innovation.[94] The book considers nine disruptive innovation cases, seven successful, one failed, and one inchoate. Successful cases examined are: American carrier warfare, Japanese carrier warfare, surface land attack warfare, composite warfare commander, amphibious warfare, air mobility, and Marine Air-Ground Task Force warfare. The one failed case is defensive sea control warfare and the one disruptive case that is still evolving is Maneuver Warfare. The case selection provides ample variation to test external independent variables (civilian intervention, interservice rivalry, intraservice rivalry, and organizational culture) as well as internal independent variables (small group(s), disguising rhetoric, product champion support, and junior officer promotion) and the dependent variable (disruptive innovation). For comparison, the study also examines five sustaining innovations including continuous aim gunfire, carrier battle group, defensive advance base force, MEU(SOC) warfare, and maritime predeployed logistics.

Table 1.1 Seventeen naval innovations: nine disruptive and eight sustaining

	Navy	Marine Corps
Disruptive	Carrier Warfare, 1918–43	Amphibious Warfare, 1905–40
Architectural Innovations	Defensive Sea Control 1970–74	Airmobility Warfare, 1948–76
	CWC – Composite Warfare Commander, 1985–88	MAGTF Warfare, 1975–89
	Surface Land Attack Warfare, 1988–90	Maneuver Warfare, 1988–99
	Japanese Carrier Warfare, 1917–42	
Sustaining Architectural Innovations	Continuous Aim Gunfire, 1899–1904	Defensive Advanced Base, 1900–42
	Carrier Battle Group, 1974–78	MEUSOC Warfare, 1976–88
		Predisposition of Logistics, 1984–89
Radical Technologies Supporting Disruptive Innovations	Aegis Radar, 1978–86	
	Tomahawk Missile, 1978–86	
	Tactical Component Network (TCN®), 1996–2003	

The study also includes three radical technologies that disrupted three sustaining innovations including the Tomahawk cruise missile, Aegis radar, and Tactical Collaborative Network (TCN®).

Building a disruptive innovation theory framework

In Chapter 2, the study uses the British and German armored warfare case study to develop an inductive disruptive innovation theory. Armored warfare development during the interwar years by the British and German armies offers an excellent set of cases to develop a framework for explaining the external and internal causes of disruptive innovation. The British invented the tank but failed to invent armored warfare. The Germans did not have tanks until the later half of the 1930s and they invent armored warfare. This discussion guides the remainder of the study.

Following this chapter, the study tests the disruptive theory using US Navy, Marine Corps, and Japanese cases. In Chapters 3–8, the study develops and analyzes Marine Corps disruptive and sustaining innovations. In Chapters 9–15, the study constructs and analyzes Navy disruptive and sustaining innovations. This includes the Japanese carrier warfare case. In Chapter 16, the study provides a comparative assessment of the different hypotheses and extends the analysis to civilian policymakers who are interested in promoting disruptive innovations in the military services.

2 Explaining disruptive innovations

This chapter focuses on the interplay between two different drivers of military innovation. The first involves external factors that describe why and when innovation happens as defined by traditional international relations balance of power theory and by that portion of organizational theory that addresses the dynamics of bureaucratic politics. As discussed in the last chapter they are civilian intervention, intraservice rivalry, interservice rivalry, and cultural factors. Over the past two decades, these theories have provided the most prominent explanations for why military innovation occurs.

The central propositions of this study, however, are drawn from the internal factors that describe the second driver, that is, *how* military leaders manage innovation. The study argues that innovation management is more than a tool for coordinating a sequence of standard actions;[1] it encompasses a myriad of skills, including the ability to foster a culture of sustaining innovations, while preventing that culture from permanently grounding the organization in the past.[2] The study also argues that the ability of naval leaders to manage internally incongruent innovations such as Rosen's disruptive innovations and Davis' sustaining innovations explains much of the variance in patterns of naval innovation.

Internal causes: *how* innovations are managed

This section develops the concepts and hypotheses that will guide the remainder of this study. The framework for disruptive innovation theory is derived inductively from British and German armored warfare experiences, which provide a 'natural experiment' for the innovation explanation advanced in this study. In particular, the successful German case provides a pattern of leadership, which involves both disruptive and sustaining managing efforts, for achieving a disruptive innovation that we can compare with the failure of British leadership to achieve armored maneuver warfare. Those areas of difference both inside and outside the German experience are helpful in isolating those factors that are necessary to achieve a disruptive innovation including the impact of managing sustaining innovations on achieving new ways of fighting.

The disruptive innovation framework is built upon Steve Rosen's intraservice competition theory. Intraservice competition model is the most useful general

Compare German's success with British failure

theory available, but recent civilian innovation theories suggest this framework should be revised to account for how modern militaries, which are investing aggressively in new technologies, resist or fail to innovate when they are confronted with disruptive innovations.[3] Intraservice innovation theory is framed solely in terms of disruptive innovation. This conception should be remodeled, however, to account for the other factors that senior military leaders consider when championing a new way of war. Although championing new ways of war is important (and the focus of this book), it is not the only consideration. Arguably, most military leaders have been promoted to senior positions because they are good warfighters as well as being good managers of people, resources, and sustaining innovations. Reported research of some business industries holds 'good management' was the most powerful reason that the best companies failed to stay atop their industries. An innovation researcher writes, '... there is something about the way decisions get made in successful organizations that sows the seeds of eventual failure [of the firm]'.[4] Is it possible that a key factor that causes disruptive innovation failure are senior military leaders who rely on sustaining management principles to champion new ways of war?

In answering this question, there seems to be no pattern in how senior leaders achieve Rosen-type major innovations that create new ways of fighting and Davis-type major ones that sustain old ways of waging war. Both models have intellectual and political processes that are similar in content. After a close study of armor warfare development, however, a repetitive pattern of successes and failures lead to the solution presented in this book that is called disruptive innovation theory. It is a reformulation of Rosen's theory (which includes parts of Davis' theory) and new factors reflected in the civilian innovation literature and derived from the historical evidence of British and German armored warfare cases.

Building a disruptive management framework

Propositions

The disruptive innovation framework is built upon five findings that can be incorporated with Rosen's two-step process innovation model. The first factor is that warfighting can be viewed as set of integrated components linked by an architecture (doctrine). This component-linkage innovation model gives rise to four types of innovation depending on the degree of change of the components and how the linkages are adjusted. These four types of innovations are managed differently, but they will usually be assimilated to an old doctrine (without changing the doctrine). The key proposition is that all military doctrine changes are architectural innovations, which are defined as linking together existing components in a novel way. Another proposition is that architectural innovation is difficult to recognize and militaries underestimate its disastrous effects. This leads to the incumbent technology failure assertion defined as incumbents who invent a new technology often do not champion a disruptive innovation because they have a difficult time recognizing the new technology linked in a novel way.

Second, architectural innovations can be either disruptive or sustaining in nature. This is important because disruptive and sustaining architectural innovations are managed differently. The majority of architectural innovations are sustaining because they usually improve warfighting (sometimes enormously) along a dimension of performance valued by traditionalists. Civilian intervention and maverick officers can cause sustaining architectural innovations. Disruptive innovations, however, are rare architectural changes that result in a new way of warfighting and are not valued by traditionalists. Civilian intervention and maverick officers cannot cause disruptive architectural innovations.

Third, the pace of sustaining technological progress often exceeds the warfighting performance demanded. For example, once the dreadnought battleship arrived, navies continued to build bigger and better battleships until such time they 'overshot' the battleship's warfighting performance requirement. Such was the case of the mammoth Japanese 18-inch gun battleships, as well as the American 16-inch gun battleships, whose usefulness was quickly diminished by the introduction of the disruptive aircraft carrier.

Fourth, when senior military leaders create and directly manage small innovation groups, the more likely new architectural linkages will emerge that will eventually define a new disruptive doctrine.

Fifth, the greater the product champions can disguise or shape the disruptive transformation as a sustaining innovation, the greater the possibility the disruptive innovation will survive.

(i) Linkages and components: architectural innovation

The negative consequences of disruptive innovations often being adopted and fully exploited first by someone other than the nation inventing the new technology has been recognized by many innovation scholars. Richard Hundley's *Past Revolutions Future Transformations*, among others, has noted how militaries that dominate one generation of technology often fail to incorporate this technology in a novel doctrine that leads to a new way of war. Attempts to explain why this occurs have eluded innovation scholars. This study argues that the roots of this failure lay in the incremental versus radical categorization of technology innovation.

The search for explanations of military transformation during peacetime could begin with a different view of innovation. Looking at innovations as incremental, radical, modular, or architectural (both technological and doctrinal), allows one to study the ways these different innovations are managed.

Historically, scholars have characterized technological innovation as being either incremental or radical. Most people have little difficulty recognizing minor changes to an existing product as an incremental innovation. Next year's model of the Ford Taurus would be an example of an incremental innovation. It refines and extends an established design, but requires no new core components or adjustments to how the components are linked (system architecture). Likewise, most people have little difficulty recognizing the emergence of a new product, such as the Marine Corps Osprey (vertical take-off/landing plane), as a radical innovation requiring new core components and system architecture.

Rebecca Henderson and Kim Clark have argued that this traditional categorization is incomplete, because it fails to capture those innovations that enhance the core components but alter the overall architecture. Such an example occurred in the mid-1970s when competitors of Xerox's large plain-paper copiers introduced a smaller and more reliable copier. As Henderson and Clark point out, the new smaller copier required little new engineering knowledge, only new architectural knowledge. Despite the fact that Xerox had invented the core technologies and had the most experience, they were unable to introduce a competitive small copier for eight years. In that time, Xerox almost failed as it lost over half its market share.[5] Henderson and Clark identified the smaller copiers as a new category of innovation, which they called architectural innovation.

The Henderson and Clark theoretical breakthrough was that they identified a new category of innovations that involve apparently modest changes to existing technology that have dramatic competitive consequences. Henderson and Clark clarify this concept:

> The essence of an architectural innovation is the reconfiguration of an established system to link together existing components in a new way. This does not mean that the components themselves are untouched by architectural innovation. Architectural innovation is often triggered by a change in a component – perhaps size or some other subsidiary parameter of its design – that creates new interactions and new linkages with other components in the established product. The important point is the core design concept behind each component – and the associated scientific and engineering knowledge – remain the same.[6]

By focusing on warfighting as a set of integrated components linked by architectural doctrine, component-linkage theory categorizes military innovation according to changes in core concept designs as well as the way in which components are linked together. The theory posits that successful warfighting innovation requires two types of knowledge. First, it requires component knowledge, or knowledge about each of the core design concepts and the way in which they are implemented in a particular component. Second, it requires architectural knowledge (or doctrinal knowledge) about the ways in which the components are integrated and linked together into a coherent whole. The distinction between the components and the links between components provides insights into the ways in which innovations differ from one another.[7] The key assertion is that because these two types of knowledge depict very different innovations, they must be managed differently. If component innovation is more common than linkage innovation, the changes are incremental and the organization remains stable. Arguably, senior military leaders spend most of their time managing these types of innovations. They are investing an enormous amount of capital that is focused on component advances.

If, however, linkage innovation is the dominant tendency, then stability is scarce, because the architectural knowledge is being uprooted, ending established routines and standard operating procedures (SOPs). Since military organizations prefer stable organizations, architectural innovation is rare. This component-linkage proposition explains why new technologies (or components) will be assimilated to an old doctrine or established architecture (without changing the doctrine). Senior military leaders are responsible for interpreting and assimilating new military technologies. As Barry Posen has argued, 'a new technology will normally be assimilated to an old doctrine rather than stimulate change to a new one'.[8] Posen derives this proposition from the empirical work of Bernard Brodie and Edward L. Katzenbach.[9] Component-linkage theory supports this proposition.

The central challenge in managing architectural innovations is being able to step back from incremental innovations and see how small improvements in legacy systems can be reconnected in new ways. That is easier said than done. As discussed in a subsequent section, Heinz Guderian as one of the champions of architectural innovation, armored warfare, faced enormous obstacles in establishing novel linkages between existing core technologies (e.g., the tank, plane, mobile infantry, and radio).[10] Similarly, Lieutenant Sims as the product champion of another architectural innovation, continuous aim gunfire, faced enormous obstacles in achieving slightly modified linkages between core components. The key point is that both sustaining and disruptive architectural innovations can face tremendous resistance from the *status quo* culture.

Generally, the same set of eyes cannot see both how to improve the way that existing systems fight and simultaneously envisage a new way of fighting them. Since an architectural innovation alters design linkages, but not components, organizations find them difficult to manage because organizations suffer from the same myopia as individuals. They tend to rely on what they know about the ways in which components are currently integrated, and fail to see how components could be connected differently.[11] Architectural innovation is all the more difficult because evolutionary change generally precedes it. Organizations get lulled into straight-line extrapolation of the future, making a (disruptive) architectural change devastating once it occurs. This was the Allied response to the *Blitzkrieg* in May 1940. Although the Allies had same components as the Germans (tanks, airplanes, radios, mobile troop carriers), they failed to seek how new linkages among these components could lead to a new way of warfare. Williamson Murray calls attention to this phenomena when he writes, 'The French also believed that the Germans *could not and would not*, in the end, perform in a radically different fashion from their own forces. To an extent this rigidity reflected an inability and unwillingness to recognize not only that their opponent possessed alternative options and conceptions, but that he might exercise those options. This was mirror-imagining of the worse sort.'[12]

Summary

By focusing on warfighting as a set of integrated components linked by architectural doctrine, component-linkage theory provides insight on how different innovations are managed. Component-linkage theory categorizes innovations according to changes in components and the way in which the components are linked together. Component innovation causes incremental changes in warfighting and is the most common innovation because linkage SOPs remain intact and the organization remains stable. This explains why new technology (components) will normally be assimilated to old doctrine (existing linkages). Linkage innovation, however, causes new ways of warfighting and is less common because it results in a less stable organization because of the requirement of new linkage SOPs.

Because organizations strive for stability, they are reluctant to alter linkages and thus they focus on incremental innovations. Consequently, militaries that invent and have the most experience with core components are least likely to change the architecture (doctrine) that links the components. This explains why architectural innovations are often adopted and fully exploited first by someone other than the nation inventing the new technology. As will be discussed, this is the case with the British who invented the tank but failed to invent armored maneuver warfare.

Warfighting performance as a measure of effectiveness: disruptive and sustaining[13]

A key proposition in this study is that all military doctrinal changes are architectural innovations because they establish new linkages among existing components (old and new weapons or systems). Although the literature on military innovation is enormous, almost all of it focuses on explaining architectural innovation.[14]

This is good because component-linkage theory can provide insights into the complex subject of military innovation. Unfortunately, not all architectural innovations lead to a new way of warfighting. For example, Rosen focused on amphibious warfare and carrier aviation and Davis focused on continuous aim gunfire and the introduction of atomic bombs into the US naval aviation strike force. According to Rosen, not all architectural innovations result in new ways of war. As discussed in the last chapter, Rosen argues that Davis' architectural innovations are not 'major innovations' because these new technologies were used to help perform existing missions better and not to change them radically.

The empirical evidence supports Rosen's claim that the majority of architectural doctrine changes do not result in a new way of warfighting. For example, as Kevin Sheehan makes clear in his study of four major army doctrinal innovations since the end of World War II including 'the pentomic' division, active defense, and AirLand Battle, and Vietnam counterinsurgency doctrine, the architectural doctrine was slightly modified, but at a deeper level, very little changed. None of these architectural doctrine changes resulted in a new way of fighting. Instead, most of them resulted in an improved performance of the established way of fighting. Sheehan concludes that most of the (architectural) doctrine changes were attempts to take advantage of new technology developments in the kind of weaponry that either the United States or the Soviets might employ in Europe.[15]

A useful study that is an invaluable tool in understanding the impact of different types of architectural innovations on organizations and how they are managed is Clayton Christensen's *The Innovator's Dilemma*.[16] In adapting the Christensen model for military use, innovations can be framed in terms of warfighting performance. Sustaining innovations are those that resulted in an improved performance along a trajectory that traditionally has been valued. This study asserts that the changes in linkages among the existing components are modest so that sustaining innovation occurs in a generally stable architecture. Stable architectures result in improvements to warfighting, not new ways of fighting. Christensen asserts that sustaining innovations can be incremental and architectural in character, while others are radical and modular in nature.

Disruptive innovations are those that resulted in an improved performance along a warfighting trajectory that traditionally had not been valued. Occasionally, an architectural innovation emerges that is disruptive, such as the *Blitzkrieg*. Disruptive innovations are the result of novel linkages among components and occur in an unstable architecture. Unstable architectures may lead to a new way of warfighting. By adding the sustaining and disruptive dimension to the component-linkage model, a two-dimensional framework is generated to view innovations. This two-dimension framework is at the heart of disruptive innovation theory. Depending upon the degree of changes among components and linkages innovations can be sorted into four groups – incremental, modular, radical, and architectural. All four types of innovation can be sustaining in nature. All disruptive innovations, however, are architectural (see Figure 2.1).

Sustaining and disruptive architectural innovations have different warfighting consequences because they require quite different organizational capabilities. Sustaining innovations reinforce the capabilities of the organization, while disruptive innovations require new skills and routines. What sustaining and disruptive architectural innovations have in common is that they are both the result of

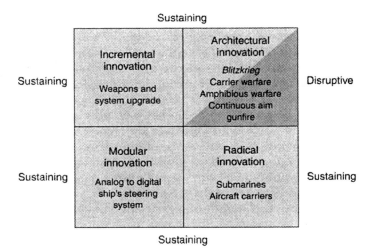

Figure 2.1 A disruptive typology for defining architectural innovations.

seemingly minor improvements in technology. The difference between them is that disruptive architectural innovations can have a disastrous effect on militaries that invent new technology, but fail to adopt and exploit it in a new and unsuspected way.

Figure 2.1 is a different view of architectural innovation typology that shows that most innovations are sustaining in nature and only a few are disruptive. One can infer from this that our senior naval leaders spend most of their time managing sustaining innovations. Although this study is focused on explaining disruptive innovation (Rosen-like innovations), one of its major propositions is that senior naval leaders spend most of their energies managing sustaining innovations (Davis-like architectural, incremental, modular, and radical), which, in turn, has a negative impact on championing Rosen-like disruptive innovations.

Disruptive innovations are unexpected because their value and application are uncertain, according to the criteria used by officers supporting the current methods of warfighting.[17] This means that the disruptive innovation almost always results in outcomes that cannot be fully predicted in advance, making them vulnerable to opponents who support the current way of fighting. Consequently, leaders must manage the results differently than they would a sustaining innovation. Disruptive innovations must be nurtured and protected from those who deem them a threat.

According to Christensen's model, if innovations are sustaining, head-to-head competition between like products continues. If innovations are disruptive, one competitor gains a significant advantage over the other and quickly grabs a bigger market share. This model can be applied to the military where new warfighting methods can be so dramatic that they can be considered disruptive rather than sustaining. In the military, most new technologies are sustaining because they foster improved warfighting performance along a performance trajectory that senior military leaders have historically valued and fits within military doctrine. There is nothing wrong with this as long as both friends and foes travel along a similar trajectory.

When confronted by a disruptive innovation, however, even a historically successful trajectory can collapse in defeat. For example, senior French military leaders during the interwar period prepared for war by building a wall of high technology – the Maginot Line. The French focused on fostering sustaining technological innovations in which they incrementally improved components of an established warfighting system. As the central component of an overall 'methodical' warfighting system, the so-called 'Maginot Line complex' – a technology wonder – worked as designed.[18] Its lethal turrets in the main line of fortifications derailed all German advances in the fighting of May–June 1940.[19] As long as the Germans fought in a predictable manner, the sustaining wall of high technology outperformed the *Blitzkrieg* in holding established positions and territory; where it fell down was in its lack of mobility – a performance dimension that was not highly valued in the decisions that led to the investments that created the Maginot Line. As the French painfully learned, employing new technologies in old ways of fighting are helpful, but seldom decisive. Rather, existing technologies employed in new disruptive ways of fighting are by comparison generally more effective.

Disruptive innovations are rare. By disruptive, they have features that only a few senior leaders value, and thus most senior leaders oppose disruptive innovations.

The irony is that militaries do not succumb to disruptive innovations because of a lack of foresight, but a lack of insight. They underestimate the role the new disruptive innovation may play in future conflicts and choose to focus on developing sustaining innovations.

For example, when the British first introduced the tank in 1917 and 1918, it was 'a slow, difficult-to-maneuver weapon of war', since it offered its crew 'minimum vision, maximum discomfort, and general mechanical unreliability'.[20] Williamson Murray writes that whatever its promise, 'the performance of those ungainly vehicles [tanks] in World War I was spotty'.[21] German senior military leaders agreed with their British counterparts that the tank was a weapon designed for one simple task: 'crossing the killing zone between trench lines and breaking into enemy defenses.'[22] In fact, by 1936 one German commentator declared that 'antitank weapons had made the tank a technology fad, that would suffer the same fate as the horse on the battlefield.'[23]

Typically, a disruptive innovation is one that appears to sneak onto the battlefield – because senior military leaders failed to recognize the threat it posed – then it outperforms the established way of fighting and defeats the unsuspecting military. Take the machine gun for example. An American invention, General George Armstrong Custer's 7th Cavalry possessed four Gatling guns, but did not feel they possessed tactical value and left them in garrison prior to the Little Big Horn campaign. The machine gun's disruptive potential was demonstrated in 1914 when the Germans employed them in an integrated fashion from a dug-in position to stop the Allied advance near the river Aisne. This event marked the beginnings of World War I trench warfare.[24] The machine gun is a classic case of a technology-driven innovation adopted and fully exploited first by someone other than the military that invented it.

The utility of disruptive innovation is frequently controversial and in doubt until the moment it is proven in battle. Initially, senior leaders reject the disruptive innovation because they cannot envision how the change will be used. An example occurred in 1936 when 'German doubters questioned whether mechanized formations could make the deep penetrations that advocates like Guderian claimed . . .'.[25] This doubt remained for the next five years as Williamson Murray states, 'there was considerable skepticism about the potential of panzer units up to the 1939 Polish campaign'.[26] 'It was not until Poland that the [German] officer corps as a whole began to grasp the potential of armored exploitation on the operational level of war.'[27] But even so, as Murray notes, 'on the eve of the 1940 [French] campaign, few of even those German officers involved in development of armored warfare during the interwar period had a firm belief that their efforts would transform land warfare'.[28]

Civilian intervention

Viewing architectural innovations in terms of warfighting performance explains why civilian intervention and military mavericks may produce a sustaining innovation but not a disruptive one. Civilian intervention may spawn a sustaining innovation because it will improve warfighting along a dimension of performance valued by the traditional senior military leaders. Ironically, there are several examples of the military being forced to buy new technologies that extend the current way of fighting well beyond the warfighting performance metrics desired by military leaders. For example, in 1994 the US Navy wanted to kill the new Sea Wolf submarine program because this oversized attack submarine was designed to fight the Soviet Navy that did not exist anymore. Civilian leadership, however, forced the Navy to build two before the Navy could cancel the program. Civilian intervention, however, cannot cause disruptive innovation because traditionalists do not value the new way of warfighting. Favoring the old way of fighting, traditionalists will 'log roll' any civilian efforts to force a new way of fighting that is not supported by the military culture.[29]

Summary

Architectural innovations can be either disruptive or sustaining in nature, but all disruptive innovations are architectural. Sustaining innovations are those that resulted in an improved performance along a trajectory that traditionally has been valued. Disruptive innovations are those that resulted in an improved performance along a warfighting trajectory that traditionally had not been valued. Disruptive innovation is more likely to fail when product champions use a sustaining managing approach. Disruptive innovation introduces a new warfighting metric that is more likely to under-perform initially the established approach, but when fully exploited is more likely to provide a significant battlefield advantage. The historical evidence supports the proposition that civilian intervention and maverick officers can cause sustaining innovations, but they do not cause disruptive innovations.

③ *Performance trajectory overshoot: sustaining innovation*

Two central propositions of the disruptive innovation framework are as follows: first, the competitiveness of different warfighting approaches can change with respect to combat over time. Second, the pace of sustaining technological progress continues well beyond the effectiveness of that approach to warfare. An illustration of both propositions is when linear warfare in the form of the Maginot Line technology surpassed the requirements of methodical warfare relative to the introduction of tanks and *Blitzkrieg* warfare. This is also an example of a nascent disruptive innovation – *Blitzkrieg* warfare – that initially under-performed linear warfare in head-to-head conflict, but fully outperformed fixed linear warfare and its installations in subsequent engagements (see Figure 2.2).

Another illustration of sustaining innovation overshoot is the US Navy's obsession of building larger battleships with bigger guns during the interwar period (see Figure 2.3). Although the British had introduced carrier aviation during World War I, the focus of the American battleship admirals was on how best to exploit new technology that would result in heavier guns with longer ranges. By

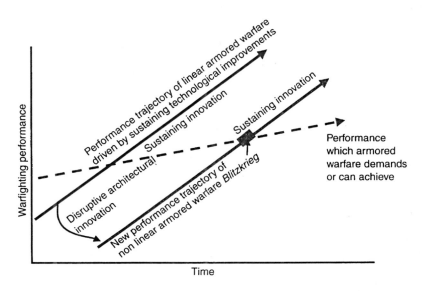

Figure 2.2 Disruptive innovation: armored warfare.

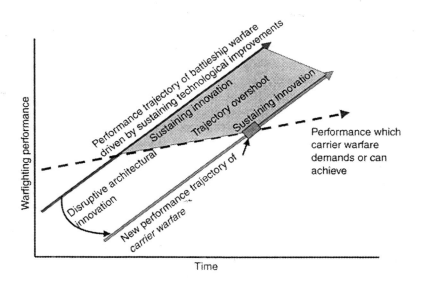

Figure 2.3 Sustaining innovation: performance trajectory overshoot battleship warfare.

the start of World War II, battleship warfare surpassed the requirements of a great fleet engagement between surface forces (as at Jutland) relative to the introduction of aircraft and carrier warfare. These events illustrate how an inchoate disruptive innovation – carrier warfare – that initially under-performed surface fleet dueling engagements, fully outperformed battleships in subsequent engagements.

Small innovation groups: disruptive innovation

Product champions, who recognize they must create a new way of war, face a major challenge. They need to build and apply new architectural knowledge effectively. Simply recognizing that a new innovation is disruptive in character does not give the organization the architecture knowledge it needs. Product champions must switch to a new mode of learning where new linkages between components can be developed. According to one hypothesis of disruptive innovation, product champions are likely to be more successful when they create small innovation groups. Because these groups are tasked with looking at new warfighting paradigms, they are more likely to generate new linkages that will define a new architecture. A related hypothesis posits the more time and resources that product champions invest in learning about the new architecture using small innovation groups the more likely the disruptive innovation will lead to a new way of war.

Clearly, how architectural and component knowledge is managed makes an enormous difference. Senior military leaders prefer, and are more likely to manage successfully, sustaining innovation. They prefer sustaining innovations because they occur in a stable architecture – an architecture that is embedded in the organization's culture and the one in which they have been successful.[30] In contrast, disruptive innovation requires a different management style than sustaining innovation. Architectural knowledge requires explicit management by senior military leaders because disruptive innovation places a premium on exploration of new linkages among components and the assimilation of new knowledge. The process of rooting out old information and creating new knowledge usually takes considerable time and is usually done through a process of trial and error. New architectures, quite simply, create different interactions between components and require innovators to build new knowledge about them.[31]

Disruptive innovation theory posits that this process is initially performed in a separate organization and is directly managed by a product champion in order to generate new linkages leading to a new way of war. Evidence supporting such a claim is provided by a study conducted by the Naval War College that concluded, 'the best way to foster innovation in a large bureaucracy is to create enclaves that can operate as small organizations'.[32] By directly managing the small innovation group, the product champion can place a premium on exploration in architectural design and the assimilation of new knowledge. New knowledge comes from wargames, simulations, modeling, and fleet experiments.

Based on considerable evidence, the study concludes that disruptive innovations must be managed differently than sustaining innovations. The danger of managing a disruptive innovation incrementally is that those in charge will underestimate its potential impact and will fail to exploit it. Sustaining innovations are generally managed by a hierarchical chain of command using centralized control and formal structures. The result is an efficient and evolutionary process. In contrast, disruptive product champions often bypass the chain of command, use informal structures and decentralized control. This arrangement inevitably leads to conflict and dissent between tradition-bound organizational units and risk-taking disruptive groups.[33] Because the power and resources of organizations are usually anchored in the sustaining innovation units, they try to ignore or subvert the disruptive groups. Thus, product champions must not only protect the disruptive groups, but also keep them separate from the main bureaucratic organization.

Summary

Because a disruptive innovation has the potential to destroy the way the organization conducts warfighting, it must be incubated in an independent organization such as a small innovation group. When product champions create and directly manage small innovation groups, the more likely new linkages will emerge that will eventually define a disruptive doctrine. The greater the visibility of the small innovation group, the greater the opposition to it from traditionalist practicing the old way of war. The more closely the product champion manages an innovation groups' outcome, the less likely opponents will be able to inhibit the development of a disruptive innovation.

(5) *Disguising new ways of war: disruptive innovation*

A most important discovery during the historical analysis of the armor warfare case study was the disguising factor. In James Q. Wilson's *Bureaucracy: What Government Agencies Do and Why They Do It*, he observed that product champions of disruptive innovation require the skill of misdirection.[34] Elaborating on Wilson's misdirection theme, this study discovered that product champions advanced their disruptive innovation by disguising it as a sustaining innovation so that it appeared merely to enhance the capabilities of the old way of fighting. In this way, senior leaders, under the cover of sustaining innovation, could promote and fully develop the new form of warfighting and set the stage for the political process.

The fact that senior military leaders generally view all innovations as sustaining is what makes disruptive innovations so devastating and unexpected. It is also the reason that disruptive product champions can successfully disguise their innovation as sustaining because the traditionalists will typically only support those efforts that improve the warfighting performance in a trajectory valued by them. By disguising, the study does not imply that experiments and fleet were exercises were held in secret. Instead, the product champion couched his efforts as sustaining the old way of fighting, which is what most military leaders spend most of their time doing.

Summary

Disruptive innovation must also have a product champion who can promote it as sustaining in nature. The greater the product champions can disguise or shape the disruptive innovation as a sustaining innovation, the greater the possibility the

disruptive innovation will survive. The chance that a disruptive innovation will survive if advertised as a new way of war is slim to none. Such an assertion has broad implications for product champions in securing resources. *budget*
fighting

Armored warfare development and disruptive innovation theory

In the sections that follow, the creation of German armored warfare, know tactically as the *Blitzkrieg*, will be examined and compared with British experiences in developing conceptions of armored warfare in the 1920s and 1930s. It is a story that suggests that reformers who followed the propositions of disruptive innovation were successful in creating a new way of war, and those who did not were not successful or simply failed to recognize the possibilities of an armored warfare disruptive innovation. It is also a story that shows that disruptive innovation is complex and can take many years to implement. In the German case it consists of General Hans von Seeckt championing disruptive maneuver warfare and combined-arms doctrine in the early 1920s. Building upon this disruptive doctrine was a sustaining innovation effort in the 1930s championed by General Ludwig Beck, chief of the general staff and the army's commander, General Werner von Fritsch, who provided the support and resources for the new form of warfare, and Heinz Guderian who lead the development of the Panzer Force. The important element about this story is that the sustaining innovation effort lead by the German armor advocates remained fully within the disruptive combined-arms framework that was championed by von Seeckt. The story of German armored warfare is a good example of how a disruptive innovation initially under performs the old way of fighting, but after an evolutionary development period of sustaining innovation it eventually exceeds or disrupts the warfighting performance of the old way of fighting. It is also an excellent case to illustrate the different ways senior military leaders champion disruptive and sustaining innovations.

British armored warfare development

At the end of World War I, the British introduced the tank. Unlike the large mobile German Panzer tanks of World War II, it was a slow, difficult-to-maneuver armored fighting vehicle. Nevertheless, during the October 1917 Battle of Cambrai the British conducted the first large-scale attempt to use a surprise tank attack to break through the first-line German positions without conducting a prolonged preliminary artillery bombardment. After seizing a few kilometers of German entrenchments, the British declared a tactical victory, which resulted in church bells in England being rung to celebrate the great triumph.[35] Within a few days, however, the Germans mounted a counterattack that did not use tanks that not only erased the British gains, but also drove the British infantry out of half of its own positions.[36]

Extrapolating from the tank lessons of Cambrai, J.F.C. Fuller, a British staff officer with the Tank Corps, created 'Plan 1919' a combined-arms scheme that called for the use of a new faster tank that could penetrate deep into the enemies rear lines. The British, however, would wait until 1932 before they began formally to examine the lessons and theories of tank warfare. Rather than conducting a critical analysis of the army's performance, the studies presented the army's performance in the most favorable light.[37] Despite these exaggerated reports, the British experiments with armored warfare between 1926 and 1934 continued in the tradition of Fuller's 'Plan 1919'.

In 1934, the tank maneuvers represented the capstone for development of the armored warfare. During these maneuvers, innovators introduced the radio and immediately recognized its importance for deep penetration raids. After-exercise reports commented on the flexibility, mobility, and firepower of armored forces, which consistently proved superior to those of more conventional forces.[38] Interestingly, a senior infantry general named Burnett-Stuart purposely designed a complex and difficult set of tactical problems as a final test for Brigadier General George Lindsay's armored forces. Afterwards, Burnett-Stuart claimed that the armored forces failed to display an advantage over conventional forces. In fact, 'many senior officers came away from the exercises with the impression that the tank arm had not lived up to the expectations of its advocates'.[39] Liddel Hart, a military pundit, immediately took Burnett-Stuart and the senior officers severely to task for misrepresenting the potential of armored forces, but the senior officers prevailed.

Ironically, Burnett-Stuart considered himself a friend to the mechanization cause. He felt, however, that the armor force needed a robust test that stressed executing logistics and difficult tasks. By design, the exercise was exceptionally difficult for the armored forces, but as any good commander should do, Burnett-Stuart wanted to see his forces perform in unfavorable circumstances in order to improve their overall capabilities. The unintended consequence was the negative impact on Lindsay's championing effort of the combined-arms concept of armored warfare. Lindsay was the most sophisticated of the Army's tank enthusiasts and was a proponent of the 'all arms concept' that was a balanced force construct, which could penetrate deeply into the enemy's rear areas to disrupt its command and logistics network. The combined armored warfare advocates never recovered from this setback, as the War Office between 1934 and 1939 did almost nothing to build on the framework of earlier experiments with tanks.[40]

After the 1934 maneuvers, the British lacked a coherent effort to pursue armor's potential. Instead of developing a separate track for armored warfare, senior military leaders tied the progress of armored development to the infantry and cavalry establishments. The cavalry arm was the most resistant to new ideas and to the tank. Field Marshal Haig's quote reflected the views of the retired officer community when he stated, 'I feel sure that as time goes on you will find just as much use for the horse – the well-bred horse – as you have ever done in the past.' The commander of the cavalry, General Sir Alexander Godley, echoed Haig when he declared, 'On the other hand, if I were asked, "Will you go to war with a mobile force composed [of] armored cars, tanks, and such-like?" I think I should refuse to go! I should say that I would not go without a force of cavalry. I should want, and should insist on having, an ample portion of mounted troops.'[41]

Perhaps the most egregious error occurred during Field Marshal Alan Brooke's leadership throughout World War II. 'In particular, he insured that *none* of innovators in armored warfare reached senior positions – division level and above – of armored forces committed to battle. Consequently, the British Army lost the hard-earned lessons of the 1920s and 1930s for almost the entire course of World War II.'[42]

Small group proposition

The consequences of not having senior military leaders serve as champions of disruptive innovation is demonstrated in the woeful British post-war efforts to develop their armor tactics. The British War Office in 1920 assigned the task of rewriting the infantry tactical manual to Basil H. Liddell Hart, a twenty-four-year-old lieutenant.[43] Although Hart introduced some innovative maneuver concepts, the War Office simply deleted these chapters from the 1920 infantry manual.[44]

It is apparent, therefore, that disruptive champions should be senior enough to create and directly manage small innovation teams to moderate the danger of allowing old architectural knowledge to remain embedded within the new architectural linkages. Creating a small innovation group is not sufficient to forge new architectural linkages among components. Generally speaking, the tendency of small innovation groups is to underestimate the impact of the innovation on embedded architectural knowledge because they are handicapped by a legacy of partially irrelevant architectural knowledge.[45] As Henderson and Clark point out, 'Since the core concepts of the design remain untouched, the organization may mistakenly believe that it understands the new technology.'[46] Tremendous effort is required to avoid relying on architectural knowledge derived from experience, which tends to blind the small innovation group to critical aspects of the new architectural innovation. This effort should come from the product champion who 'protects' his vision. Partial blinding will inevitably lead to an underestimation of the innovation's potential as demonstrated by the British underestimation of armor maneuver warfare. In sum, innovation champions must provide direct oversight of the small innovation groups in order to create a disruptive innovation.

In 1926, British Lieutenant Colonel Frederick Pile carried out a stunning 25-mile tank maneuver that brought an entire exercise to a halt. Two years later, similar results were received with an experimental armored force during armored maneuvers. Instead of supporting the development of new linkages, the general in charge 'emphasized the negative impact that the new formations were having on the traditional branches by their success in exercises; therefore, he argued that it was wrong to create a force equipped with new armament. Rather, he suggested, mechanization and motorization should take place throughout the whole army'.[47]

As Williamson Murray notes, 'Despite such inhibiting factors, the British experiments with armored warfare between 1926 and 1934 contributed to a considerable extent to the creation of the German panzer forces after Hitler came to power.'[48] Under the direction of senior military leaders, such as General Ludwig Beck, chief of the general staff, innovators were allowed to conduct extensive linkage experiments among components. Beck, in fact, circulated an extensive lessons learned document about British tank maneuver warfare.[49] Evidence suggests that German military leaders learned more in the long run from these experiments than did the British, since they watched the exercises with great interest and disseminated the results widely.[50] As a result, the German officers understood the principles of mobile, armored war long before they received their first tanks.[51] German small innovation groups, which were manned with the best people and directly supervised by the product champion, appeared to be the critical difference between the Germans and the British. Ironically, his seniors ignored Liddell Hart, the junior officer the British selected to update doctrine manuals, but the Germans did not overlook his writings.[52] In fact, Murray argues that Hart's forward thinking about the mobile tank exercises 'formed much of the basis of the German Army's analysis of British experiments with armor'.[53] General Oswald Lutz, the chief of armored development in Germany until 1938, told Sir John Dill during the latter's visit to Germany in 1935 'with considerable pride that the German tank corps had been modeled on the British [armored experiments]'.[54] Murray echoes this assessment by stating, 'the British experiments with armored warfare during the interwar years contributed to a considerable extent to the creation of the German panzer force'.[55]

The evidence of Lieutenant Basil Hart introducing many innovative maneuver concepts that were ignored by his superiors supports the proposition that disruptive champions should be senior leaders. The advantage of being a senior leader champion is that they can form small innovation groups to protect and incubate an inchoate disruptive innovation, which has the potential to destroy the way the organization conducts warfighting.

Incumbent technology failure

The belief that militaries underestimate the disastrous effects of disruptive innovations constructed with modest technological change lies at the heart of disruptive innovation theory. According to this view, militaries that introduce new technologies into warfighting are more likely to be totally shocked when confronted by competitors who use these technologies in a disruptive way.

Disruptive theory provides us with another view of why disruptive innovation is so difficult to recognize.[56] Senior military leaders may identify a disruptive innovation or detect another military doing so, but they may fail to act because they underestimate the innovation's impact on warfighting. A disruptive innovation can at first appear to be nothing more than a sustaining innovation, but not necessarily a better, way of performing traditional missions. In other words, the embedded knowledge of the existing doctrine serves as a powerful lens for viewing new architectural arrangements. Such a lens may filter out the new linkages leaving the old linkages with which the organization is familiar.

The belief that disruptive innovation can be accommodated within old frameworks helps explain why someone other than the military inventing the new design often adopts and fully exploits it as a disruptive innovation. For example, the British invented the tank and successfully used it in the stunning Allied victory of 8 August 1918, and they understood the need to continue to develop tank expertise following the war. After the war, senior military leaders failed to recognize any need for disruptive architectural changes in the ways in which tanks and other core components were linked. As Stephen Rosen argues, 'By 1918, the British army was in a position to learn how to use tanks against a reactive enemy and had developed all of the intellectual bases for blitzkrieg warfare.'[57] Yet as Williamson Murray notes, 'the British Army's deep-rooted regimental system and disdain for the serious examination of the last war [World War I] certainly stacked the odds against development and adoption of mobile armor warfare'.[58] Put simply, a continued reliance on the infantry-artillery paradigm derived from the World War I experience blinded the British to critical aspects of armored warfare.

While British engineers were able to push the limits of the component technology (building better tanks), they had great difficulty emulating what the Germans accomplished in their development of armored warfare. Several interesting implications emerge. Factors that emphasize component knowledge over architecture knowledge tend to increase the rigidity of linkages between components. In the British case, they expended their primary resources constructing better tanks, whereas they expended few resources examining new linkages. One of the primary venues for exploring new architectural linkages is the willingness to experiment with new concepts and ideas in annual maneuvers, experiments, and wargames.[59]

Paradoxically, one of the few key individuals capable of contributing to disruptive innovation was J.F.C. Fuller, Britain's product champion for armored warfare. Fuller, however, refused to take command of the experimental tank unit because his conditions for taking command were not met.[60] Consequently, without a product champion to revise component linkages despite their successful experiments with mobile tank warfare, British military leaders underestimated maneuver warfare's potential impact and failed to see how it would give them a decisive advantage.[61] This failure made the German exploitation of armored warfare all the more devastating.[62]

Because the German's tank experience occurred later and was much more limited than that of the British, they were more open to exploring new uses for the tank. Negative experiences, such as the lack of effective communication that resulted in the loss of two tanks to friendly artillery fire, convinced them there was a better way to conduct tank warfare.[63] If these experiences were insufficient to break old linkages, the Treaty of Versailles, which banned tanks in the German army, made sure that the Germans were going to have a clean slate once they did begin thinking about tank warfare. The Germans began by using tank mockups to test a new doctrine that emphasized maneuver, speed, and combined arms.[64] The key difference for the Germans was that, compared with the British, they were new entrants to tank warfare. While the British emphasized building component knowledge, the German's product champion of mobile warfare, Hans von Seeckt, focused on building architectural knowledge. Under the direction of Heinz Guderian, one of the product champions of armored warfare, senior military leaders approved a wide array of experiments to evaluate the possibilities of the tank's use in mobile warfare.

What the British case shows is that some of what an organization knows is not only useful, but may actually handicap product champions. Disruptive theory posits that recognizing what is useful and what is not, and acquiring and applying new knowledge when necessary, is more difficult for the organization that invented the technology because it typically assimilates it as sustaining innovation. Hence, others may exploit the innovation's potential much more effectively, since they are not handicapped by a legacy of partially irrelevant architecture knowledge.

A central assertion for both incumbents and new entrants is that a considerable amount of time is necessary to flush out the subtleties in a disruptive innovation. Results from wargaming and exercises that might warn the organization that a particular innovation is disruptive may be screened out by senior military leaders who mistakenly believe that they understand how the new technology should be employed in a sustaining way.

French incumbent technology failure

The French faced a similar situation in that proponents of the old method of fighting prevailed. They were not men opposed to change. Quite the contrary, they embraced new technology and tried to incorporate it into their doctrine. Using the Maginot Line again, it is an excellent illustration of the incumbent technology failure proposition. Rather than couching the Maginot Line as notorious metaphor for bungling, we should instead attempt to understand it as a sustaining innovation of ingenious engineering and technological accomplishment. While French engineers were able to push the limits of component technology – developing impregnable fortresses, as well tanks that were superior in protection and armament to the Germans – a reliance on their old frameworks blinded them to critical aspects of maneuver warfare.[65] Consequently, they underestimated maneuver warfare's disruptive potential and failed to see how new interactions in component development would give them a decisive advantage. This disruptive failure by the French was one of the factors that led to the German to exploit armored warfare.[66]

As one of several sustaining technology investments in the interwar years, the Maginot Line's shortcomings derived not from execution, but from the inability of its proponents to anticipate the advent of a disruptive innovation – mobile armor warfare.

The British and French are examples of why incumbents of new technology usually do not champion a disruptive innovation because they have a difficult time recognizing the new technology linked in a novel way.

Disguising proposition

In contrast to the German case, British advocates of armored warfare directly attacked senior leaders for not transitioning to a new way of war. 'Both Fuller and his compatriot Liddell Hart launched increasingly vitriolic attacks on the British Army's leadership in the 1930s. Neither Fuller nor Hart attempted to disguise armored warfare as a sustaining innovation. The consequence was that they exercised a decreasing influence over the thought processes of the army itself.'[67] Williamson Murray continues, 'To a great extent, the result of their strident advocacy served to exacerbate the split between innovators (by far the smaller group) and the great mass of professional soldiers and assured their ideas played a decreasing role in the preparation of British ground forces for war.'[68]

German armored warfare development

The popular account that most scholars have come to believe is that the German success in armored warfare was the result of the Germans, reacting to defeat in World War I, developed a revolutionary approach to war, one that emphasized maneuver and armored war as a means to escape the strategic stalemate of trench warfare. Political scientist Barry Posen refined this argument by incorrectly giving primary credit for the *Blitzkrieg* innovation to Hitler, but he also acknowledged the role of Heinz Guderian, who acted as a military 'maverick' by providing Hitler with military expertise that he lacked.

Few military historians credit this explanation any more as the picture that has evolved since Posen's 1986 *The Sources of Military Doctrine* has substantially altered the traditional view. German military historians including Williamson Murray, James Corum, and Robert Citino have specifically responded to Posen's civilian intervention analysis and have collectively offered an alternative explanation that armored warfare was an evolutionary, rather than revolutionary, development based upon a combined-arms concept that began during World War I and was codified by Hans von Seeckt immediately after the end of the war.[69] As an illustration of these views, Williamson Murray writes, '... the Fuhrer [Hitler] played little or no role in the development of army doctrine in the tactical and operational fields'. Murray does argue, however, that Hitler did provide funds to provide the military forces. What is striking in the current literature is the absence (or mention) of Hitler's influence on the innovation process.[70]

Civilian intervention

The findings from this study disagree with Barry Posen's findings. Posen characterizes von Seeckt's post-World War I innovation efforts as sustaining and the subsequent innovation efforts of Heinz Guderian buoyed by Adolf Hitler as a disruptive. Posen contends that von Seeckt's innovations were sustaining because his reforms were evolutionary and not revolutionary. Posen states, 'The postwar Reichswehr [lead by von Seeckt] did not pioneer either the weapons or the principles of *Blitzkrieg*.'[71] Posen elaborates on this observation by pointing out that he finds himself 'hard put' to call the measures introduced in the early 1920s by the commander of the German Army, General Hans von Seeckt, as an innovation. Instead, he considers von Seeckt a 'reformer' not an 'innovator', because Seeckt preferred to superimpose some new tactics and technology on old doctrine, rather than innovate doctrinally to exploit more fully all the offensive potential of the new technology.[72] Although inventive, Posen argues that von Seeckt's focus on combining motorized transport rather than armored fighting vehicles with infiltration doctrine was not a disruptive innovation.[73]

Instead, Posen argues, incorrectly, that Hitler was the primary creator of *Blitzkrieg* military doctrine.[74] The agent of Hitler's disruptive innovation was the military maverick Heinz Guderian who was a proponent of daring, high-speed, deep armored thrusts.[75] Posen bases this analysis on the observation that Guderian's armored warfare concept was opposed within the German Army. Furthermore, Hitler's emphasis on the newest weaponry was also resisted by senior military leaders.[76] Posen argues that Hitler may not have fully conceptualized the *Blitzkrieg*, but Guderian did. Once Hitler understood Guderian's armored warfare concept, it was Hitler's intervention and unabashed support for the *Blitzkrieg* 'that brought the doctrine to operational fruition'.[77]

In contrast, this study of the origins of the *Blitzkrieg* purports that instead of viewing Guderian's efforts as disruptive and von Seeckt's as sustaining, the opposite is true. That is, Guderian's championing of the new Panzer divisions was merely a sustaining innovation of the disruptive innovation principles that von Seeckt had developed.[78] From this perspective, von Seeckt championed a disruptive maneuver warfare doctrine that emphasized combined arms and independent action by commanding officers at all levels.[79] This disruptive doctrine was spelled out in Army Regulation 487, *Leadership and Battle with Combined Arms*, and largely institutionalized by 1926 when von Seeckt retired.[80]

Over the next decade General Ludwig Beck, General Freiherr Werner von Fritsch, and Guderian worked to sustain von Seeckt's disruptive doctrine by championing experiments and new technologies such as tanks and tactical bombing aircraft. Although Guderian faced 'considerable skepticism among senior officers' about his vision of armor formations conducting deep penetrations of enemy positions, his sustaining efforts eventually culminated in a Panzer Force that could execute the disruptive maneuver warfare doctrine championed by von Seeckt, which together is known as armored warfare. As with many sustaining innovations, the critical support of civilian leaders, most notably Hitler, contributed to achieving the innovation. This is not meant to detract from Guderian's role, but rather to underline the tenet that sustaining innovations can be as almost difficult to achieve as disruptive innovations. Civilian intervention, however, did not have an impact on influence von Seeckt's successful effort in championing the disruptive principles of armored warfare.

Early development of German armored warfare:
small group proposition

The selection of von Seeckt as the commander-in-chief of the army following the war was a key factor in creating the *Blitzkrieg*. Seeckt was the product champion of mobile warfare. During his tenure he produced the first edition of *Leadership and*

Battle with Combined Arms in 1923. By 1933 the Germans possessed a military doctrine that fully took into account the lessons of the last war, which included the principles that all officers were expected to display – initiative, exploitation, and maneuver. This doctrine spelled out the conceptual framework that determined how the organization would fight and within this conceptual framework the development of the German Panzer Force took place.[81] It was a doctrine of mobile warfare – a tactical system of maneuver and close cooperation among all arms, which in today's vernacular is called combined arms warfare.[82]

Von Seeckt championed mobile warfare by creating several small innovation groups immediately following World War I to collect and analyze the war experiences and create a new body of military doctrine.[83] Seeckt appointed 109 officers and former officers to serve on a total of 57 architectural groups that he directly managed and the result was the 'extraordinary Army Regulation 487 *Leadership and Battle with Combined Arms* 1921, and 1923'. This doctrine served as the framework within which the development of the German Panzer Force took place.[84] Williamson Murray considers *Leadership and Battle* the key factor in the development of the German *Blitzkrieg*.[85]

The challenge for small innovation groups is to build new architectural knowledge (doctrine) without eliminating relevant old architectural knowledge.[86] Given the tendency of architectural knowledge to become embedded in the organization, separating the relevant from the irrelevant is like separating rocks from mortar. Consequently, this requires explicit senior management and attention.[87] Von Seeckt accomplished this by issuing specific directives and questions for each group and by seeking those officers with the best reputations to serve on the groups.[88]

Disguising proposition: maneuver warfare

This study argues that disguising a disruptive innovation eases its acceptance by the organization. This is the case with von Seeckt who disguised mobile warfare. After reporting to the Allies that the German General Staff had been disbanded, Seeckt worked to maintain a disguised German General Staff in defiance of the terms of the Treaty of Versailles. He also made secret arrangements with Soviet Russia that allowed the construction of a tank school at Kazan and a flying school at Lipetsk.[89]

As noted earlier in this chapter, von Seeckt also created small innovation groups that lead to the disruptive innovation of mobile warfare. In creating mobile warfare he used dissimulation. He disguised his efforts to create mobile warfare from the Allies. Because the disarmament provisions of the peace prevented tanks, warplanes, and armored vehicles von Seeckt disguised his continued use of these tools to create mobile warfare. Believing that superior mobility would enable the army to wage offensive warfare, von Seeckt in October 1921 championed maneuvers of motorized units in the Hartz Mountains. This exercise was followed by several secretive experiments involving dummy tanks made of cardboard mounted on automobile chassis. By 1923–24 von Seeckt began experimenting with a new doctrine linkage between motorized ground forces and air forces.[90]

Sustaining innovation of armored warfare:
Panzer Force development

As discussed, Seeckt was the champion of the disruptive mobile warfare innovation. The next task for the reformers was to sustain the disruptive efforts of Seeckt and they did this by closely following the British experiments with tanks. The Germans read the British press reports quite carefully regarding British tank maneuvers and the Germans concluded that tanks 'be allowed to break through repeatedly' to attack the enemy's command post and rear guard units. From the British they developed their own conceptions of penetration and exploitation.[91] In fact, Guderian admitted in his memoirs that the Germans had translated the current British field manual on the employment of armored fighting vehicles and used it as the basic primer for developing armored warfare.[92]

After several years of using von Seeckt's disruptive innovation to conduct experiments in how armored war should be waged, German reform leaders faced the second major challenge of disruptive innovation: the need to apply the architectural knowledge – the linking of existing components, the tank, airplanes, mobile troops carriers – throughout the army.[93] Simply recognizing that von Seeckt's innovation was disruptive in character does not provide the architectural knowledge that was needed to achieve change.[94] Von Seeckt's disruptive combined-arms doctrine emphasized speed, surprise, mobility, decentralization, and exploitation for the entire army, not just for tank doctrine. Although challenging, the transition from mobile infantry to mobile tank tactics did not represent an insurmountable hurdle.[95]

In 1932, Generals von Fritsch and Beck, the army's future commander-in-chief and chief of staff, reworked von Seeckt's 1921 *Leadership and Combat of Combined-Arms Forces* in a manual called *Die Trüppenführung* that effectively crystallized the entire theoretical groundwork for the *Blitzkrieg*. Fritsch and Beck's sustaining innovation guided the use of the tank in the attack and pursuit as a combined-arms weapon, not as an individual wonder weapon.[96] As significant as what Fritsch and Beck wrote about tanks and the emphasis on decentralized tactics and rapid exploitation, the German army did not yet possess a single such machine. Clearly, German armored tactics were simply the conscious adaptation of tanks and radios to a disruptive conceptual framework developed by von Seeckt and sustained by Beck and Fritsch.[97]

One of the key sustaining innovators in championing the *Blitzkrieg* was Heinz Guderian, the tank pioneer and proponent of armored warfare. Posen concludes, again incorrectly, that Guderian developed a tactical doctrine at odds with the opponents of armored warfare who, in turn, embraced von Seeckt's mobile warfare doctrine. On close inspection, Guderian was one of the champions of armored warfare, along with Beck and Fritsch.[98] Armor warfare was a sustaining innovation because Guderian, Beck, and Fritsch were operating within the bounds of mobile warfare that von Seeckt had established. These reformers successfully achieved a sustaining innovation by transforming the tank from a support weapon for the infantry to the main weapon of the attack.

Many German officers opposed this transformation, especially many of the infantry and cavalry officers. Yet, the German military leaders did not consider

Guderian a radical who was pushing some revolutionary development. They all accepted von Seeckt's disruptive doctrine of mobile warfare and combined arms. At issue was the role the tank was to play. The infantry was very much inclined to view the tank as a support weapon for the infantry. As a signals officer during World War I, Guderian had the credentials that infantry officers would respect. In the beginning, Guderian endorsed the tank as a support weapon for the infantry, but he quietly sought support from von Fritsch and General Beck who provided resources and protected and promoted Guderian's tank innovations.[99] Beck and Fritsch willingly allowed Guderian to experiment and allotted considerable resources to von Seeckt's relatively unproven disruptive doctrine.

There is little question that Hitler's support in the late 1930s eventually altered the balance of resources in favor of the tools of armored warfare such as tanks, armored vehicles, and bomber planes. The lack of new technologies, however, was not the fault of senior military leaders. From 1918 on, German military leaders had been interested in developing and building the technologies they needed to acquire to defend themselves. Preventing them from doing so was the Treaty of Versailles, which prohibited Germany from owning tanks, armored vehicles, and warplanes. Thus, rather than opposing Hitler's order to ignore the Treaty of Versailles and rearm, senior military leaders welcomed it.

The important point to note is that real tanks replaced dummy canvas tanks when they became available and they were incorporated successfully and rapidly into German Panzer units. This occurred not because of Hitler's intervention, but because von Seeckt had laid a sound and intellectual mobile warfare foundation for the use of this new technology. As two historians have recently noted, 'Thus, in the 1930s German officers understood the principles of mobile, armored war long before they received their first tanks ... In 1935 Beck conducted a general staff tour on how a panzer division might be employed, and by the next year the general staff was examining the potential of a panzer army.'[100] Thus, the new Panzer divisions represented the sustaining innovation efforts of Guderian to extend to the tank and armored vehicles the disruptive principles of mobile warfare von Seeckt had championed. Thus disruptive innovation efforts (by von Seeckt), not intervention by civilians or military 'mavericks', followed by steady sustaining innovation development (by Guderian, Beck, and Fritsch) with some support from within the military (with some outside support from Hitler), explains the *Blitzkrieg* disruptive innovation.

In sum, Williamson Murray writes, 'Hitler played little role in development of the panzer forces except to make a few favorable remarks to Guderian in 1934.'[101] This study has no disagreement with Posen's analysis that Hitler had extraordinary goals, which were nothing less than the subjugation of the entire continent. But as Murray writes, Hitler's goals 'did not lead to a demand for some specialized new form of warfare'.[102] Instead, it was von Seeckt's influence in promoting mobile warfare and Guderian's efforts in introducing the tank into mobile warfare that allowed the Germans to innovate in such devastating fashion.

There is strong evidence that Hitler's intervention and support of Guderian in 1938 lead to the General's selection as the first commander of both Panzer troops (motorized force) and Mobile troops (cavalry force). Guderian initially refused the promotion, however, as he regarded Hitler's 'suggested innovation as a step in the wrong direction'.[103] Guderian explained to Hitler that coupling tanks with the cavalry was not a prudent idea. In fact, Guderian makes the argument that such a move appeared to support the theory of the tank as an infantry support weapon, rather than the tank being the main weapon.[104] Hitler listened patiently to Guderian's argument for about 20 minutes and then ordered Guderian to take command of both tank and cavalry arms. If anything, Hitler actually retarded the development of *Blitzkrieg* warfare as Guderian complained after the war that he should have been focusing on solving the logistics problems of the *Blitzkrieg* rather than commanding the cavalry.

Disguising armored warfare

In the 1930s Guderian played a central role in the development of the Panzer Force. He was not, however, the sole creator of armored forces and doctrine.[105] Armored warfare development in the interwar period was largely championed by a group of officers that included Guderian. Few would quarrel with this characterization, but some would question if they used dissimulation in promoting armored warfare. This study argues that disguising is a prerequisite for disruptive innovation, but not necessarily so for a sustaining one. There is not a sharp demarcation line where von Seeckt's disruptive mobile warfare innovation and the armor champion's sustaining innovation begins. Part I of von Seeckt's *Leadership and Battle with Combined Arms* (Army Regulation 487) appeared in 1921 and Part II in 1923. Institutionalizing mobile warfare doctrine was an evolutionary process that took some time. Many of the armor champion's effort took place shortly after the mobile warfare doctrine was published, which meant that some of the senior leaders were still traditionalists who held on to the tenets of linear warfare. It is consistent with disruptive theory that traces of disguising might be found among the armor champions as they began to experiment with incorporating tanks with mobile warfare doctrine.

As one of the sustaining champions of the armored warfare, Guderian's efforts are worth detailing to find evidence of the disguising thesis. Guderian saw the tank not as a mechanical horse, but rather the means for a new way of war. The evidence shows that in championing the Panzer Division, Guderian couched the tank innovation as a sustaining change that did not threaten the core interests of mobile warfare. At issue was the role the tank was to play. The organizational culture of the German Army was very much inclined to view the tank as von Seeckt had viewed it – as an infantry support weapon. As Matthew Cooper concludes in *The German Army, 1933–1945*, 'Infantry divisions would remain the deciding factor of the strategy of decisive encirclement, and the motorized infantry and armor would be subordinated to their needs.'[106] While endorsing the tenets of mobile warfare, which initially limited the role of the tank, Guderian quietly promoted the idea of armored warfare. At first he argued that the tank could substantially aid the infantry in pushing their way through enemy defensive positions and making possible infantry exploitation. While doing so, Guderian simultaneously promoted the idea of the Luftwaffe providing close air support of armored warfare.[107]

Guderian's beginning in mechanized warfare dates to 1922, when as a captain he was assigned to work for the Inspector of Transport Troops of the Motorized Transport Department. During this time, Guderian began reading Liddell Hart's writings about using the tank as something more than an infantry support weapon. Guderian was impressed with the idea of combining Panzer tanks with motorized infantry units.[108]

In 1924, Guderian was responsible for a series of exercises to explore mobile warfare. He used the exercises to test the possibility of transforming his motorized units from supply troops into combat troops. While doing so, he also tested the employment of the tank as a reconnaissance vehicle in connection with the cavalry reconnaissance units. Here is a good example of Guderian working with the cavalry, while simultaneously testing future armored warfare concepts. His actions did not appear suspicious nor did he have to disguise them because he did not have any tanks to experiment with as they were banned by the Treaty of Versailles. Similarly, Guderian states that he had frequent opportunities to test his ideas in tactical simulated wargames, and again he did not threaten anyone because the culture of the German Army permitted a wide latitude and freedom to those developing doctrine.

Over a period of eight years starting in 1926, the Germans received accurate reports on the progress that the British were making in their tank experiments. They learned that technological improvements in the speed and maneuverability of tanks meant that the armored vehicles were no longer tied to the pace of infantry. This meant that tanks could become a strike force capable of rapidly exploiting a breakthrough.[109] In 1928 Guderian began teaching tank tactics while assigned to the motor transport troops.

In 1929, Guderian became convinced that 'tanks working on their own or in conjunction with infantry could never achieve decisive importance'.[110] Instead, he theorized that tanks, supported by mobile infantry, must play the primary role, and that other weapons should be subordinated to the requirements of armor. During the summer field exercises, Guderian based all the exercises on the tank supporting other arms with the exception of one exercise. In this exercise, he successfully employed an imaginary armored division and became convinced he was on the right track. However, his immediate superior, now General Otto von Stulpnagel, believed Guderian's tank concept was a utopian dream and restricted Guderian's 'magical' employment of tanks to units equal to or smaller than regimental strength.[111]

In 1931, Guderian was given command of a motorized battalion that consisted of a company of dummy tanks. Now he was free to conduct unrestricted exercises with his units out of sight of the Inspector of Transport Troops who had little faith in the employment of the tank. By this time, Guderian was seeking ways to 'persuade' the other arms of the service and the Commander-in-Chief of the Army that he should be allowed to form Panzer Divisions and Corps.[112] As long as Guderian was commanding motorized troops, who were only 'service troops', few senior leaders of the other arms took his innovation seriously. As Guderian states, '…no one then believed that the motorized troops…were capable of producing new and fruitful ideas in tactical and even the operational field'.[113] After all, the other arms, particularly the infantry and cavalry, regarded themselves as the most important parts of the army. The result is that these arms were quite prepared to have Guderian's armored units support the infantry, but they would not agree to the concept of the tank as the principal weapon.

Guderian's chief opponent, however, was the cavalry arm. He shrewdly took advantage of the cavalry commander, General von Hirschberg, when he announced that the cavalry was training to fight their own battles. Guderian and his superior convinced Hirschberg to hand over the job of operational reconnaissance to the motorized troops. Guderian immediately began to train a Panzer Reconnaissance Battalion for this task. In doing so, he continued to gain valuable expertise in developing armored warfare. To the extent that Guderian used dissimulation in promoting his sustaining innovation, it ended at this point. His subsequent efforts to promote armor warfare were not disguised as he sought to replace the horse with the tank.

Both the senior leaders of the infantry and cavalry did not understand the extent of Guderian's progress in armored warfare. Until 1933, they did not take tank warfare too seriously as the infantrymen found it amusing that they could defend themselves from the canvas walls of the dummy tanks by throwing sticks and stones. This changed, however, in 1933 when General Lutz organized mobile warfare exercises that included armor-plated reconnaissance cars mounted on the chassis of a six-wheel lorry. Immediately, the cavalry objected to the use of armored vehicles in the exercises. As long as Guderian was using dummy vehicles made of canvas the cavalry leaders did not consider Guderian's concepts a threat. But now, with the advent of the genuine armored vehicle on the battlefield, they did.

Also in 1933, the newly appointed War Minister, General von Blomberg, favored Guderian's tank ideas and arranged for him to demonstrate to Hitler a platoon of Panzers supported by a motorcycle platoon. Guderian records Hitler's response as 'That's what I need! That's what I want to have!'[114] After this, Guderian was convinced that with Hitler's newly-found support he would be able to create the Panzer Division. Guderian was sorely disappointed. He writes, 'The rigidity of procedure in our army, and the opposition of the persons in authority over me – the General Staff Officers who stood between Blomberg and me – were the principal obstacle to this [Panzer Division] plan.'[115] The chief obstacle to creating the Panzer Division continued to be those who favored the role of the tank as being employed primarily as infantry support weapons or as a separate large tank unit. Consequently, senior military leaders limited Guderian to brigade-size Panzer units.

Beginning in 1934 Guderian shifted his focus from advocating that armored warfare's focus of effort was to support the infantry to advocating that it was large enough to attack the enemy on its own. General Oswald Lutz, the armor pioneer, and Guderian, emphasized that infantry should not be ignored, but motorized so they could become a part of the combined arms Panzer Division of motorized engineers and signal troops, as well as tanks.[116] In 1935 senior military leaders formed three Panzer Divisions. In 1937, under the instructions of General Lutz, Guderian prepared a book that was published as *Attack Panzer*, which described his disruptive innovation in detail. Guderian was then promoted Major General and commanded a Panzer Division.

The empirical evidence is clear that there were a number of young General Staff officers who promoted armored warfare as an innovation that did not have to be tied to the infantry. Naturally, there was some reaction in the General Staff from those who believed armored warfare was as an innovation that should support the infantry. Guderian and the other tank enthusiasts, however, were allowed to experiment and did not have to disguise these armored warfare maneuvers as sustaining the infantry. In fact, von Seeckt from 1920 to 1925 stressed the importance of motorized transport for combat troops.[117] Once the British tank experiments warned the Germans that technological improvements had severed the tie to the pace of the infantry, Guderian and Lutz demanded a mobile force that could rapidly exploit gaps and attack enemy critical vulnerabilities.[118] Until Guderian had fully developed the armored warfare concept, he would conduct exercises where the tank would support the infantry, but Lutz would permit Guderian to slip in one additional exercise that would test his idea of the tank operating independently from the infantry.

Generally speaking, most of the senior officers were aware of Guderian's efforts to promote armored warfare. It is apparent that champions of sustaining innovations, such as armored warfare, did not have to disguise their efforts. What this also shows is that institutionalizing sustaining innovations can be just as challenging as championing a disruptive innovation. 'As Rundstedt, who led Army Group A's drive (and the panzer forces) through the Ardennes in May 1940, commented to Guderian at the end of an armor exercise in the late 1930s: "All nonsense, my dear Guderian, all nonsense." Nevertheless, Rundstedt's skepticism did not prevent him – and officers like him – from recognizing that tanks might extend the infantry's capacity to exploit tactical situations on the battlefield.'[119] With tacit approval to experiment from senior leaders, Guderian continued to develop armored warfare.

German Blitzkrieg *logistics failure: failure of disruptive innovation to evolve*

Once an organization succeeds in reorienting itself to implement a disruptive innovation, building new architectural knowledge still takes time and resources. As noted earlier, the discovery process as well as the process of creating new information and rooting out the old is difficult.[120] The more valid that senior military leaders view old linkages, the more likely that irrelevant old linkages will find their way into and handicap the new architecture. For example, the Germans were successful in establishing new linkages between tanks, mobile infantry, and planes. But the *Blitzkrieg* concept never evolved as a 'total' warfighting system because the Germans never questioned the validity of their linear logistics system for supplying their forces. With relatively short supply lines into Poland and France, linear logistics proved adequate. In the Russian campaign, however, long supply lines proved inadequate for supporting the *Blitzkrieg*. As Williamson Murray notes, 'If tactical battlefield innovations provided the Germans an initial advantage, these were not sufficient to overcome the gross mistakes they made in logistics... largely as a result of their military culture.'[121]

External causes of disruptive innovation: why and when

The study identified four hypotheses for the external causes of disruptive innovation. The evidence from the armored warfare cases found to a great extent that interservice, intraservice, and cultural factors explained the why and when of disruptive innovation. The study showed, however, that civilian intervention to assist military 'mavericks' was not the means that produced innovation in the German armored warfare case. Instead, what has proved more useful in this study is the argument that civilian intervention is effective only to the extent it can support a steady doctrinal development of sustaining innovation and protect product champions. This type of intervention differs from the model proposed by Posen because the initiative for the reform came from within the military, not from an external source and the civilian intervention came in support of senior officers who did not see themselves as hostile to the dominant values of their service. These officers were not, in fact, mavericks.[122] Since this conclusion is in conflict with the civilian intervention proposition, the study will elaborate on its findings.

Civilian intervention

The study found little evidence supporting Posen's claim that, 'the *Blitzkrieg* doctrine is another innovation that required civilian intervention'.[123] Posen argues, incorrectly in the judgment of this study, that Seeckt's mobile warfare doctrine was not an innovation. Posen states 'The fundamental aim of German doctrine remained the annihilation of enemy forces; the preferred maneuver remained the single or double envelopment. In contrast to the *Blitzkrieg* doctrine that emerged as a competitor within the German Army by the late 1930s, the doctrine of the von Seeckt army was only an incremental change from that which preceded it.'[124]

Posen states that the *Blitzkrieg* differed from mobile warfare because it aimed 'directly at the adversary's command, control, communications, and intelligence functions (C^3I)'.[125] 'Attacking the enemy's brain, his C^3I', Posen argues, 'is the characteristic that fundamentally distinguishes the Blitzkrieg from earlier German military strategy [Seeckt's mobile warfare].'[126] Posen claims that Guderian recognized the value of disrupting the adversary's ability to control its organization and designed the armor warfare doctrine to attack the enemy's brain.[127]

The historical evidence does not support Posen's claim that Guderian parted from Seeckt's mobile warfare doctrine and designed a new tactical doctrine that attacked an opponent's C^3I. As James Corum argues, 'the tactics of *Blitzkrieg* warfare in the era between 1939 and 1941 originated in the military doctrine and training of the 1920s Reichswehr [von Seeckt's army]'.[128] 'As a result of the high command's attitude, the mobile-war doctrines of the 1920s were gradually transformed into the *Blitzkrieg* concepts of the 1930s.'[129]

Supporting the claim that by 1918 the Germans sought to win battles by shattering the enemy's cohesion through a variety of rapid and unexpected actions are historians who have focused on the German Army during the last two years of World War I.[130] Timothy Lupfer in *Dynamics of Doctrine: The Changes in German Tactical Doctrine During the First World War*, documents the origin of the *Blitzkrieg* tactics in the German attack doctrine of 1918. *Attack*, which became the basic document for the German offensives of 1918, described an attack-in-depth to be accomplished by combined-arms penetration relying on surprise. Lupfer writes, '*Attack* noted that the strategic breakthrough [of Allied lines] was the ultimate goal of the penetration. In order to achieve that goal the attack had to strike deeply into the enemy position.' Lupfer continues, 'Acknowledging the impossibility of destroying all enemy forces in such a deep penetration, the German tactical doctrine did not require complete destruction. Instead, disruption of enemy units and communications was essential. Throughout the doctrine, keeping the enemy off balance, pressing the attack continuously, and retaining the initiative received great emphasis.'[131]

To conduct the attack, the German infantry depended upon speed and organizational depth as the means of securing their flanks and rear. 'Speed to keep the enemy from reacting in time to the attack, and depth to provide and prevent the follow-up units, which would isolate by bypassed pockets of resistance and prevent there remnants from interfering with the continuation of the attack.'[132] Described by the Allies as 'infiltration tactics', the German tactics go beyond individual units bypassing resistance and pushing forward as far as possible. Infiltration connotes individual or small squad movement. Lupfer points out that infiltration as practiced by the Germans, however, emphasized the combined-arms movement of an entire division.[133] In the German 1918 offensive principles, which Seeckt adopted and refined after the war, 'the Germans did not aspire to achieve total destruction at the thin area of initial contact; they used firepower and maneuver in a complementary fashion to strike suddenly at the entire enemy organization'.[134]

As the historical record shows, the new German infiltration tactics did not solve the problem of firepower and maneuver beyond the range of supporting artillery fire. Without motorized support, breakthroughs in 1918 had scant opportunity to reach deep into enemy rear lines and shatter enemy cohesion. From these lessons, however, Seeckt created a mobile doctrine. Not a doctrine of annihilation as Posen claims, but a doctrine of maneuver designed to attack the enemy's cohesion.

The historical evidence is clear on the following observations: Hitler did not order the disruptive champion of the *Blitzkrieg*, General Hans von Seeckt, to create small innovation groups to create the doctrine of combined-arms mobile warfare. Nor did Hitler order Seeckt to secretly create tank and airplane bases in

Soviet Russia to train according to the mobile warfare doctrine. Furthermore, Posen's explanation tends to downplay the early role played by Guderian, the second product champion of the *Blitzkrieg*, in developing armored warfare in the mid-1920s and early 1930s, before civilian intervention. This study agrees with Posen that many senior officers opposed Guderian's concept of employing the tank as the key part of mobile warfare. But Posen ignores the official history when he argues that senior military leaders opposed the use of tanks and that Guderian created armored warfare out of a vacuum after Hitler's intervention. First, senior military leaders did not oppose the use of the tank, only how it was to be used. The arguments arose over whether it should be used as an infantry support weapon or as the main weapon itself. Second, the historical record shows that Guderian placed the sustaining innovation of armored forces within the framework of Seeckt's mobile warfare doctrine.[135]

The assistance Hitler provided Guderian tends to support the argument in this study that civilian intervention is effective in spurring sustaining innovations and in disruptive innovation it is effective only to the extent it can support or protect product champions. The case of German armored warfare seems to contradict the theory of civilian intervention on behalf of mavericks as the cause of disruptive innovation.

Organizational culture

The study found evidence supporting Elizabeth Kier's claim that culture is a determining factor in achieving a disruptive innovation.

German culture

The success of the Germans in the forging of Panzer Forces was a result of the German high command allowing wide latitude and freedom to those who had conceived and were developing armored forces. The roots of the Army's success are cultural. Senior leaders demonstrated open-mindedness in examining the lessons of World War I and a receptiveness of new ideas that could forge a disruptive innovation.[136]

French culture

Murray notes that 'the French high command maintained a stranglehold over the operational and tactical development of the mechanized forces in 1930s'.[137] This approach may have been sufficient if Germany had followed its World War I doctrine of static positions, overwhelming firepower, and stalemate. Unable itself to define a non-linear role for armored forces, the French high command made sure its forces were ready to fight the last war.[138] Constrained by its World War I defensive approach that had established the invulnerability of large fortresses, the French army ensured they would be unprepared for the German disruptive armored warfare innovation.[139]

British culture

Following 1918, the British Army shifted its focus from war-torn France to the demands of its empire in India, Ireland, and Southeast Asia. Consequently, with a steadily diminishing defense budget, experimentation was limited. Nevertheless, it did establish an independent tank corps for developing tank tactics.[140] J.F.C. Fuller and Liddell Hart's writings on mobile armored warfare fueled the experimental tank force exercises from 1926–35, which was an important factor in converting theory to doctrine.[141] Ironically, the Germans, rather than the British, received the maximum benefit from the British tank experiments.

Progress between 1935 and 1939 in British armor warfare development stagnated as both Fuller and Liddell Hart fell from grace by angering senior leaders to the point that their views had minimal effect on development of armor warfare. Viewed by many as a zealot, Fuller advocated the tank replacing infantry and cavalry instead of the tank being a key component of a combined arms force. To his credit Hart realized that armor could transpose the German infiltration tactics into a mobile armored force.[142] Hart's perceived influence on the Cabinet and Secretary State for War, however, resulted in bitter feelings from senior leaders who felt Hart did not follow the chain of command.[143] Admittedly, the British Army's culture supported tank experimentation that yielded extraordinary results. Part of the explanation for failing to achieve armored warfare, however, was this same culture was still grounded in a World War I tactical doctrine that did not emphasize speed, exploitation, initiative, and drive. In sum, culture was a root cause in preventing the British from developing armor warfare.[144]

Conclusion

The central proposition of this thesis is that the ways senior military leaders manage disruptive and sustaining types of innovation explains much of the variance in patterns of military innovation. A related proposition is that although senior military leaders spend most of their time managing sustaining innovations, this management approach will not lead to a disruptive innovation.

Driven by broad structural changes in the security environment, military reformers consider the need for disruptive innovation. During this intellectual process, they create small innovation groups, which they directly manage to develop and incubate new visions of war. Once the character of the new warfighting way is conceptualized, the politics of effecting disruptive change are eased by using a dissimulation strategy for presenting it to the organization as a sustaining innovation. Dissimulation of disruptive innovations is not a deterministic social science prediction, since the presence of a disguising strategy does not guarantee success. Also required is a political process where the product champion can successfully protect and promote junior officers in the new way of war. By incorporating the disguising and small group factors that create favorable conditions for transformation, disruptive innovation theory provides a better explanation of innovation than does Rosen's intraservice rivalry theory.

3 US Marine Corps innovation

The development of amphibious warfare

Before 1916, the US Marine Corps' traditional missions included performing as naval infantry on board ships and naval stations and behaving as colonial infantry guarding American interests in the Caribbean and Far East. United States naval forces had conducted unopposed landings at Santiago, Cuba, and Manila, Philippines (1898), but World War I experiences – in particular the British debacle at Gallipoli – provided strong evidence that an opposed landing could result in untenable losses and even failure. By 1940, however, the Marine Corps had transformed into a Fleet Marine Force, a potent new naval warfighting arm capable of assaulting and defending advance naval bases.

The creation of amphibious warfare in the US Marine Corps is not only a study in the complete transformation from conventional naval infantry into a landing strike force, but also one of the major developments of World War II.[1] J.F.C. Fuller, the British tank theorist and later military historian, commented that it was 'in all probability...the most far-reaching tactical innovation of the war'.[2] From identification of the need for advance base operations to the first amphibious assault of World War II, the entire innovation process took some 37 years.[3]

The innovation question is difficult to address in the case of the US Marine Corps. Although there was almost no connection between the Marine Corps performance in Europe and the development of amphibious doctrine, some have argued, incorrectly, that Marine Corps innovation strategy was a natural evolution after the defeat of Germany. General Charles C. Krulak, former Commandant of the Marine Corps (1995–99), states, '[The veterans of Belleau Wood] could see the incredible potential in amphibious assault when all the self-proclaimed "experts" considered it futile in light of the 1915 debacle at Gallipoli.'[4] In the 1951 Princeton study of the Marine Corps, the authors argue that the Corps 'took on its more formidable job of developing an up-to-date doctrine of amphibious warfare shortly after the close of World War I. ... The Marines in 1940 had a sound amphibious doctrine...[that] underwent no basic change during World War II.'[5] Generally speaking, this perspective sees the Marine Corps as having transitioned easily to an amphibious assault force from a force that had manned naval ships as security guards and fought small wars in Asia and the Caribbean and large war in Europe.

On the other hand, some Marine Corps historians, such as Lieutenant Colonel Jon Hoffman (deputy director of the Marine Corps Historical Center), argue that the postwar Marine Corps brought about amphibious warfare with no clear vision and only after considerable infighting as to whether the amphibious assault mission was the service's *raison d'être*.[6]

How are these two positions to be reconciled? This study proposes to do so by investigating the amphibious warfare phenomena using a disruptive innovation perspective. From this view, amphibious warfare can be understood not as a single innovation led by a single product champion, but as two innovations woven together by two product champions – John Lejeune and John Russell, both of whom served as Commandant of the Marine Corps.

Defensive and offensive operations were the two architectural innovations that emerged during the interwar years. The defensive innovation was sustaining in nature and was a natural evolution of the traditional missions the Marine Corps had been practicing. The offensive innovation was disruptive in nature because it introduced a new warfighting metric that was not accepted previously by the Corps as a mission. Although they occurred simultaneously, the defensive innovation was caused more by changes in the security environment, and the offensive innovation was caused more by interservice rivalry. In sum, to unravel the complexity of managing the amphibious warfare innovation, the study examines it through the lenses of sustaining and disruptive innovation.

Sustaining innovation: advanced base force – occupying and defending bases

The Fleet Marine Force (FMF) had two primary responsibilities at the outbreak of hostilities with Japan. The better known was the amphibious assault mission played in the Pacific and highlighted by the flag raising on Iwo Jima's Mount Suribachi. The second task was advanced base defense, as at such places as Midway and Wake in the opening days of World War II.[7] Although it has received less attention from historians, from 1900 until late 1939 advanced base defense had equal or even greater significance in naval strategy than did amphibious assault.[8] Before World War I, the assault mission did not exist in war plans, but advanced base defense did. In fact, the advanced base force originated in theory in 1898 and evolved into the defense battalion in 1939.

Engine of change: why and when

Defense battalions can be traced back to 1894, when Congress assigned to the Marine Corps the mission of providing forces that could occupy and defend advanced naval bases.[9] In 1898, the Marines produced the first advanced base force battalion, which successfully occupied and defended an advanced base at Guantanamo Bay, Cuba, which, in turn, enabled Admiral William T. Sampson to keep his fleet in the area.[10] Subsequent victory in the Spanish-American War propelled the United States into the role of major colonial power, with new possessions and a concomitant need for the defense of new outposts. As a result, the Marine Corps' role in naval strategy was transformed.[11]

As early as 1900, Japan had become, from the naval viewpoint, a serious poten-tial enemy. Naval planners realized that a network of bases was needed to support the fleet operating in the western Pacific, but they did not envision attacking defended beaches to secure advanced bases. Changes in the security environment, however, would trigger an offensive innovation for occupying advanced bases.[12] The Japanese seizure of German-held islands in Micronesia during the opening months of World War I was a potential threat to US supply lines to the Philippines.[13] To fulfill the strategic requirement of dominating the seas, the Navy's general board charged the Marine Corps with creating a force that could occupy and defend the advanced bases. These bases would provide secure refueling and repairing facilities for the fleet as it progressed across the ocean.

Another trigger was intraservice (Navy and Marine Corps) and interservice (Army and Marine Corps) rivalry. In December 1908, Navy Captain William Fuller convinced President Theodore Roosevelt that Marines should be removed from naval vessels. After Roosevelt's executive order to that effect, the *Washington Post* ran a feature declaring that the Army was to get the Marines by transfer to the Army infantry.[14] Fortunately for the Navy, Admiral George Dewey, president of the Navy's General Board, in a letter to the House Naval Affairs Committee, convinced Congress the Navy needed the Marines within the Department of the Navy because of the requirement for an expeditionary force to assist the fleet in seizing and defend-ing advanced bases.[15] Three months later the Senate overturned Roosevelt's order.

Throttle of change: how

Intellectual process

A full explanation of the success of the advanced base force must go back to the immediate aftermath of the Spanish-American War. At that time, Admiral Dewey, head of the Navy's General Board, assigned the defensive advanced base mission to the Marine Corps.[16] In making this decision, the General Board in 1900 noted that the Marines would be 'best adapted and most available for immediate and sudden call'.[17] Following the Board's advice, the Secretary of Navy directed the Marine Corps to establish a unit capable of defending advanced bases. Almost a year later, the Corps had established an emergency defense battalion of four companies of approximately 100 men each. In 1902, the defense battalion sailed for Culebra, an island a few miles from Puerto Rico, where they practiced an unopposed landing and then setting up a defense for the island.

There was, however, little intellectual progress made over the next decade in developing the advanced base concept as senior Marine leaders debated the Corps' primary roles and missions. These included the traditional service on board ships and naval stations, as colonial infantry serving in small-scale actions short of war, defense of advanced naval bases, which the fleet needed for coaling stations, and service with the Army in major land campaigns.

There was little agreement within the Corps regarding the priority of these missions, asserts Marine Corps General Victor Krulak. In his view, many advocates of the traditional roles fell within the group that focused on creating the *Small War Manual*, while the other general group focused on creating defensive amphibious doctrine. In his memoirs, Krulak notes the greatest difficulty the offensive assault group faced was that, 'until the 1920s, there was no real institutional dedication in the Corps to the idea of an assault landing attack against organized defenses'.[18]

Late in 1909 the General Board urged the Marine Corps to begin 'serious' consideration of the advanced base problem as its primary mission. At that time, Major John Russell, among others, was advancing the theory of advanced base defense, in fact submitting three studies to the board on how to establish such bases.[19] His efforts provided the intellectual scaffolding for understanding amphibious operations, which Major Pete Ellis would build on in the early 1920s in predicting that Japan would strike first and outlining how Marine forces should conduct both defensive and offensive amphibious operations.[20] Russell observed that when the fleet was operating at a distance from permanent bases it should carry with it 'a sufficient force and material for seizing and defending' an advanced base in the theater of operations.[21]

The next year, General George Elliot, then Commandant of the Marine Corps, submitted to the Secretary of Navy a copy of the proposed course of instruction for the advanced base school. It would take several years, however, before the General Board would decide what would be the proper defense force. That same year, an aide-for-inspection for the Navy, Captain Fullam, delivered a devastating report against the Marine Corps stating that the advanced base operations were a failure. He attributed these failures to the Marine Corps and its organization and recommended removing Marines from shipboard duty so they could concentrate on the advanced base problem.[22] The General Board elected not to follow this advice, stating, 'this action...may eventually cause the loss of the Marine Corps to the Navy and its absorption by the Army'.[23]

As a result of this internal bickering, the General Board in 1914 requested the Marine Corps give practical instruction in several areas of defending advanced bases. Attempting to spur the Marines to embrace this crucial mission, the board also requested they conduct advanced base exercises each year.[24] Whether the Marines would have become fully engaged in the advanced base mission is a moot point because the service's expeditionary roles in Mexico, Haiti, Santo Domingo, and France prevented any further training for advanced base operations until 1922.

The intellectual redefinition of Marine warfare from guarding ships and naval bases and fighting small wars to defending advanced bases was the work of General John Lejeune. Prior to his tenure as Commandant (1920–29), Lejeune was one of the few Marines who saw the need to secure base facilities in the Pacific.[25] Krulak writes that Lejeune 'had been disappointed with the inability of Commandants Heywood (1891–1903) and Elliott (1903–10) to grasp the relationship between the global needs of the Navy and the creation and defense of overseas naval bases. Their view was that the century-old Marine Corps role of providing ship's guards and security for naval stations should still be foremost, that to commit Marine resources to advanced base force duty was an imprudent diffusion of effort'.[26]

The 'Young Turks' of the Marine Corps, led by Lejeune and two other future commandants, led a quiet revolt. Disappointed over the Corp's reluctance to embrace the advanced base concept, they formed the Marine Corps Association and its publication *The Marine Corps Gazette*.[27] One of the central articles in the first edition of the *Gazette* was written by the product champion of amphibious assault, Major Bill Russell. In 'A Plea for a Mission and Doctrine', Russell argues that the Marine Corps was rudderless without a general mission and suggests the Corps look to become a striking arm of the Navy. As to doctrine, Russell argues that the Corps should develop a *Tentative Doctrine* to achieve the greatest efficiency in cooperating with the Navy.[28]

The first intellectual step toward defensive amphibious operations was made when Lejeune established a small innovation group to examine the problem. In December 1920, Lejeune created a new planning section in the Division of Operations and Training (DOT), with Pete Ellis as intelligence officer. Within DOT, Lejeune had assembled the Corps' most influential officers, all of whom knew and admired Ellis and could be counted on to promote his career.[29] With the tacit approval of Lejeune and the direct approval his immediate senior, Brigadier General Feland, Ellis immersed himself in monastic fashion and produced two definitive works on Pacific naval strategy – 'Advance Base Operations in Micronesia' and 'Navy Bases: Their Location, Resources, and Security'.[30] Although Ellis envisioned offensive assault against Japanese-defended bases, he also recognized that once the Marines captured an advanced base, they would have to defend it against counterattack.[31] The evidence is convincing that Ellis, as well as Russell and others, promoted defensive and offensive advanced base operations.[32]

The next intellectual step was Lejeune's approval and acceptance in total of the Ellis study, which he renamed Operation Plan 712D, part of the Orange Plan. There were two segments of Plan 712D – defensive operations in defending an advance base and offensive operations in seizing one. Surprisingly, Ellis devoted the most pages to the older mission of defense, although his work on the conduct of seizure of defended islands, especially his prophetic description of ship-to-shore movement, is one of the remarkable military documents of the twentieth century.[33] It is fair to say that the intellectual transition from a force with several competing missions to one with two primary missions appears to be the work of one man, Major Earl (Pete) Ellis.

Lejeune chose to focus on the defensive concepts of amphibious operations. Not surprisingly, he still considered small wars as a primary mission of the Corps, and the evidence is strong that he did not consider offensive amphibious operations as described by Ellis as a priority mission. As an illustration, General Holland 'Howlin' Mad' Smith, the Marine general who trained both Army and Marine units in offensive amphibious warfare, stated that when he arrived at the Marine Field Officers course at Quantico in 1926, 'I was appalled to encounter there the same degree of outmoded military thought as I had found at the Naval

War College.... I found myself deep in difficulties because I objected to the emphasis placed upon defensive tactics.'[34] It would not be until 1932 that the Marines, in Smith's opinion, engaged in the first large-scale amphibious operation 'with Japan actually in mind'. Supported by the fleet, the Marines conducted a ship-to-shore movement that the general considered a 'dismal exhibition'. He notes, 'I realized that we had a great deal to learn before we approached anything like efficiency in amphibious warfare.'[35]

Political process

As Rosen notes, the intellectual redefinition of advanced based warfare would have been futile unless product champions gained power within the officer corps and succeeded in promoting and protecting officers focused on the new way of war.[36] As the product champions for defensive amphibious warfare, Lejeune and his allies succeeded in this struggle.

Following World War I, the Marine Corps was in search of a mission, as it felt duplicating tasks traditionally performed by the Army was unwise. The Army, being in the business of landmass warfare, however, was not interested in seizing beaches, especially considering the British experience at Gallipoli.[37] To give the advance base mission higher priority, newly appointed Commandant John Lejeune reorganized the officer fitness report card to include professional qualification comments in the area of 'Advance Base Work'.[38] By doing so, he ensured that officers who obtained skills in defensive warfare would be more likely to be promoted than officers who had not.

The next step in the political process was to overturn the 1920 Russell board promotion results. Then-Colonel Russell (future product champion of offensive amphibious warfare) chaired the promotion board on which wartime officers eventually would receive permanent commissions. One reason Russell and the other members of the board were selected was because they had not served in France with the 4th Marine Brigade. Russell instructed board members that they were selecting the next generation of Marine leaders and that they should apply prewar standards and not just reward combat experience. His intent was to build a new officer corps with the intellectual capacity to solve hard problems, such as the offensive amphibious warfare question.[39]

Lejeune convened a new board that would moot the results of the Russell board. By taking such action and failing to provide a permanent structure that would allow Marines to study offensive amphibious warfare, Lejeune ensured that the disruptive innovation of offensive warfare would remain merely a theoretical concept. Without a permanent standing fighting organization, developing offensive amphibious warfare would be almost impossible.

For a great part of his tenure as Commandant, Lejeune remained focused on developing defensive amphibious warfare, fighting small wars in Central America, and monitoring events in China. His apparent inaction in pursuing offensive assault warfare as the Corps' primary mission has confused scholars. Admittedly, Lejeune ensured Ellis's study on both defensive and offensive amphibious operations was adopted into War Plan Orange, but as Rosen notes, he 'gave the whole subject of advanced forces two paragraphs in his 1930 memoirs, and made no mention of amphibious assaults at all'.[40] Rosen posits that perhaps Lejeune was reticent about assault operations because of a perceived need for secrecy.

Disruptive theory provides us with a different explanation. Using this approach, it is reasonable to posit that Lejeune did not recognize the disruptive nature of offensive amphibious warfare. Therefore, he managed the offensive mission by letting it be an outgrowth of the sustaining defensive mission, on which he placed a higher priority. More than likely, he saw offensive amphibious operations as merely a sustaining innovation that, in time, could be achieved along the same performance metric as the defensive amphibious operations he was developing; that is, if you understand how to defend advanced bases then by working in reverse you understand how to attack them. Offensive operations would consist of little more than dashing across a beach with a company of Marines, much as had been done during fighting at Belleau Wood.[41] Such a perspective would explain why Lejeune did not comment on offensive operations in his memoirs – he was focused on defensive operations and offensive operations would be just a subsequent development.

Lejeune's behavior is reminiscent of that of his British Army counterparts who had returned home victorious from France. British intellectual maverick Liddell Hart was the equivalent of Pete Ellis – both achieved intellectual breakthroughs (Hart in mobile tank warfare) to redefine military tasks and missions. Yet their senior leaders failed to achieve the political task of transforming the officer corps to achieve the disruptive innovation. They failed because they attempted to manage a disruptive innovation using a sustaining approach. Clearly mobile tank warfare was not merely an extension of linear tank warfare, or offensive amphibious innovations a subset of defensive amphibious operations.

A great part of Lejeune's failure to recognize the disruptive nature of offensive amphibious operations can be traced to the influence the prestigious Army War College and his experience fighting alongside the Army in France. The only Marine to attend the 1907/08 Army War College class, Lejeune adopted Army procedures and by his own assertion was for all intents and purposes an Army officer during this period.[42] As a consequence, Lejeune as Commandant approved a new curriculum for the Marine Officers' Training School that was strongly influenced by the teachings and methods of Army.[43] The field officers class was designed after the Army Command and General Staff course at Leavenworth, Kansas.[44] In addition, Lejeune started a new Correspondence School in 1926 that adopted Army correspondence courses as the basis for instruction.[45] Lejeune's Army emphasis caused some difficulties. There was a question, for example, of whether the school should develop a course of study geared to the fleet, or a curriculum that favored Army methods and land warfare – an uncertainty not clarified until Russell organized the Marine Corps around the Fleet Marine Force some years later.[46]

As disruptive theory predicts, Lejeune had difficulty recognizing offensive amphibious operations as disruptive because the Army architectural knowledge embedded in the Marine Corps way of fighting blinded him to critical new aspects of this disruptive innovation. As an illustration, Army Tables of Organization were used for conducting Marine assaults from the sea and they did not work. As discussed in the next section, Russell finally solved this problem by discarding all Army courses and teachings for conducting offensive operations.[47]

As product champion for offensive amphibious operations, Russell demonstrated that such operations would require not only a new organization – the Fleet Marine Force – but also a new architecture in the form of doctrine. With Russell as Commandant, the Corps' highest priority shifted from defensive advanced beach operations to offensive beach operations, although defensive operations continued to evolve.

By 1939 Lejeune's vision of advanced base theory had evolved into the defense battalion.[48] On the eve of Germany invading Poland, the General Board reported to the Secretary of Navy its final assessment of naval forces' ability to fight a major war. Less than optimistic, the report detailed several deficiencies and several recommendations that listed the offensive amphibious warfare mission as secondary to defending fixed naval bases.[49] At the outbreak of World War II, the Marines had defense battalions at several locations, including Wake and Midway, where they performed superbly against Japanese attacks. The offensive mission of the Marine Corps, however, eventually would supplant the defensive mission in the final assault on Japanese territory and homeland.

In sum, Rosen is correct in stating that Lejeune did not reorganize the Corps around amphibious warfare after Ellis had achieved the intellectual breakthrough. Arguably, this was not because Lejeune was trying to keep offensive warfare a secret, but because he was using a sustaining method for managing this disruptive innovation. The strongest evidence found for this explanation is Russell's disruptive product champion General Breckinridge's comment about Lejeune's and Russell's innovation efforts: 'We [Lejeune, Russell, and Breckinridge] have been traveling on a certain track, unavoidable at first, but always more or less suitable to date; but the suitability is steadily decreasing. In a manner of speech we have arrived at a turntable, upon which we shall place ourselves, and pick up the new track [offensive warfare] upon which we must travel for the future as far as we can see it.'[50] The new track of offensive warfare required a product champion willing to break with the past and create an entirely new way of thinking about warfare and the doctrine to support it. This product champion would be General Russell.

Disruptive innovation: Fleet Marine Force and amphibious assault warfare

As discussed in the section, offensive amphibious warfare did not evolve from defensive amphibious warfare. Its intellectual roots are in the last third of the Ellis's famous report for Lejeune. Its organizational roots are in Russell's revolutionary offensive unit – the Fleet Marine Force. By creating the FMF in 1933, Russell had provided a permanent cauldron for testing new recipes for conducting assaults. In Holland Smith's words, the creation of the FMF 'was the most important advance in the history of the Marine Corps, for it firmly established the Marine Corps as part of the organization of the US Fleet, available for operations with the Fleet ashore or afloat'.[51] For purposes of this study, a most important aspect of Russell's strategy for effecting his disruptive innovation was his skill in initially disguising the revolutionary aspects of the FMF as a *status quo* organization. In fact, he was willing to end the careers of zealots who supported his new way of fighting because he did not want them drawing attention to his hidden revolution.

Engine of change: why and when

Amphibious warfare resulted from the naval services' efforts to deal with changes in the strategic security environment in the Pacific after World War I and with intense interservice rivalry. The new Republican administration of 1921 favored isolationism and during the so-called Washington Conference secured a limitation on naval construction among several of the big naval powers. Besides curtailing naval construction, the United States also agreed not to fortify its Far East possessions, including the Philippines, Guam, Wake, and the Aleutians, if Japan agreed not to expand naval holdings in Formosa or the mandated islands in the Central Pacific. All else being equal, the treaties signed in Washington left the US naval bases in the Pacific marooned and indefensible. In its rush to achieve disarmament, the nation unwittingly had given Japan naval supremacy in the western Pacific.[52] Immediately afterward, a major revision to War Plan Orange began at the Naval War College as planners anticipated the requirement to occupy and defend advanced bases in the Marshalls or Carolines.[53]

During this time, the Marine Corps was focused on developing a defensive doctrine for defending advanced naval bases. This emphasis changed abruptly in 1933, when General Douglas MacArthur (then chief of staff of the Army), openly antagonistic to the Marine Corps, proposed that the Corps be transferred to the Army.[54] When he failed to gain support, the general proposed 'that at least the bulk of the Corps be transferred to the Army, leaving the Marines with only base defense and seagoing detachment functions'.[55] Marine Corps General Victor Krulak in his memoirs writes, 'The substantial influence wielded by MacArthur impressed Commandant Ben H. Fuller (1930–34) with the gravity of the threat, and gave his assistant, General Russell, the opportunity he sought to drive the [offensive] amphibious subject to the surface.'[56]

Finally, in the early 1930s senior political leaders decided to pursue an isolationist policy and avoid if possible sending Marines worldwide to fight small wars. The result was that most Marines returned stateside and were available to train and exercise in amphibious assault.

Throttle of change: how

Intellectual process

Although his work remained a secret document for many years, the intellectual breakthrough to redefine Marine Corps tasks and missions as offensive in nature had been made by Ellis in Lejeune's architectural innovation group. In 1931, the next intellectual step toward offensive amphibious operation was made by Commandant Fuller when he selected Brigadier General Randolph C. Berkeley as the first commanding general of Marine Corps schools. With Fuller's approval Berkeley established an small innovation group under the direction of Colonel Charles Price to start work on a tentative text for 'Marine Corps Landing Operations'.[57] A key member of this group was Navy Lieutenant Walter C. Ansel, selected because Fuller realized the importance of addressing naval issues such as gunfire support. Although this small innovation group did not write a formal publication, it did make some strides in translating Ellis's vision of offensive warfare into new tasks.[58]

The next intellectual step was the most important one. Russell persuaded senior leaders in the naval services to establish the FMF. He needed an organizational mechanism to transform the Corps into an offensive assault force and the FMF would serve this purpose. Although Russell had been a proponent of specialized training for offensive amphibious operations since 1916, he did not have the resources to carry out such training. With regard to manpower, the Corps was stretched thin to meet its heavy expeditionary obligations in Nicaragua and China. As an illustration of the severity of the problem, while Russell had strongly encouraged the East and West Coast Expeditionary Forces (formerly known as the advanced base force) to train at amphibious operations, as Holland Smith writes, 'expeditionary duty in China and Nicaragua so reduced the Marine Corps as to prevent participation in fleet exercises... In 1933, there were again no units available for extensive landing exercises'.[59]

In 1933, three reasons prompted Russell to recommend creating a permanent organization, the FMF, for the study and practice of amphibious warfare that would be recognized by the Army and the Navy and consist of an independent command ashore.[60] First, he was encouraged by senior Navy leaders who were beginning to demonstrate a real interest in developing amphibious assault warfare. Second, the planned withdrawal of the Marine brigades from China and Nicaragua would free Marines to train for base seizure.[61] Third, General MacArthur had attempted to have the Marine Corps transferred to the Army. In creating the FMF, General Victor Krulak notes, Russell 'may well have exerted greater influence in rationalizing and regularizing the amphibious assault than any other single individual in the Corps'.[62]

Three aspects of the FMF of 1933 made it as novel within the framework of American arms: (1) it was singly and openly organized, equipped, and trained for landing operations incident to naval campaigns; (2) it was never skeletonized to the extent that it was not capable of rapid embarkation in useful combat units and movement by sea; and (3) realizing how garrison duty can sap the combat training of any tactical unit, the Marine Corps from the outset drew a sharp distinction between FMF units and the post troops needed for normal garrison, security, maintenance, and similar duties. At each post where FMF troops were stationed, additional post troops also were maintained so that Fleet Marine Force training could proceed unhampered. Individual Marines were rotated between FMF and non-FMF duties so that all members of the Corps were fully trained in combat roles.[63]

The next intellectual step occurred in 1934 when General Russell, became Commandant of the Marine Corps. As Rosen notes, Russell was able now to combine the intellectual tasks with the political organization tasks needed to bring about a genuine disruptive innovation.[64] Russell turned his focus to his architectural innovation group, the Marine Corps schools at Quantico. Realizing that without a new offensive doctrine to guide its training, the FMF would not be able to achieve the capability for opposed assault, Russell directed that Marine Corps schools be suspended for six months to prepare a formal amphibious doctrine.[65] He had several committees formed to address different issues of the new doctrine. The result was the 'Tentative Landing Operations Manual', published in 1935, which formed the basis of both Navy and Army doctrine for World War II.[66]

Disguising strategy

With the internal intellectual breakthrough to redefine the Marine Corps' central mission as offensive amphibious operations made and the organizational mechanism for making it happen, Russell faced the external intellectual task of managing the disruptive innovation as a sustaining innovation. The common account of Russell creating the FMF then moving effortlessly on the next challenge is popular myth. Creating the FMF was not without risk and opposition. Russell's move to create a Marine strike unit, which the senior fleet admiral could tactically control in naval campaigns, was not popular among many of senior Marines. Victor Krulak, a strong supporter of Russell, writes that Russell's eloquent case for amphibious assault task as the Marine Corps primary mission was seen as 'almost heretical'.[67] There is a stern reality in Krulak's portrayal of what Russell faced: 'Only a few, a very few, visionaries were willing to attack the formidable, tactical and material problems associated with the modern amphibious assault landing.'[68] Krulak sums up Russell's FMF development as being 'the results of unusual and brave actions'.[69]

Russell's opponents were those officers who felt the Marine's core interests were threatened by the magnitude of the change he proposed. Krulak accused these officers as being 'proponents of a retrospective philosophy that went back a hundred years'.[70] Russell's primary resistance, however, came from the pro-Army Marine officers. Many of these officers had fought in France, and to them the land campaign mission was the most important and the one they should be prepared to do. They believed that the Marines' future lay in winning battles such as Belleau Wood, where the enemy is not only defeated, but the Corps wins the public relations battle as well at the Army's expense.[71]

The history of the Marine Corps Schools confirms that the land campaign mission was the primary one as the Marine Corps used all Army manuals, doctrine, and training techniques to teach Marine students. Furthermore, the evidence shows that Marine instructors were sent to Army schools to learn how to teach Army methods to Marines. The strong link to Army schools came from General

Lejeune. A graduate of the Army War College, Lejeune was the Commandant most ambitious to send Marine Officers to Army schools. In fact, he was so impressed with Army schools that he intended that Army school completion should be regarded as part of an officer's fitness report for special assignment. Marine Historian Allan Millett writes that the Army school completion 'might serve as a moral equivalent of promotion and the key to rapid advancement if the Corps went to war again'.[72]

In creating the FMF, Russell sought to create an organization that performed differently from the Army. In 1932, Russell had stated to the head of Marine Corps Schools that that the courses had not placed sufficient emphasis upon amphibious assault missions. The head of Marine Corps schools told Russell that 'our reorientation will require a mental wrench to part from the universal leadership of Army Schools'.[73]

With this opposition in mind, Russell disguised the true intention of the FMF from his opponents by suggesting that it was 'essentially' a name change from Marine Expeditionary Force to the Fleet Marine Force.[74] A close analysis of Russell's memorandum to the Chief of Naval Operations would seem alarming to opponents of the amphibious assault. In fact, Russell never mentioned the phrase 'amphibious assault'. To further belay the fears of his opponents, Russell gave the CNO the option of naming the new force as either 'Fleet Marine Force' or 'Fleet Base Defense Force'. Thus, it appeared that Russell's FMF was merely another name for a defensive force. Shortly after the CNO had agreed to not sending the FMF on expeditionary missions, the Navy Department ignored the agreement and sent a Marine expeditionary force to Cuba. Indirectly, this helped Russell disguise the FMF's *raison d'être* as amphibious assault as it appeared to Russell's opponents that missions had not changed.

The next intellectual step is perhaps the most interesting and important to the disruptive innovation story. After the FMF was created, Russell tasked Brigadier General James Breckinridge, commandant of the Marines' officers' school system, to develop the doctrine and training necessary for the FMF to conduct amphibious warfare. The driving force was Colonel Ellis Bell Miller, who as the head of the steering committee, created the *Tentative Manual for Landing Operations, 1934*. Miller used the staff and students to help him produce the doctrine. Numbers wise, they were only 15 students attending the Field Officers course and only 30 students attending the Company Officers' School.[75] Miller was a zealot advocate of amphibious warfare. Krulak used the phrase 'apostolic fervor' to describe Miller's personality, and described him as 'demanding, intolerant of any dissent, and impatient with those who could not maintain the pace'.[76]

Russell understood that he still faced strong opposition from Marine traditionalists who opposed the amphibious assault as the Marines' new primary mission. As a result, he needed to exercise misdirection in artfully moving the Marine Corps toward defining the amphibious assault as its central mission. During this fragile transformation period of sorting out the new task and doctrine, Russell did not desire undue attention to his efforts at Quantico. On 12 May 1934 he received a spirited and emotionally written memorandum from Colonel Miller.

Miller described his diatribe as a way to 'lay our cards on the table' to discuss his misgivings about Russell not taking on the Army-trained Marines. Miller writes, 'we feel that we are like the prophet without honor in his home town. We feel that Marcorps [*sic*], deep down in their hearts, do not accord us the respect and recognition that is given by them to other similar military institutions'. Miller then attacked the Marine officers who were Army school graduates that he had to work with, writing, 'Captain Collier, a graduate of Leavenworth [Army school]...and the one who attempted to write the artillery section of the Manual...had no concept of the employment of an FMF in defense of a base,' and when Miller questioned him about it replied, 'I [Captain Collier] admit it. I never had a picture of it until I saw the rehearsal for N.W.C. problem a few days ago.' And yet for four months he had been paraphrasing Army school literature and trying to build a structure of a false foundation, <u>without knowledge that his foundation was false</u> [underline in original].' Miller continues, 'We are not anti-Army, but are pro-Marine...Do we want to build up the Fleet Marine Force with officers who are prepared to face Naval-Marine problems or Army problems?... How can we be recognized as worth while, when even our own headquarters don't believe we are a good school.'[77]

Russell in a memorandum dated 14 May 1934 commenting on Miller's charges writes, 'A careful reading of the attached memorandum [Colonel Miller's] leads me to the conclusion that the detachment of this officer, which occurs in the near future, is for the best interest of the Marine Corps School and the Marine Corps. While the enthusiasm of the officer in his work is to be commended, it is nevertheless, apparent his vision of the Marine Corps and the obstacles under which it is operating, is decidedly limited. At times the language and inferences are not all that might be desired but this is accounted for by his intense enthusiasm. He is an excellent officer but is very apt to be carried away by his interest in the subject. He needs a balance wheel.'[78]

Here is strong evidence that a product champion such as Russell, who was attempting to disguise his disruptive innovation as a sustaining change, has no use for a zealot subordinate. Disruptive theory predicts that this would be the case. A zealot would tend to counteract the disguising efforts of the product champion and as such would be counterproductive. As discussed in Chapter 8, however, in the Navy's continuous gunfire case, Lieutenant Sims, a zealot product champion, is useful in achieving a sustaining innovation as the zealot is merely trying to introduce new technologies that will help perform existing missions better.

The next intellectual step came in 1934 when Russell, a close personal friend of President Franklin Roosevelt, arranged for the President's son, James, to be commissioned as a Lieutenant Colonel in the Marine Corps reserve. Cunningly, Russell assigned the President's son to advocate the FMF.[79]

The next intellectual step came with Russell's successor, General Holcomb, who also deserves credit for managing the nascent disruptive assault mission to a mature science. Holcomb continued shifting as many Marines as possible into the FMF to gain crucial offensive assault training. With the FMF at less than one-third its planned wartime strength of 25,000 Marines, he desperately needed more Marines, but Congress would not fund offensive 'interventionist' forces. As Millett notes, 'The political chances for increasing the FMF were not promising. Still Depression-anxious, Congress avoided financing military modernization with the exception of some shipbuilding. With isolationism at high tide, Congress saw the FMF as provocative and interventionist.'[80] As a result, Holcomb disguised the disruptive innovation as a sustaining innovation – defensive advanced operations – that Congress would support. As Millett writes, 'Holcomb agreed with Chief of Naval Operations William D. Leahy that it was politically wiser to stress the FMF's base defense role for American possessions rather than the seizing of enemy bases, for Congress would not fund "interventionist" forces.'[81]

Political process

Russell pushed through several key political processes in the areas of promotions and training to ensure his new way of war would survive. At the time he created the FMF, the Marine Corps had too many senior officers. Many were too old for their grade and many would not welcome a new way of fighting. Through the Selection Bill, Russell lobbied successfully to streamline the top-heavy officer corps. The bill allowed the Corps to retire 'overage' officers promoted through seniority. Another significant change Russell instituted to support amphibious doctrine was the introduction of the officer promotion selection system, which set up officer promotion boards where promotions would be based upon merit and skills developed in the FMF rather than promotions based on time in service. This helped promote amphibious warfare as those officers skilled in this new way of war would be the ones being promoted.

Russell also focused on how the FMF would train in the new way of fighting. The struggle for the 'heart and soul' of the Marine Corps was waged at Quantico and within the Corps' professional educational system. Russell emphasized development of an offensive amphibious mission for the Marine Corps, leading to the seizure and not just the defense of advanced naval bases as part of a naval campaign.

Consistent with disruptive theory, Russell did not have difficulty recognizing offensive amphibious operations as disruptive. Just before his retirement, Russell in 1936 wrote in *Naval Proceedings* that 'lacking bases for distant operations, the fleet would have to seize them'. He noted that assaulting defended beaches was more complicated and required a different approach and organization to accomplish. 'You can no longer hit the beach with some navy landing gun, some push-carts, and your rifles and bayonets'; special equipment and capabilities also would be needed.[82] Russell had created the FMF that would serve as a component in the new way of fighting. What he needed next was a new architecture that would link the components in different ways.

Russell's small innovation groups at Quantico would solve the problem of new linkages. The next step was to root out Army influence from the Marine training system. The problem Russell faced at the Quantico schools is summed up by the school's executive officer, who stated, '[The] prolonged used of this Army material, not taken from all of the Army schools, has so saturated the entire Marine Corps Schools system that its foundation is still resting on Army principles, Army organization, and Army thought...[To succeed] it will require a natural wrench to part from the universal leadership of the Army schools. But, we have been on our own feet for twelve years and we no longer need to be lead by anyone'.[83] The natural wrench was Russell. With his approval, courses were infused with naval thought. Gradually the Corps shifted from the defensive mode of defending bases, which was associated with the Army, to the offensive mode of assaulting defended bases, which was associated with the Navy.

Within two years of the organization of the FMF, funds were available to conduct the first FMF landing at Culebra. This was carried out in 1935 and conducted every year after until the beginning of World War II.

Technological innovations: landing craft development

The offensive operations envisioned by Russell required several accompanying technological innovations, such as converting traditional low-trajectory armor-piercing naval ordnance designed for attacking ships to amphibious high-trajectory ordnance for attacking soft shore targets often located on the far side of a sloped beach. But the most immediate problem was developing landing craft, which could be used in the ship-to-shore movement phase of the assault.[84]

Another intellectual breakthrough was needed to solve the material problems of amphibious warfare. The intellectual step for accomplishing this was the establishment of an architectural innovation group under the direction of product champion Assistant Commandant of the Marine Corps General Russell.

Engine of change: why and when

As Marine General Victor Krulak writes, 'The amphibious assault, with its early need for heavy weapons, vehicles, and equipment, could never come into its own until the need for ship-to-shore transportation was met'.[85] In a series of landing problems called Fleet Landing Exercises (FLEX) from 1934 to 1941, amphibious forces tested the doctrine in the tentative manual. During the FLEX 1, participants again argued that existing landing craft, some still towed to shore, for the ship-to-shore movement were inadequate.[86]

Throttle of change: how

Intellectual process

In 1933, Russell realized that ship-to-shore movement, which could deliver heavy weapons, vehicles, and equipment, was the critical challenge in conducting amphibious assaults. In light of this concern, he established an 11-member architectural innovation group called the Marine Corps Equipment Board. It was the first professional body in the United States to devote all its time and study to the development of material suitable for use of troops in amphibious warfare. For the next several years, the board worked closely with the Navy to develop the needed landing craft for amphibious assault.

The intellectual transition from a force that could land heavy equipment dockside to one that could land such equipment in shallow water beachside appears to be the work of one man, First Lieutenant Victor Krulak. Krulak, who was Ellis' counterpart in amphibious assault equipment development, was serving in Shanghai in 1937 when he observed a Japanese landing assault on Chinese positions at the mouth of the Yangtze River. He took pictures of an odd-looking craft with a ramp in the bow that was designed to negotiate the surf for shallow beach landings.[87] Krulak realized this was the big break the Marine Corps was looking for, 'sturdy, ramp-bow-type boats capable of transporting heavy vehicles and depositing them directly on the beaches. What we saw was that the Japanese were light years ahead of us in landing craft design'.[88]

Political process

The Navy's Bureau of Construction and Repair was responsible for designing and developing all ships and boats, but in the 1930s the development of landing craft was at the bottom of their agenda.[89] The Navy's prewar assumption for creating amphibious shipping was converting merchantmen and liners to military service. The Marine Corps was unsuccessful in convincing the Navy to adopt an alternative program that would build amphibious troop and landing ships.[90] Only through the persistence of the Marine Corps architectural group did the Navy's bureau consider a request for building ship-to-shore movement craft.

As an illustration, Krulak sent his pictures and report to the bureau of ships, where it sat for two years in a file with the note, 'some nut out in China'.[91] Krulak retrieved his report and presented it to Commandant Holcomb, who showed it to the Secretary of Navy. In 1941, Major Ernest Linsert, a member of the architectural innovation group, showed Krulak's pictures of a Japanese landing craft with a ramp in the bow to Andrew Jackson Higgins, a civilian boat builder. Higgins, in turn, developed the landing craft vehicle, personnel (LCVP), a ramp-bow craft used extensively throughout World War II. It is fair to say that this pattern of the equipment architectural innovation group pioneering and sponsoring equipment ideas and presenting them to the Navy's bureau of shipbuilding was how the Marine Corps developed the now-renowned amphibian tractor or alligator, which carried troops and equipment across every beachhead from Tarawa to the Rhine. Of note, the alligator concept came from Colonel Ellis's 'war portfolio', which called for a reef-crossing vehicle to deal with the coral-ringed atolls.

The story of the landing ship, tank (LST), the workhorse of the Pacific campaigns, appears far less an American innovation than a copy of Japanese efforts. In 1937, British and American observers watched the first LST, the Japanese *Shinsu-maru* operating near Shanghai. The ship was a significant development in amphibious operations as it carried landing craft in a well deck, could land on a beach, and its bow opened up to allow troops and equipment to depart.

The British were the first to copy the design, and the US Navy subsequently copied both the Japanese and British versions of the ship.[92]

Wartime innovation

Generally speaking, the amphibious tactics employed to defeat Japan and Germany were fundamentally the same, based on the amphibious doctrine the US Marine Corps developed during the interwar years. The European and Pacific areas, however, presented different problems. Variances in geography, the nature of enemy defenses, and the personalities of forces and commanders resulted in different doctrines in the European, North African, Central Pacific, and South Pacific theaters. As Marine Corps Historian Jon Hoffman notes, 'By the time of Iwo Jima in February 1945, United States and Allies can successfully execute amphibious assaults and ferocity that Marine planners in 1933 probably never dreamed of [conducting]. Parts of the 1933 doctrine were still being used, but much was new and many other things had been tried and discarded along the way.' The Marine Corps deserves enormous credit for creating the doctrine for amphibious assaults, but it was extremely tough for the Marines of the 1930s and 1940s to get it right.[93]

Southwest Pacific: MacArthur's amphibious navy

In the South Pacific, amphibious operations generally were conducted on large land masses where much of the coast was undefended or only lightly defended. This did not mean, however, that these amphibious forces did not face severe challenges. The amphibious forces commanded by General Douglas MacArthur in New Guinea suffered from a dearth of big amphibious transports and the smaller but faster APDs – converted from World War I destroyers for use by raiding parties.[94]

As a result, the Seventh Amphibious Force relied mainly on beaching-type ships, such as LSTs. Because air cover was provided by temporary loan of Pacific aircraft carriers, MacArthur (and Admiral Barbey) altered the amphibious doctrine developed by the Marine Corps during the interwar years. Enemy strong points were bypassed and Allied troops landed on lightly-defended beaches of their own choosing. Once a beach was chosen, Admiral Barbey used multiple narrow landing points to seek out enemy weakness and exploit them by shifting forces. Limited by the lack of amphibious shipping and unsure of Japanese defensive concentrations, Barbey became an expert at landing in limited visibility across lightly-defended landing points, rapidly reinforcing success and evacuating quickly if there were failures.[95] During World War II the MacArthur and Barbey team conducted 36 successful assaults in which their ships carried close to one million Allied troops and a million tons of supplies from Australia, with a casualty list of but 272 men.[96]

Central Pacific: General Holland Smith and
Admiral Richmond K. Turner

The landing challenges faced by those operating in the Central Pacific, with its small islands and atolls, were much different. These limited land masses offered few spots for landing, and these beachfronts usually were heavily defended. Considering these defenses, training emphasized techniques for storming such positions, including high coordination between the Marine Corps and Navy to maximum fire support. Tactical surprise was forsaken in exchange for extensive preparatory bombardment to achieve maximum destruction prior to a landing. Consequently, these assaults were conducted in daylight. Because of the contested ship-to-shore movement, Major General Smith devised a three-echelon landing concept. In the first echelon light mobile forces seized commanding terrain to protect the ponderous main forces of the second echelon as they built up combat power ashore. The third echelon consisted of reserve and support forces.[97]

North Africa and Europe: Admiral Connelly

Political considerations and the existence of friendly underground forces affected the landings in North Africa and Europe. The shores on which landings were made were not necessarily hostile, nor did those landings always require an assault. The enemy in most of these cases chose to rely on mobile reserves to strike the landing forces soon after they hit the beach, rather than take up defensive positions at the water's edge.[98] As a result, night and surprise landings were feasible and generally used, as parachute and airborne troops could be employed to support the flanks of the advancing assault force. There was little likelihood of enemy naval intervention, but air supremacy at the objective area could rarely be counted on.[99]

Summary of amphibious assault warfare

Perceptions of structural changes in the security environment, theoretical developments involving new military capabilities to meet these threats, and interservice rivalry appear to be at the heart of the intellectual development of amphibious warfare.[100] A few 'Young Turks' within the Corps argued for offensive amphibious warfare as their service's priority mission. After several years, one of these Young Turks, Russell, became Commandant and actively supported the new offensive vision of warfare. As product champion, he created three important assets for advocates of the new mission: (1) the Fleet Marine Force, a critical component in the new offensive architecture; (2) a set of incentives in the form on new opportunities for promotion; and (3) small innovation groups in which new doctrine was written and new equipment developed. Perhaps most important, in writing the new doctrine Russell was not blinded by the Army knowledge embedded in the Marine Corps manuals. In managing this disruptive innovation, he rooted out Army doctrine that was not relevant to amphibious operations and replaced it with new Marine Corps doctrine that supported the mission. By the end of 1934, the Corps was on its way to being transformed. It is true, however, that both Russell and Holcomb were able to protect the disruptive innovation by disguising it as a sustaining innovation. By 1941, when the United States entered the war, the Marine Corps was ready to put its new doctrine to the test.[101]

An interesting discovery in the disruptive innovation case is the role of the zealot. As demonstrated by Russell, when he encountered a zealot in Colonel Miller, he did not want him in his small innovation group. Disruptive theory supports Russell's decision as Russell does not want his disguising efforts countered by a zealot who is busy attacking his opponents, the Army-trained Marines.

With this said, we can apply the same logic to Major Pete Ellis, the Marine Corps' most famous 'maverick'. Lejeune approved of Ellis' scholarly work, which focused mainly on the sustaining innovation of defense of advanced bases. According to disruptive theory, Lejeune would approve a zealot promoting a sustaining innovation, but would not approve of a zealot promoting a disruptive innovation. Perhaps this explains why Lejeune does not mention offensive amphibious operations in his memoirs. According to disruptive theory, amphibious warfare was not a priority for Lejeune, who seemed more focused on teaching Marines to defend advanced bases and fight along side the Army in land campaigns. A more controversial finding using disruptive theory is that although Ellis did spend the last few pages of this famous report describing the future of amphibious warfare, his mysterious mission to Micronesia was not about gathering intelligence to support his offensive plan of an island-hopping campaign, but rather it was to gather intelligence to support a defensive plan for island-hopping. That is, Ellis was trying to locate the islands that were not occupied by the Japanese that the Marines could occupy using defense advanced base tactics. Ellis, simply put, wanted to be known as the defensive warfare prophet, not the offensive one, which he is known as today.[102]

4 Post-World War II Marine Corps disruptive innovations

(I) Helicopter warfare

The pattern of innovation in the two cases which follow is remarkably similar to that of the two interwar cases. As before, it is possible to identify individuals who consciously redefined critical military tasks in response to changes in the strategic environment or in technology and who had a strategy for gaining control of the officer corps in order to implement the new way of war.[1] Product champions created and managed architectural innovation groups to develop new ideas about the ways future wars would be fought and how they might be won. In managing the intellectual process, these reformers initially promoted air mobility in the Marine Corps by reaching out to fixed-wing aviators and nonaviators and convincing them that the change would sustain the established trajectories of performance improvement that senior military leaders historically had valued. Only after product champions had pioneered the innovation successfully did the change become disruptive and outperform the existing way of fighting.

As an illustration, during the early stages of development, helicopters were able to lift just slightly more than their own weight – roughly three infantrymen. A typical World War II landing craft could easily deliver seven times as many troops in half the time. As disruptive architectural theory predicts, the early helicopters (a disruptive innovation) underperformed the established landing craft in the traditional, ship-to-shore mode. They were unreliable, relatively slow, and unable to carry large cargoes. In time, however, technological advances in helicopter lift and speed resulted in the airmobile mode being superior to the surface mode.

The process of implementing these innovations also has showed considerable consistency. Senior military officers acting as product champions create a new set of operational tasks relevant to the new military capability and open a new promotion pathway for young officers to follow.

In the period after World War II, the US Marine Corps transformed itself from a force that conducted ship-to-shore amphibious assault to the first force capable of air mobility and able to conduct a vertical and surface amphibious assault. In the same period, the Corps also transformed into true combined arms operating force in the form of a Marine Air-Ground Task Force (MAGTF) (see Chapter 5).

Helicopter warfare: airmobility and vertical envelopment

Thanks to the foresight and imagination of a few officers at Quantico, the Marine Corps was the first service to employ helicopters in combat – during the Korean conflict. As Lieutenant General Gerald C. Thomas stated, 'Indeed, the helicopter gave clear evidence, from its first tactical employment [in Korea], that a major advance in combat was at hand.'[2] Building on the successful use of helicopters to transport troops in combat, the airmobile program gained interest and momentum as it presented the means to make amphibious operations viable at the dawn of atomic warfare.

The introduction and employment of helicopters into the aviation arm of the Marine Corps for use in the warfighting concept of vertical envelopment was a disruptive architectural innovation.

Engine of change: why and when

The process leading to the development of helicopter aviation in the Marine Corps appears to be due more to technological advances and interservice rivalry rather than to changing commitments resulting from changes in the security environment. The atomic bombs that ended World War II ushered in a new age. Uncertain of the weapon's potential for destruction if used against the fleet, the Navy organized Operation Crossroads in the summer of 1946. In the Bikini Lagoon in the western Marshalls, the Navy detonated two atomic bombs, one in the air, the other underwater, in the midst of a fleet of obsolescent target ships.[3] Damage to the ships indicated that World War II amphibious techniques needed to be altered drastically to cope with the new threat.

The results of the Bikini test gave the Army the ammunition it needed to take another shot at the viability of the Marine Corps' primary mission – amphibious warfare. In an off-the-record speech in 1947, an Army Air Force brigadier general stated, 'Now, as for the Marines, you know what the Marines are. They are a small, fouled-up Army talking Navy lingo. We [the Army Air Force] are going to put these Marines in the regular Army and make efficient soldiers out of them.'[4] By 1949, senior Army officers argued that amphibious assault no longer was a practicable tool in America's joint toolkit. General Omar Bradley, who had commanded the US First Army in Normandy during Operation Overload, startled the naval services with his statement, 'I am wondering whether we shall ever have another large scale amphibious operation. Frankly, the atomic bomb, properly delivered, about precludes such a possibility.'[5] The most radical suggestion, however, came from Secretary of Defense Louis Johnson, who in December 1949 stated, 'There's no reason for having a Navy and Marine Corps. General Bradley tells me that amphibious operations are a thing of the past. We'll never have any more amphibious operations. That does away with the Marine Corps. And the Air Force can do anything the Navy can do nowadays, so that does away with the Navy.'[6]

For their part, the Navy and Marine Corps argued that the atomic weapon did not alter amphibious doctrine, although it did have a profound effect on amphibious techniques. Motivated by interservice rivalry and the desire to preserve the Corps' amphibious mission, Marine planners began immediately to devise a new amphibious technique that would be workable in the atomic age. After pursuing several possibilities, senior military leaders adopted helicopter assault. The important point to note, however, is that only after product champions of airmobility succeeded in redefining amphibious assault around helicopters and building new craft and helicopter ships did amphibious warfare become viable in the atomic age.

Throttle of change: how

Intellectual process

The intellectual process that would drastically change the character of the Fleet Marine Force following World War II began with Lieutenant General Roy Geiger. As the senior Marine Corps representative at the atomic test, he wrote a stunning document about the need for the Marine Corps to reconsider its approach to amphibious warfare. Motivated by the possibility of a future enemy possessing small atomic weapons, Geiger argued for a smaller expeditionary force that would be highly trained, lightly equipped, and transported by air or submarine. He also argued that ship-to-shore movement should be conducted with a greater degree of surprise and speed and that the expeditionary force should consider dispersing over a much wider front than used in past operations. Geiger concluded his report, 'With an enemy in possession of atomic bombs, I cannot visualize another landing such as executed at Normandy or Okinawa.'[7]

After receiving Geiger's report, the Commandant of the Marine Corps, Alexander Vandegrift – the commanding general for the Guadalcanal assault – created a small disruptive innovation group led by Major General Lemuel Shepherd and including members of the aviation arm and schools branch.[8] Vandegrift's group was to create a new way of amphibious warfighting for the atomic age. In forming the small innovation group, Vandegrift made two major assumptions: (1) the amphibious force mission still was valid, and (2) the amphibious tactics and techniques developed prior to and during World War II had to change if the amphibious mission was to succeed in an era of greater firepower on the battlefield.[9] Both Vandegrift and senior Navy leaders believed that increased dispersion of naval air components and surface ships would improve protection while allowing these two naval arms still to perform their tasks. The problem, however, was that dispersing the landing striking arm would dilute its strength and impair its ability to attack swiftly and in overwhelming force at the point of landing. In simplest terms, dispersion for the Fleet Marine Force (FMF) meant risking the consequences of decreased firepower against a superior enemy.

Shepherd's small innovation group explored several ways to achieve mobility. From the beginning, they ruled out airborne operations by glider, parachute, and land plane transport as not suitable for the Marine Corps. Helicopter airborne landings, submarine landings, and large flying boats all were considered, but the group placed the most faith in helicopters as they believed the airmobile amphibious assault technique provided the speed, flexibility, and dispersion needed to counter atomic weapons.[10] With this decision, the small innovation group achieved an intellectual breakthrough – a new tactical construct that eventually would be called vertical envelopment. A Marine Corps historian explained the breakthrough as follows:

> What the new concept [vertical envelopment] envisaged, in brief, was an assault landing without concern for reefs, beaches, beach defenses, and surf; a landing from the air but free of the inflexibility, tactical disorder, and disorganization of parachute operations; an airborne attack independent of airfields and airheads; a landing force that could be launched from ships widely dispersed and under way miles off shore.[11]

Once the Shepherd group convinced Vandegrift that helicopter mobility was the direction to take, two questions remained: How would amphibious warfare incorporate helicopters, and what would the new warfare look like? Vandegrift decided on a two-step intellectual process to advance the airmobile theory and put its elements into practice. First, in a move reminiscent of Russell's 1930s initiative, in the summer of 1946 he directed the establishment of a vertical envelopment small innovation group at Quantico schools to develop a new tentative doctrine for ship-to-shore movement by helicopters. The schools immediately began work on an airmobile concept of amphibious operations, which would evolve into vertical envelopment doctrine.

Second, in 1947 Vandegrift set the wheels into motion to establish an experimental squadron (HMX-1) of 12 helicopters of the first available type.[12] He placed HMX-1 under his direct operational control via the Commanding Officer of the Marine Corps Air Station at Quantico. The squadron would test and perfect the helicopters the Corps would need and would assist the small innovation group in developing doctrine for the tactics and techniques of the employment of helicopters in amphibious operations.

Following the script of the interwar years, Vandegrift's small innovation group at Quantico did not wait for the development of larger and more powerful helicopters to work out the shape of the new way of war. As the craft itself was perfected, in 1948 the group prepared a tentative doctrine on the employment of helicopters called *Amphibious Operations – Employment of Helicopters (Tentative)*.[13] Although not all the components existed to make the new way of war work, the innovation group had forged the new warfighting architecture, which, it is important to note, did not challenge any of the central tenets of amphibious doctrine hammered out by the Marine Corps in the 1930s.

The intellectual task of going beyond a general understanding of the potential utility of an innovation to a description of how that innovation would be implemented was spearheaded by Lieutenant Colonel Keith McCutcheon, who would assume command of HMX-1 in 1950.[14] Prior to McCutcheon's arrival, the squadron had had some significant accomplishments. For example, in May 1948, with five helicopters, it had conducted the first amphibious vertical assault from the escort carrier USS *Palau*.[15] The primary lesson learned from this first airborne amphibious assault was the need for a larger helicopter. In 1949, Marine Captain Wallace Blatt conducted the first military helicopter rescue during the US withdrawal from China.[16]

McCutcheon, vertical assault's intellectual equivalent of Ellis, saw great potential for helicopters in combat. He couched his vision in three basic assertions. First, 'helicopters are a tool to help the ground forces (or naval or air) but primarily infantry. They alone can not wage effective combat let alone win a battle. They are merely a tool to assist the overall commander.'[17] Second, helicopters belong to the fixed wing aviation arm, not the infantry arm. Third, individuals controlling helicopter assets should have the requisite technological expertise and tactical experience.

During the Korean conflict, McCutcheon was responsible for introducing the first helicopter unit trained and organized for combat duty.[18] During the Chosin operations in November–December 1951, the Marine Corps helicopters provided the only liaison among isolated commands. McCutcheon was instrumental in pioneering these vertical tactics under combat conditions.[19] Another first was the use of McCutcheon's helicopters to evacuate and transport wounded Marines from the front to a hospital ship.[20] Finally, in Operation Mousetrap, McCutcheon provided short-notice helicopter lift to troops carrying out anti-guerrilla operations in the Korean countryside, an operation that would be carried out as a matter of routine in Vietnam a decade later.[21]

After several operations in Korea, McCutcheon, the product champion of vertical envelopment, wrote, 'A military without helicopters in the future will be as obsolete as a cart without a horse. They will give to the military a new style of cavalry with the all important characteristics of mobility, speed, and dispersion.'[22]

The next intellectual step occurred in 1956 when the Commandant of the Marine Corps, General Randolph Pate, created a small innovation group named after Major General Robert E. Hogaboom.[23] McCutcheon, the only member of this group with experience in command of a helicopter squadron, and 15 other officers were tasked to recommend the components, such as composition and equipment, and the architectural structure for the FMF that would yield the optimum force to perform vertical envelopment.[24] The central issue was whether ship-to-shore movement in the future should become an 'all helicopter concept' or be combined with 'traditional' crossing-the-beach operations. The board recommended that a mixed technique be used and that the FMF be redesigned in a way that 'would facilitate and be consistent with the requirement for the projection of seapower at any selected point in the world littoral without the necessity of direct assault on the intervening shoreline'.[25]

Although the Korean experience demonstrated that air mobility was a viable concept, architectural problems of how to integrate the air and ground components remained to be solved. The fashion in which product champions managed this process coincides with the predictions of disruptive innovation theory.

Intellectual process

Although helicopters had proved of value in the Korean War, their long-term use depended on product champions such as McCutcheon convincing senior Marine leaders that there should be a mix of helicopters and fixed-wing aircraft in the service's aviation arm. McCutcheon, like Russell in the early 1930s, recognized that the best strategy was to get both fixed-wing and helicopter pilots on his team.

The debate between the fixed-wing and helicopter communities was centered on mix, not mission. Helicopter advocates pushed the idea that it was in the best interest of the Corps to be able to transport an amphibious assault division entirely by air. In reply, the fixed-wing pilots argued that they must have enough fighter aircraft to provide adequate air cover of an amphibious landing.[26] Fixed-wing aviators also argued that the airmobility concept had limited utility and should not be expanded to the point where fixed-wing aircraft would be sacrificed.[27] They were disgruntled by the number of helicopters that had to be added to the aviation arm under the new airmobile amphibious concept. Apparently, the helicopter community had severely underestimated the ability of helicopter manufacturers to deliver the numbers of helicopters required for vertical envelopment,[28] and senior helicopter pilots had begun exploring the possibility of procuring other types of helicopters from different manufacturers.

McCutcheon ensured that helicopters would remain a viable part of the air arm by forging a compromise during the first rounds of negotiations and settling for a fixed-wing to helicopter ratio that fell short of the number of helicopters required to conduct an all-vertical-envelopment assault. McCutcheon also was able to convince the architectural board and Shepherd that all fighter squadrons should be outfitted with aircraft capable of providing close air support as a secondary mission, thus reducing the need for more attack squadrons.[29] In the long term, dual-mission fixed-wing aircraft would result in fewer planes, which would create the space needed for the anticipated increase in heavy lift helicopter squadrons.[30]

McCutcheon had strong opposition from other corners. In addition to mollifying the fixed-winged opponents in his own air arm, he had to convince infantry officers that the helicopter mission would not interfere with the way they fought. Naturally, not all ground commanders believed him. The airmobility concept also faced resistance from the artillerymen, who thought the helicopter pioneers eventually would try to form an alliance with the close air support adherents in an attempt to rob the Marine Corps of its field artillery.[31]

Disguising strategy

To ensure that helicopter warfare survived within the aviation arm, McCutcheon presented the innovation as sustaining rather than disruptive. He understood that fixed-wing pilots were not concerned that helicopters would replace them as close air support or fighter craft. In fact, they had little interest in competing for the airmobile concept because they did not have the technology to accomplish this task. So airmobile product champions attempted to assuage fixed-wing aviators by couching the innovation as sustaining in nature, asserting that close air support operations and the vertical assault missions were inseparable.[32]

Once the helicopter advocates had implemented their airmobile strategy, however, they turned their attention to the problem of fixed-wing aircraft's inability to provide ground-fire suppression in the landing zone. During off-loading, helicopters and infantry units were vulnerable to enemy ground fire, and fixed-wing aircraft could not offer adequate protection because they could not deliver their heavy ordnance accurately and could not stay on station for prolonged periods. Helicopter advocates began to look for answers, eventually developing an armored helicopter that could compete with the fixed-wing units in the tasks of close air support and air to air combat, which led to the development of the Cobra gunship.

In the classic pattern of disruptive innovation, helicopter advocates did not compete head-on with fixed-wing pilots, but rather sought to dominate a task in which fixed-wing pilots had no interest. Only after they had the assets to perform the airmobile mission did they turned their sights to fixed-wing missions by introducing the Cobra gunship.

Political process

The effort to promote air mobility by reaching out to fixed-wing aviators who had established their legitimacy according to traditional Corps values began with McCutcheon, the product champion of helicopters and air mobility. A fixed-winged aviator with a master's degree in aeronautical engineering from Massachusetts Institute of Technology, during World War II he had developed the doctrine for close air support.[33] Although not a qualified helicopter pilot, in July 1950 McCutcheon was assigned as commander of the experimental helicopter squadron, HMX-1.[34] It was here that he engineered much of his important work in promoting airmobility.

Advocates of air mobility began with a large handicap – there was a serious shortage of helicopter pilots. To remedy this, fixed-wing aviators were transferred to helicopters. Unfortunately, many senior Marine Corps aviators were not committed to the helicopter program because they considered it less prestigious than flying fixed-wing craft. Consequently, many new pilots coming to the helicopter program from fixed-wing squadrons were disenchanted by the prospect. McCutcheon therefore recommended that only officers who were majors and above and who were totally committed to becoming career helicopter pilots be allowed into the program.[35]

McCutcheon did not become or act like a maverick or a zealot in promoting the airmobile concept with the ground commanders. He was keenly aware of the traditional military standards of infantrymen, and he worked assiduously to create an airmobile community that respected the infantry arm. As an illustration, during the

Korean conflict, he repeatedly urged his pilots not to engage in flights of convenience that might distract from their true mission because he feared that this type of flying might spoil relations with the ground commanders to whom they were trying to 'sell' helicopter assault.[36] McCutcheon stated, 'Our hardest battle is to eliminate or keep to a minimum the number of unnecessary, worthless, or misuse hops, which are made more for convenience for personnel than anything else. We are not ready to replace feet or wheels 100 percent yet!'[37]

Shortly after taking over as Commandant in 1952, General Lemuel Shepherd, in an effort to begin the process by which young aviation officers (including helicopter officers) would be promoted to the highest ranks in the Marine Corps, proposed elevating the senior air commander in the FMF to the grade of lieutenant general. McCutcheon strongly agreed with promoting aviators, but he worried that this proposal might create the embarrassing situation of having the senior aviator outrank his immediate superiors in the FMF. He countered that a better way to create a three-star position for an aviator was to create a vice commandant billet with the requirement that its occupant be an air officer if the commandant was an infantry officer or an infantryman if the commandant was an aviator.[38]

With these efforts the product champions of helicopter warfare created new career paths for junior helicopter officers, which eventually resulted in a change in the distribution of power as helicopter pilots achieved flag rank.

Summary of air mobility

The dawn of the atomic age changed the character of the security environment and provided the intellectual basis for the development of helicopter warfare in the US Marine Corps. Marine General Geiger, after watching the Bikini atomic test, alerted the Corps that the techniques of amphibious warfare must transform if it were to remain the service's primary mission. Immediately, the Commandant created the Shepherd small innovation group, which recommended increased dispersion of and mobility for the striking force. Subsequently, senior Marine leaders called on the Marine schools to establish a small innovation group to explore the Shepherd's group's conclusions. Although naval air warfare and surface warfare could be conducted in an atomic attack with increased dispersion of the fleet, the challenge for the FMF would be to provide protection for a scattered landing force.

Aviation advocates emerged, arguing that helicopters should be considered as a means of accomplishing the ship-to-shore movement of an amphibious assault. This airmobility concept fit nicely with 'dispersion theory'. A key member of several small innovation groups as a mid- and senior-grade officer was McCutcheon, who emerged as the Corps' most prominent airmobile expert. In the Korean War, he demonstrated the potential of helicopters as a mode of transport during combat. Afterward, he successfully spearheaded the effort to develop new components and architecture needed to make 'vertical envelopment' a workable tactical doctrine.

In managing the helicopter warfare innovation, McCutcheon created new aviation billets within Headquarters Marine Corps that would ensure senior aviators were equal in rank and responsibility to infantry officers. By doing so, he created incentives not only for aviators but also for helicopter pilots, who were considered by the aviation arm as second-class citizens for many years. Next, McCutcheon successfully sold the airmobile concept as a sustaining innovation that would pose no threat to either the fixed-wing pilots or artillery officers. He did this by siding with fixed-wing pilots, who favored a mixed air and surface assault force rather than an 'all-helicopter concept'. In the end, however, once helicopters had been established as the technology of choice for air mobility, McCutcheon shifted gears and managed the development of the Cobra gunship as a disruptive innovation, which eventually substituted for fixed-wing aircraft and replaced artillery in the close air support mission.[39] When the Navy began building helicopter carriers at the end of the Vietnam War, it was clear that McCutcheon had successfully managed and protected one of the most important disruptive innovations in Marine Corps history.

5 Post-World War II Marine Corps disruptive innovations

(II) MAGTF warfare – combined arms operations

Occurring concurrently with the vertical assault innovation was the combined-arms innovation of the Marine Air Ground Task Force (MAGTF). At the organizational level, the heart of the MAGTF concept was the integration of the aviation arm (both fixed-wing and helicopter) into the infantry assault divisions. Much slower in development and implementation, the MAGTF concept can be traced to the earliest days of Marine Corps aviation, when aviation units deployed with expeditions to the Caribbean and China between the world wars, but it did not evolve into a true combined arms force until senior military leaders were able to adapt and integrate aviation technology into the FMF.[1] MAGTF warfare is a disruptive innovation that took almost 37 years to develop and is analogous to the German *Blitzkrieg* combined-arms innovation in its length and sustained effort.

Engine of change: why and when

As a Marine Corps historian writes, 'Integration of Marine air ground forces might not have proceeded further had it not been for the deployment of atomic weapons at the end of the war.'[2] Believing that Marine aviation and ground units should operate as an integrated unit so that aviation units could provide close air support, the Marine Corps achieved tacit recognition of the concept in the National Security Act of 1947 by incorporating the following statements:

> The United States Marine Corps...shall include land combat and service forces and such aviation as may be organic thereto.

> The Marine Corps shall be organized, trained and equipped to provide Fleet Marine Forces of combined arms, together with supporting air components, for service.

Senior Marine leaders also emphasized the bond between air and ground arms to protect the air arm from interservice rivalry and Air Force proponents who wanted to absorb Marine air once the MAGTF arrived in theater.

Throttle of change: how

Intellectual process

The Marine air arm as a component of MAGTF warfare was the vision a young lieutenant, Alfred Cunningham. In 1909, the Marine Corps did not have airplanes and had no apparent desire to buy any. During World War I, however, Cunningham organized a Marine aviation unit for France, and through his lobbying efforts with Congress, the Marine Corps in 1919 was able to establish aviation as a permanent combat arm.

Intraservice rivalry grew quickly between 'real infantry' Marines and the 'aviator' Marines that almost killed the concept before it got off the ground. For the infantry, two issues dominated: (1) Marine aviation's failure to support Marine ground forces in France, although no fault of their own, created considerable tension; and (2) Marine aviation did not seem to support the primary mission of the Corps – amphibious warfare.[3]

With Russell as Commandant, however, Marine Corps schools produced the *Tentative Manual for Landing Operations* and, most important, Marine aviation played a significant role in gaining air superiority over landing beaches. At Guadalcanal, for example, Marine pilots performed superbly in air-to-air and close-air-support missions. It was not always possible, however, to employ aviation and ground units according to the tenets of the integrated air-ground team. As one Marine Corps historian notes, 'the concepts/techniques necessary for integrated ground-air teamwork were not fully developed early in the war'.[4] Navy and Marine Corps aviation grew more rapidly than the carrier-building program. Many Marine Corps aviation units were sent to the Pacific to operate from land bases, while Navy units were given most of the carrier spaces. When Marine landings were conducted beyond the range of land-based aviation, carriers and Navy units generally supported them.[5] Tactical nuclear weapons portended the end to Marine air supporting the landing force as planners argued over the viability of amphibious operations in atomic warfare.

The next intellectual step was the rebirth of the MAGTF doctrine. The MAGTF is a task organization tailored to accomplish a specific mission(s). Composition may vary, but the force normally will include a command element, ground combat element, aviation element, and combat service support element. The MAGTF organization is designed to accomplish combined arms – the integration of arms such that to counteract one, the enemy must become vulnerable to another – to maximize combat power. A combined-arms force poses the enemy with a no-win situation.[6] An example is as follows:

> We use assault support aircraft to quickly concentrate superior ground forces for a breakthrough. We use artillery and close air support to support the infantry penetration, and we use deep air support to interdict enemy reinforcements that move to contain the penetration. Targets which cannot be effectively supported by artillery are engaged by close air support. In order

to defend against the enemy attack, the enemy must make himself vulnerable to the supporting arms. If he seeks cover from the supporting arms, our infantry can maneuver against him. In order to block our penetration, the enemy must reinforce quickly with his reserve. However, in order to avoid our deep air support, he must stay off the roads, which means he can only move slowly. If he moves slowly, he cannot reinforce in time to prevent our breakthrough. We have put him in a dilemma.[7]

Although the MAGTF concept had been in existence for 25 years, its evolution as a disruptive innovation was based on and influenced by the advent of great tactical mobility (helicopters) and firepower (atomic weapons).[8] Since the introduction of air power, the Marine Corps had considered aviation assets as complementary to the ground forces, and air support was an integral part of the amphibious operation[9] – but helicopters and atomic weapons introduced dispersion theory and thus control and composition of the FMF had to be reconsidered. General Lemuel Shepherd in 1955 stated that with or without nuclear weapons, the most effective employment of the FMF was in the form of MAGTF in which both air and ground elements were responsible to a single commander.[10]

The product champion of MAGTF warfare was Marine Corps General Keith McCutcheon, who developed the first close air support doctrine for the Marine Corps.[11] As military historian James Ginther notes, 'The close cooperation necessary to deliver close air support to ground forces became one of the foundations of the post-World War II MAGTF concept.'[12]

After returning from Korea, McCutcheon was assigned to the Marine Corps Equipment Board. In addition, he served on two small innovation groups whose efforts resulted in the restructured FMF and Marine air wings in light of the evolving MAGTF concept.[13] McCutcheon developed the command and control principles in which helicopter assets could interact with ground forces to ensure that the force as a whole could adapt continuously to changing requirements. He believed that helicopter transport units should be under the control of a Marine force commander who had control of both air and ground operations. Contrary to the Air Force's insistence on control of all aviation assets, McCutcheon argued that if Marine helicopters were operating with Air Force units, they would remain under the operational control of the Marine landing force commander to ensure they were used to support Marine ground units.[14]

The next intellectual step in occurred in 1952 when the new Commandant, General Lemuel Shepherd, created the small innovation group called the Marine Corps Advanced Research Group at Quantico. This group of ten colonels met for a year to develop recommendations on how the MAGTF should evolve structurally to meet the challenges of atomic warfare and new technologies such as helicopters and high-speed aircraft. Notably, eight of the ten colonels would become general officers.[15]

The intellectual breakthrough on how to redefine Marine Corps tasks around the MAGTF concept was achieved by this group. Its recommendation was the 'single weapons system concept' – a new doctrine focused on the idea that Marine

Corps air and ground forces be viewed and employed as a single integrated weapon system. Based on this recommendation, the group argued for the total integration of air and ground units, including their command staffs. History had demonstrated that the operational separation of air units from their ground counterparts had poisoned the air–ground relationship. Thus, the group's intent was to prevent such separation of Marine air and ground units.[16] Shepherd generally agreed with the single weapons system concept, and for the first time in Marine Corps history, headquarters viewed Marine aviation as a distinct but coequal entity within the Corps and not simply as a subordinate arm.[17] The single weapons system concept became the central doctrine of the Marine Corps when in December 1955 Shepherd issued his memorandum, 'Marine Corps Air-Ground Task Force [MAGTF] concept'.[18]

The next intellectual step occurred in 1956, when the Hogaboom small innovation group recommended the further development of the 'Marine-Air-Ground Task Force' architectural concept. During the Vietnam conflict, the MAGTF concept allowed the Marine Corps to use the helicopter extensively both as a gunship and as a mobile transport for infantry.

Political process

The Corps emerged from the 1950s with a MAGTF concept that was little more than a working theory.[19] From his position as Director of Marine Aviation, McCutcheon worked diligently to provide the equipment and organizational structure to make the MAGTF concept operational.

In 1956, a move by the Helicopter 'Young Turks' threatened McCutcheon's efforts to develop the nascent helicopter concept into an established part of the air arm. Led by Lieutenant Colonel William Mitchell, they felt that the lack of standard operating procedures threatened the helicopter program and contributed to the troubled relationships between air and ground unit commanders over control of Marine helicopter operations. Without such guidance, rules were improvised, and because most MAGTF units were dominated by senior ground officers with insufficient understanding of aviation capabilities and problems, the 'Young Turks' felt the helicopter mission was being throttled. Mitchell noted that the senior infantry officers often were unwilling to defer to their more junior aviation colleagues. Furthermore, he accused the senior fixed-winged officers of being unwilling to join in development efforts and provide procedures for helicopter operations.[20] His solution was to create an independent helicopter board to deal with these problems.

McCutcheon opposed an independent board because he felt that any distancing of the helicopter program from fixed-winged aviators and ground officers would mean disaster. He believed that if the helicopter was to be an organic part of the MAGTF, problem solving had to include representative of all the armed branches.[21] Instead, he placed a few extra aviators and infantry officers on the doctrine board who could focus on the SOPs of helicopters operating within the MAGTF.[22]

As operations chief of staff of Commander-in-Chief, Pacific (CincPac) and later as commander of the 1st Air Wing during the Vietnam War, McCutcheon fought off interservice doctrinal differences that threatened the integrity of the MAGTF concept. During this process, he earned respect as an equal from Marine ground officers as they saw him as an aviator who was willing to fight for causes that helped the infantry fight better.[23]

McCutcheon also argued that aviators should be allowed to serve alongside infantry officers in Headquarters and FMF general staff billets. As it stood at the time, only infantry officers could hold these positions. McCutcheon stated, 'It is my sincere considered belief that this aircraft [the helicopter] will prove to be the binding link that really makes us the air-ground team [MAGTF] we think we are.'[24] Perhaps more than any other event, McCutcheon's warding of the attack of the Air Force solidified the integration of Marine air and ground forces.

During the last half of the 1960s, McCutcheon skillfully used his position as the Commandant's Deputy Chief of Staff for Air to entrench the MAGTF concept as the Corps' fundamental operational and organizational doctrine.[25] As Ginther notes, 'Through astute political maneuvering, keen insight, and force of character, he was able to protect the MAGTF structure from influences that threatened its demise.'[26] Aviators and especially helicopter pilots, McCutcheon noted, should serve in all positions that infantry officers could serve in.

> The ground [infantry arm] is enthusiastic about aviation's latest contribution and there is no reason why we should cease development [of helicopters and tactics and techniques for their employment]. It is possible that command of joint air-ground task forces will be given to aviators sometime in the future so it is desirable to be on the ground floor and not wait until so many ground rules have been made that we would have trouble breaching their defense.[27]

During his tenure with the helicopter program, McCutcheon carefully couched the capabilities of vertical envelopment as a sustaining innovation that would permit the Corps to continue to perform the amphibious assault mission in the era of the atomic weapon.

Summary of MAGTF warfare

The desirability of close coordination and teamwork between aviation and ground elements was widely recognized from the introduction of aviation units. As with air mobility, however, it was the advent of the atomic weapon that spurred the development of the MAGTF concept. Perhaps more important, interservice rivalry between the Air Force and the Marine Corps drove the Corps toward a combined-arms concept. Specifically, the Air Force on more than one occasion attempted to have Marine Corps air come under Air Force control. MAGTF product champions countered these Air Force maneuvers by noting the Marine Corps infantry's need for organic air for close support and to achieve air supremacy over the amphibious objective area during the crucial ship-to-shore movement.

Although McCutcheon did not invent the concept of the integrated air-ground team, as product champion of the MAGTF he did more than anyone to ensure its intellectual and political process development. He was able to accomplish the combined arms feat by focusing on creating greater opportunities for aviators and by increasing their status in the ranks, ensuring that the promotion system permitted Marine Corps pilots to take their place as equals to their infantry counterparts. Again, product champions of the MAGTF concept used the small innovation groups to intellectually develop the doctrine. Over and over again, especially in the Vietnam conflict, McCutcheon successfully waged the battle to ensure the MAGTF stayed an integrated air-ground team.

6 US Marine Corps inchoate disruptive innovation

Maneuver warfare

In 1989, the Marine Corps attempted to adopt a new philosophy on warfighting named maneuver warfare.[1] As with any disruptive innovation that requires a massive change in an organization's cultural framework, it is going to become a long-term effort by its very nature. Presently, the Marine Corps is in the sustaining innovation stage of maneuver warfare where it continues the process of institutionalizing the disruptive innovation throughout all levels of command. It is apparent, however, that with its performance in Operations Desert Storm and Iraqi Freedom, the Marine Corps is responding positively to maneuver warfare in terms of doctrinal refinements, acquisition programs, and force structure modifications. Also, the evidence suggests the development of maneuver warfare is following the pattern of a successful disruptive innovation.

In reviewing this pattern, disruptive theory argues that two general propositions are necessary for disruptive innovations to be successful. At the intellectual level, senior military leaders assess the security environment and perceive a need to innovate. This leads to an intellectual process that translates the perceived need for innovation into new concepts of military operations. At the practical level, the disruptive innovation depends on a political process whereby senior military leaders disguise the disruptive innovation as sustaining, attract other officers with traditional military credentials, and then make it possible for younger officers to rise to positions of command while pursuing the innovation.[2]

If these propositions are true, then the absence of any one of the specified factors or the presence of the civilian intervention factor should derail or delay the development of maneuver warfare. An examination of the inchoate Marine Corps disruptive innovation in shifting from the French model of methodical battle to German-style maneuver doctrine suggests that civilian intervention in the early stages may have retarded maneuver warfare development. It is worth noting that some of the civilian advocates of Marine Corps maneuver warfare incorrectly believe that the disruptive innovation failed.

As the civilian product champion of maneuver warfare, Bill Lind, writes, 'The Marine Corps formally adopted maneuver doctrine in the 1980s, but nothing was done to implement it beyond rewriting a few field manuals (which themselves are seldom read and less often understood).' He continues, 'All that changed were a few "buzzwords". The shift in institutional culture that makes maneuver doctrine real – a shift from centralized, inward-focused, imposed-discipline culture to one that is decentralized, focused outward on the enemy and the situation, prizes initiative and depends on self-discipline – simply never happened.'[3] Lind concludes 'the Marine Corps made a genuine effort to reform itself. But an attempt should not be confused with a result. The attempt was laudable; the results were minimal.'[4] The historical evidence does not support Lind's thesis, however.

FMFM-1, *Warfighting*, published in 1989, reflected the maneuver warfare philosophy of its product champion, General Alfred Gray, 29th Commandant of the Marine Corps. Not since 1934, when the *Tentative Manual for Landing Operations* began to guide the newly established Fleet Marine Force, had a single doctrinal publication promised such great change in the Corps. Marine Corps Lieutenant General Bernard Trainor (retired) notes that in adopting the maneuver warfare doctrine the Corps was 'returning to its roots in warfighting learned in the Banana Wars, but shelved in the set piece battles of the Pacific and Korea'.[5] He goes on to state 'The "maneuver warfare" emergence [by General Gray] was a reflection of our Vietnam experience...and the demoralizing effect of Beruit and its aftermath.'[6]

During General Mundy's term as the 30th Commandant, some progress was made in experimenting and developing maneuver warfare. Mundy did establish the Marine Corps Warfighting Center, designed to experiment with the maneuver warfare doctrine to develop tactics, techniques, and procedures. An examination of the Mundy tenure, however, suggests that it was primarily dominated by personalities, both civilian and military, who debated the merits of maneuver warfare. Lieutenant General Trainor recalls that the debate centered on Bill Lind trying to convince the Marine Corps that 'maneuver itself would cause an enemy to collapse without the need for combat'.[7] Many Marines disagreed with Lind, including Trainor who responded to Lind's thesis with an article in the *Marine Corps Gazette*, 'which pointed out that maneuver of itself is not enough and that at some point battle must take place'.[8] Another criticism of Lind was his apparent belief that the Marine Corps was committed to attrition warfare as the sole vehicle of warfare. Again, Trainor notes, 'this was complete nonsense'.[9] General Charles Krulak, the 31st Commandant, followed Mundy and he was a strong proponent of maneuver warfare who updated and revised several of Gray's manuals.

While Gray can be compared to the German General Hans von Seeckt, who championed the disruptive innovation doctrine of the *Blitzkrieg*, Krulak can be compared to Generals Ludwig Beck and Werner von Fritsch, who sustained the disruptive doctrine innovation by championing the mechanisms for experimentation. In doing so, Krulak focused on experimenting with new ideas to discover what could be done with new technologies and combinations of new technologies and what military tasks could be done differently using the maneuver warfare construct. Besides updating maneuver warfare doctrine, Krulak focused his efforts on the Marine Corps Warfighting Lab, which he hoped would be the mechanism by which new technologies could be translated into maneuver warfare innovations. Succeeding Krulak was General Jim Jones, who was a mentor of General Gray. Both Krulak and Jones had powerful sources of support both inside and outside their service and they effectively institutionalized maneuver warfare as a philosophy for thinking about and in war.[10] In the end as demonstrated by Operation Iraqi Freedom, maneuver warfare doctrine did change the way the Corps fought.

Disruptive innovation: maneuver warfare

Spurred by the imagination of General Gray, then commanding general of 2nd Marine Division at Camp Lejeune, the Marine Corps in the 1980s began experimenting with a style of fighting called maneuver warfare. Bill Lind, a civilian aide to Senator Taft and then Senator Gary Hart and the leader of the military reform movement,[11] coined the term, and in his 1985 *Maneuver Warfare Handbook*, laid out the concept of maneuver warfare, which he 'addressed primarily to Marines'.[12] Lind claimed that only the Germans and Israelis have institutionalized the practice of maneuver warfare in recent times. It was his goal that the Marine Corps adopt and institutionalize maneuver warfare as well.

Engine of change: why and when

The process leading to the development of maneuver warfare doctrine in the Marine Corps appears to be due to changing commitments resulting from shifts in the security environment and civilian intervention. In the early 1980s, strategists began urging the Army and Marine Corps to adopt the doctrine of maneuver warfare, which they claimed could be used to defeat the numerically superior Warsaw Pact on the plains of Europe. Proponents noted that Israeli Defense Forces had adopted this German approach to war as a way out of Israel's own strategic dilemma. Although the evidence does not support this thesis, they argued that surrounded by numerically superior enemies it could not match, Israel turned to maneuver warfare and in four conventional conflicts, defeated their stronger Arab opponents in a matter of weeks.[13] This suggested that maneuver warfare might be the prescription for success for the Marine Corps.

Some senior Marine leaders found maneuver warfare an appealing alternative to the concentration of superior firepower that was at the heart of attrition warfare. First, it rejects the attrition mind-set that contributed so much to the 'body count' mentality that led to disaster in Vietnam.[14] Second, besides providing an answer to fighting in Europe, it would be useful in a potential conflict in the Persian Gulf, against the larger, half-million-man armies of Iraq and Iran. General Gray writes, 'On its own merit, I believe that maneuver warfare is a superior way to fight. But I don't think that this fact alone can account for the interest that has been generated in this approach. It appears to me that the genesis of the interest is the recognition of the fact that the potential enemy in a major war is likely to have superior raw combat power to pit against US forces, and particularly against a deployed MAGTF.'[15] He continues, 'You don't defeat such a force by relying primarily on fire power, frontal assault, and attrition. Rather you defeat him by superior technology, maintaining the initiative with intelligent, purposeful movement, by attacking his most vulnerable point and through the application of firepower. Such is the essence of maneuver warfare.'[16]

Interest in maneuver warfare throughout the 1980s can also be traced to Army senior leadership, who were struggling with correcting the problems of the Vietnam War, and to the military reform movement, who were supported by the strong advocacy of civilian defense analysts familiar with German military history. Foremost among these advocates were William S. Lind, a civilian defense intellectual, and Gary Hart, the senator who established the Congressional Caucus on Military Reform to reform the Army and Marine Corps from the outside. Eventually known as the Military Reform Movement, the group claims it was successful in causing the Army to shift to its 1984 Air-Land Battle doctrine of maneuver warfare. This is not quite true. Although there was some insignificant debate between the Army and defense reformers, the Army created the Air-Land Battle doctrine on its own and was not spurred by the military reform movement. After the Army promulgated its Air-Land Battle Doctrine, Lind, president of the reform movement, became the central crusader in advocating the Marine Corps shift to maneuver warfare. In his word, 'Congress gave us the "bully pulpit" we needed.... The Marine Corps was especially sensitive to anything coming from Congress.'[17] Lind and his fellow reformers believed they had a ready-made Marine audience for proselytizing maneuver warfare, but Lieutenant General Trainor disagrees with this thesis. He credits 'Lind as catalyst for a movement that was already underway in many quarters within the Corps emphasizing decentralization, flexibility, and mission type orders. Two who were pushing in that direction were Colonels Mike Wyly and John Studt.'[18]

Throttle of change: how

Intellectual process

The intellectual process that would drastically alter the doctrine of the Fleet Marine Force (FMF) began with Major General Alfred Gray. As a colonel in the late 1970s, Gray heard retired Air Force Lieutenant Colonel John Boyd talk on the OODA cycle (observe, orient, decide, act) theory of war fighting.[19] The theory was completely different from the old Marine fighting doctrine, based on 1930s French military texts, which emphasized wearing down the enemy's strength. Attrition warfare was enshrined as an article of faith among older Marines.[20]

Under Gray's command, the 2nd Marine Division adopted maneuver warfare as doctrine. Like Pete Ellis, Gray was an unschooled military genius who had intuitively arrived at many of the tenets of maneuver on his own. While developing a smarter way to fight than that he had witnessed in Korea and Vietnam, Gray met Bill Lind and was impressed with his understanding of the German model of warfighting. Both were impressed with Lieutenant Colonel Boyd, who was in the midst of developing his theory for maneuver warfare.[21]

While the Army was actively debating Lind on the merits of maneuver warfare, the Marine Corps was more open.[22] Lieutenant Colonel Wyly and Colonel Gray both were interested. In 1980, Wyly, a key 'Young Turk' of maneuver warfare, became head of the tactics department at the Amphibious Warfare School. In 1980, Wyly invited Bill Lind, then legislative assistant to Senator Gary Hart and prominent member of the military reform movement, to speak to his class. Afterward, two of his students, Captain G.I. Wilson and Captain William Woods – who would become members of the 'Young Turks' maneuver warfare movement – asked Lind to continue his lectures on an informal basis, which he agreed to do. During this time the 'Young' Turks were introduced to General Gray, the product champion of maneuver warfare.[23]

In June 1980, Gray assumed command of 2nd Marine Division at Camp Lejeune, North Carolina, and immediately set a course to transform the division's warfighting philosophy from one based on attrition to one based on maneuver warfare. He sent the following message to his Marines:

> Realizing that many of our potential enemies could bring superior numbers of men and good equipment to bear against us in a distant theater, it would be foolhardy to think about engaging them in firepower-attrition duels. Historically, maneuver warfare has been the means by which smaller but more intelligently led forces have achieved victory. It is, therefore, my intention to have us improve upon our understanding of the concepts behind maneuver warfare theory and to train our units in their application.[24]

He established an small innovation group, the Maneuver Warfare Board, to help spread the concept throughout the division. Gray selected Wilson and Woods as the board's first members.[25] As Wilson recalls, the maneuver warfare 'Young Turks' generated considerable opposition and resentment from the old guard within 2nd Division, especially if Bill Lind, John Boyd, or anything German were discussed.[26] The group did, however, generate Gray's *Maneuver Warfare Notebook*, which consisted of Gray's general tenets of maneuver warfare and several maneuver warfare readings.

Eventually, Gray appointed assistant division commander Brigadier General Milligan to chair the board to consider how books and articles might be chosen for distribution within the division to stimulate interest in and knowledge of the fundamentals of effective maneuver warfare. Gray states, '[The Maneuver Warfare Board] fills our need for a professional forum on this subject and helps to institutionalize our training and education objectives.'[27]

The board spawned several of the most famous maneuverists in the Marine Corps. Colonel Tony Zinni, commanding officer of 2nd Battalion, 8th Regimental Marines brought Bill Lind to Camp Lejeune to discuss the finer points of maneuver warfare. Attending these meetings were future maneuverists such as Captain Scott Moore, Lieutenant Colonel Ray Smith (future maneuver hero of Grenada), and Lieutenant Colonel James Myatt (future leader of a Marine Division in Desert Storm).

Maneuver warfare, as practiced by Gray's division, sought to replace the traditional firepower approach of the Marine Corps with maneuver warfare procedures, tactics, and strategy. General Gray also orchestrated a series of field exercises at Fort Pickett, Virginia, to give Marines the latitude to experiment with new tactics and techniques. After each problem, Gray wanted to know 'if their scheme of maneuver was well thought out, logical, and supportive of the commander's intent. We [Gray and his senior leaders] discuss the question of whether or not the maneuver brings decisive, positive results'.[28]

For these free-play exercises, Gray established a second small innovation group led by Lieutenant Colonel K.D. Schreiber. Schreiber was born in Germany prior to World War II, and his father, a German officer, died fighting on the Russian front. After the war, Schreiber moved to the United States, where he enlisted in the Marine Corps and received the Navy Cross for his service in Vietnam. After the war, the Marine Corps sent him to the German General Staff College (formerly the famous *Kriegsakademie*) for two years. It was there that he learned maneuver warfare from the surviving German generals of World War II. Gray tasked Schreiber to the plan a series of maneuver warfare exercises at Fort Pickett. Schreiber not only secretly approached the Army's 82nd Airborne to conduct a 03:15 landing in the rear of blue forces during one of the exercises, he also secretly conferred with a visiting senior German Army officer on the finer points of the maneuver warfare exercises. After one such exercise, congressional observer Newt Gingrich was so impressed that he, Gray, and Schreiber spent more than two hours discussing the implications of maneuver warfare for the Marine Corps. It was during this period that Gray's two small innovation groups translated the theories of maneuver warfare into routine operational tasks.

During the height of the movement, the Junior Officers' Tactical Symposium in the 2nd Marine Division worked to understand maneuver warfare.[29] These 'Young Turks' considered Boyd, Wyly, and Gray as the product champions and Lind as the intellectual, political, and academic component. Lind was why the product champions were able to galvanize the 'Young Turks' into translating theory into practical application and war fighting.

The next intellectual step was the publication of the *Maneuver Warfare Handbook* in 1985. Although Lind is credited as its author, two Marines, Colonel Mike Wyly and Captain Scott Moore, translated his maneuver warfare concepts into tactics and procedures. Lind's purpose in writing the handbook for the Marines was to clear up the confusion caused by maneuver warfare.[30]

In June 1987, General Gray became Commandant of the Marine Corps. He immediately set up an small innovation group with Captain John Schmitt as its central member.[31] Schmitt translated Gray's views of maneuver warfare into a concise, readable manual, FMFM-1 *Warfighting*, which Gray published in 1989. This 98-page manual clearly explained the new Marine Corps approach to warfighting.[32] In April 1997, Charles Krulak released the first revised edition of *Warfighting*, and in the foreword he stated, '[*Warfighting*] has changed the way Marines think about warfare.... Very simply, this publication describes the philosophy which distinguishes the US Marine Corps. The thoughts contained here are not merely guidance for action in combat but a way of thinking.'[33]

Also in 1987, Gray directed that The Basic School Nonresident Program be rewritten. He established the Jackson small innovation group, led by Captain

Tim Jackson, which developed the new program, the Warfighting Skills Program, to be more in line with the tenets of maneuver warfare. A decade later, Gray would write that Jackson's Warfighting Skills Program was an extraordinary accomplishment and 'it is entirely possible that this program has had a even greater impact on instituting maneuver warfare than the more widely known *Warfighting* doctrinal manual'.[34] Jackson developed the course without input from the Schmitt small innovation group. The only members common to both groups were Lind and retired Air Force Colonel John Boyd. Jackson writes, 'Bill Lind introduced me to maneuver warfare. ...an alternative approach that made a lot of sense. It wasn't difficult to convince me that maneuver warfare was a superior approach and major change in the Marine Corps' approach to warfare.'[35] In developing the course Jackson relied on Colonel Boyd's lectures, *Patterns in Conflict*, and 'Bill Lind's *Maneuver Warfare Handbook*, particularly the tactics appendix written by Colonel Wyly'.[36]

General Charles Krulak, as 31st Commandant of the Marine Corps, was also a product champion of maneuver warfare and did much to push the ideas of Boyd and maneuver warfare.[37] He turned to the Commandant's Warfighting Lab at Quantico, Virginia, and within it, the small innovation group led by Colonel Gary Anderson and later Colonel Tony Woods, who were indirectly supported by Lieutenant General Paul Van Riper. The lab and Krulak's small innovation group were to 'develop new operational concepts, tactics techniques, technologies, and procedures to prepare Marines for combat in the 21st Century'.[38]

Krulak's warfighting lab would function as a looking glass into potential maneuver warfare operating methods. For example, senior officers took field trips to the New York Commodities Exchange to garner insights from traders as to how the Marine Corps might improve its command-and-control procedures and decision-making techniques. The lessons learned from traders' rapid decision-making in an environment of information chaos and overload are applicable to maneuver warfare.

The next intellectual step for Krulak was to eliminate Lind from the maneuver warfare movement. Implicitly, he realized that as long as Lind was pushing maneuver warfare from Capitol Hill, the Corps would never accept it and maneuver warfare would slowly fade away. With Gary Hart no longer in the Senate, Lind had less protection, and, wisely, Krulak banned him from Quantico. This turned out to be a key decision by Krulak because many Marines hated Lind for trying to change the way the Marine Corps fought, and they balked at embracing maneuver warfare because to them it meant embracing Bill Lind. With the real (or perceived) civilian intervention removed from championing maneuver warfare, Krulak could begin positioning key maneuver warfare champions within the Corps to institutionalize the concept. One of the first places he did his was in the teaching schools at Quantico.

Civilian intervention

Bill Lind states that the Military Reform Movement started in 1976, the year Senator Robert Taft Jr. released his *White Paper on Defense* and he himself wrote the critique of the Army's forthcoming FM 100–5 that he believes initiated the debate over maneuver warfare.[39] Interestingly, the model the reformers claimed to be following was the Scharnhorst reforms, which changed the way the Prussian army thought and acted.[40] The Prussians instituted a continuing, self-generating process of reform from within that, with some ups and downs, lasted until 1945.[41]

Lind and his fellow reformers focused on Capitol Hill as their center of gravity for inducing changes in the Marine Corps (as well as the Army). They had tried before to achieve military reform through legislation, but with little success, so they focused this time on the Hill because it afforded them the 'bully pulpit' they needed to get attention. Using speeches by senators and congressmen, language in committee reports, and requests for studies by the Department of Defense, Lind aimed to 'get someone who could influence DoD's budget – the only thing it cared about – to raise an issue or a point reflecting reform thinking, and thus make a connection between resources and thinking about war. That, and only that, could get the services to pay any attention to reform ideas, such as maneuver warfare.'[42]

Regarding reform, Lind noted that the Marine Corps was something of an exception. 'A significant number of Marines still regarded ideas about war as something important for their own sake, and that included some Marine generals. When they were in positions of influence, we [the reform movement] did not need to use the defense budget connection to gain the Corps' attention.'[43] Also, the Corps realized the Hill was its base, so it was extremely attentive to developments there. As Lind writes, 'If a number of Members and staff were interested in some aspect of war, such as maneuver warfare, the Marine Corps was going to be interested as well.'[44]

The reform movement also focused on creating interservice competition to gain support for maneuver warfare. Essentially, the strategy was one of threatening budgets, but on occasion they chose other things, such as roles and missions. Lind notes, 'We played the Army off against the Marine Corps this way over the intervention mission, to gain attention to the things we were interested in, which were primarily not budgetary.'[45]

Lind began writing articles about maneuver warfare in the *Marine Corps Gazette* in December 1975. Retired Marine Colonel John Greenwood, the editor of the *Gazette*, played a crucial role in defining the intellectual debate. As Lind notes, 'Without its willingness to publish on the subject, maneuver warfare would never have gotten going in the Corps.'[46] As Trainor writes about these debates, 'The greatest contribution the Lind "movement" made was in articulating the focus on the center of gravity, mission type orders, flexibility and relegating tactical decision-making to the lowest level.'[47] Many of the *Marine Corps Gazette* writers were young officers who actively participated in the maneuver warfare debate. Trainor believes the derivative of this debate is the willingness of 'today's young officers to challenge the "school solution" with those of their own. As a result, Marine Corps schools are noted more for education than training'.[48]

In the late 1970s Lind began to visit Marine Corps exercises at 29 Palms, the center of Marine Corps experimentation with mechanized forces. His purpose was to report back to Senator Hart on progress in the Corps' understanding of maneuver warfare concepts. In 1978, Lind reported, 'If the United States develops a maneuver concept of mechanized warfare, it will do so through the Marine Corps, not the Army.'[49] He concluded, '29 Palms is well on the road to becoming the *Panzer Lehr* of the Marine Corps, and indeed of the United States military.'[50] In 1979, Lind reported that although the maneuver concept was spreading and deepening and interest in new ideas remained high, 'the basic concept of maneuver warfare is still not fully understood'.[51]

By 1981, Lind still believed 29 Palms had great potential, but he reported to Hart that, unfortunately, General O'Donnell, commander of the Pacific FMF, had abolished the Mechanized Training Force (MTF). The MTF was an experimental prototype mechanized regiment that consisted of two maneuver battalions plus some artillery units. Lind argued that without the MTF or some similar structure, there could be no ongoing basis for experimenting with mechanization. He wrote, 'While congressional involvement does not appear necessary yet, it could become so, depending on whether or not the Marine Corps acts on its own initiative to address the problem.'[52]

In 1984, Lind incorrectly assessed that maneuver warfare development in the Marine Corps for the past two years was backing off. The new commander of 2nd Division was not sympathetic, and teaching maneuver warfare at the Quantico schools had been banned. 'Little done with doctrine – no equivalent to [Army's 100–5]. *[Marine Corps] Gazette* has been substantially throttled. While Headquarters USMC won't say anything publicly against it, [the Commandant] seems to be opposing internally or giving the impression it is opposed, putting commanders who are opposed in key billets. [The Commandant] has reduced it to semi-covert "young officers movement." '[53] The evidence does not support Lind's perception, however. Trainor believes that maneuver warfare continued to emerge as the Marine Corps was actively examining alternatives to fighting Lind's so-called 'attrition warfare'. What Lind did experience, however, was a less than enthusiastic embracing from the majority of Marine Corps officers who held that the professional military may well regard a civilian's input to fight differently as being outside the legitimate authority of civilian leaders. Rosen argues that officers typically believe that, 'Military professionals, not civilian politicians, are supposed to be the repository of expert knowledge on how to fight.'[54]

In 1985 the next intellectual step for Lind was to publish his handbook on how Marines should fight. This was followed by a visit the Naval War College to espouse maneuver warfare and the reform movement. The Deputy Secretary of the Navy wrote a scathing critique of Lind's performance to the Secretary of Navy, stating that Lind used the occasion as a 'forum to hold forth on what he thought was wrong with the Navy (it's brain dead), the Defense Department (it's obsessed by attrition warfare), and the United States (it's riddled with yesmen and "courtiers" who can't think for themselves and won't tell their bosses what's wrong)'.[55] The deputy secretary continued, 'Lind works from a closed intellectual system in which the tactics and organization of the German Army are confused with and substituted for the larger questions of strategy, and in which he has a neat, pat, programmed answer for every topic imaginable.' He concluded his memo, 'Lind's genuine hostility to the Navy which results from his small carrier fixation, and his diesel submarine fetish, both of which are fueled by the disingenuous posturing of his boss [Senator Gary Hart]. I urge you to remember, should there be any future requests for cooperation from this quarter that granting them will produce no benefit at all for us, and will only clothe with the respectability of experience the arguments of this turbulent liberal gasbag.'[56]

The next intellectual step for the civilian product champions was, in 1985, to accuse the Marine Corps of displays of incompetence in at least a dozen peacetime field exercises. This veiled attack was supposedly to remind the public that the Marine Corps had not yet altered its practice of simply following cookbook recipes and formulas and to teach the Corps to outthink its opponents.[57] At issue, for the civilian reformers, was the notion that new Marine Commandant, General Paul X. Kelly, would dismantle in top-down fashion the maneuver reform that his predecessor had allowed to grow from below. Led by Bill Lind and Jeffrey Record, the reformers were convinced that Kelly, who had given the impression that he was sympathetic to the need for change, was now, two years later, downplaying maneuver warfare in the schools and refusing to adopt maneuver warfare as official Marine Corps doctrine.[58] Lind writes, 'The current commandant [Kelly], in speeches to Marines, has repeatedly denounced "little groups that meet in people's basements in Washington", i.e., the groups of Marine officers that have sought to explore and spread the maneuver concept. The commanding general at Quantico has gone so far as to forbid subordinates to invite a number of civilian spokesmen for maneuver warfare – including one of the authors [Lind] – to the base.'[59] What Lind and Record feared was that the maneuver reform efforts by the 'Young Turks' of the Marine Corps would be crushed under Kelly. Lind and Record write, 'It is time for Marines and friends of the Corps to warn [Marine] Headquarters of the potential consequences of its [antimaneuver warfare] policies.'[60] The leverage the civilian reformers planned to use was interservice competition and the threat of moving resources from the Marine Corps to the Army.

In an effort to stir that interservice competition, Lind and Record noted, 'Not only has the Army abandoned its earlier opposition to maneuver warfare, it has adopted it as doctrine and issued a first-rate new field manual to promulgate it. ...The Army's internal reform movement is by no means certain of success at this point, but the effort is real. If it succeeds while the reform movement in the Marine Corps is crushed, the result my well be an increasing gap in military effectiveness between the Army and the Corps, a gap in the Army's favor.' They conclude by threatening the Marine Corps, 'Although issues of military effectiveness traditionally have been overshadowed in Congress and the press by the politics of defense budgets, politicians will eventually sense that something is seriously wrong, especially if the Marines suffer another military debacle [such as the Marine Barracks tragedy in Lebanon] resulting from professional lethargy and incompetence. The strong support for the Corps on Capitol Hill that has saved it from past takeover bids could disappear.'[61]

Intellectual process

Once at the helm, Gray immediately began to give rudder orders to steer the Marine Corps toward maneuver warfare. The problem, however, was that he did not ensure his orders were being carried out. Marine Captain Chuck Leader, who was aware of the problem, suggested that a 'change' or 'transformation manager' was needed to implement maneuver warfare, but Gray said no.[62] Without a mechanism for tracking compliance, there was a tendency for subordinates to say they were complying without actually changing established organizational routines. Trainor writes, 'Al Gray was an innovator who threw a hundred balls in the air in hopes that some of them would fly.'[63]

For example, Gray created changes in the Marine Corps officer education system, including formation of the Marine Corps University, that he thought would promote maneuver-warfare-oriented officers. The problem was that there was no guarantee that maneuver warfare instructors could be identified or would teach in the Marine Corps University system. Admittedly, Gray understood this.[64] The problem as identified by Major Mike Peznola was that the three primary officer development schools – The Basic School, Amphibious Warfare School, and Command and Staff College – did not have a mechanism for selecting maneuver warfare instructors,[65] although Gray had considerable input into who commanded these institutions, for example appointing Colonel (and later Lieutenant General) Paul Van Riper, a maneuver warfare intellectual, to head the Quantico schools. After Gray's retirement, however, the faculty changed, the intellectual energy diminished, and new instructors lost touch with his intent.[66] For example, by the mid-1990s The Basic School (TBS) had returned its focus to 'imparting knowledge' at the expense of 'developing tactical judgement'. Peznola writes, 'The Lieutenants need to know the basics, "Tactics, Techniques, and Procedures" (TTP) before they should be talking theory and reading about Austerlitz, Napoleon or [Robert E.] Lee was the common rationale used. Use of historical example was reduced at the expense of teaching "the techniques".'[67] According to Peznola, many of the TBS instructors from support specialties had not practiced tactics since leaving TBS and yet were expected to teach maneuver warfare concepts. Since infantry tactics are the tool by which the school develops decision-making and judgment, it is difficult for instructors who do not practice tactics all the time to teach them at TBS (let alone teach the skills of maneuver warfare).[68]

Disguising proposition

Disruptive innovation theory posits that senior leaders manage a disruptive innovation by disguising it as a sustaining innovation. In disguising maneuver warfare as a sustaining innovation Gray argued, 'it won't be hard to recognize that it [maneuver warfare] is a style that many Marines have employed over the years and that it has been at the conceptual core of some of our most successful amphibious operations. Inchon comes to mind immediately'.[69] Gray goes on to say that he is far from the first Marine to employ the tenets of maneuver warfare, as 'the World War II Pacific Island campaigns offer other examples'.[70] He continues, 'The concepts that we employ in maneuver warfare have been around for a long while. I find interesting something Bill Lind said in a television interview last April – that there are probably no new ideas on war that have been introduced since the eighteenth century. The concepts that we are promoting in the Second Division have been around for centuries and Marines from Generals Harry Lee to Puller to Walt to Barrow all have employed them.'[71] Gray contends that what he was attempting at 2nd Division was not new. 'What is new,' however, 'is the process of codifying it in our manuals, training for it in our exercises and in our approach to leadership.'[72]

Trainor generally agrees with Gray's assessment that maneuver warfare was not something radically new or different. He writes that Gray's maneuver warfare doctrine, '*Warfighting* simply stated the obvious as reflected in the way the Marines fought in Grenada, Panama, Desert Storm, and Iraqi Freedom...The problem was that some of Gray's successors believed it was something new.'[73]

Gray agreed with Lind's definition of maneuver warfare as a style of warfare opposed to firepower attrition that seeks to destroy the capability of the enemy to wage war.[74] However, in an effort to root maneuver warfare in some of the traditions of the Corps, he embellished Lind's definition with General Trainor's emphases on the importance of seizing and maintaining the initiative, which Trainor espoused as a foundation to his new 'thoughts on war'.[75]

In regards to training, Gray disguised maneuver warfare concepts as merely a sustaining innovation. He stated that maneuver warfare training builds on basic skills and training that Marines learn at boot camp and TBS. 'I want to emphasize that there is nothing we are proposing under this concept that is alien to the fundamental training, operations or administrative routines of the Marine Corps. The training and experience our Marines get in our depots, centers and schools equip them to participate in and contribute to the maneuver warfare objectives of this Division. ... What we are doing neither contradicts nor replaces those basic skills Marines bring to this Division from our schools and other operating units.'[76]

Gray was quite aware that Lind's naming of maneuver warfare was disruptive and controversial. In 1983, the Marine Corps Liaison Officer at the Army's Infantry School drafted a memo to Gray suggesting the Marine Corps avoid using Lind's term. The important point is that 'both Rommel and Patton called their brand of battle – Mobile Warfare. A study of their methods and a review of the pronouncements of the maneuverists would indicate there is no difference between the two titles'. The liaison officer recommended that 'maneuver is not equal to but rather a part of mobile warfare and should not be treated as a separate and distinct entity'.[77]

In 1984, General Gray admitted, 'I'm not certain that it [maneuver warfare] is the proper title. The term carries with it a lot of questionable baggage. To many it implies a heavy reliance on mechanization, to others it connotes an exclusively ground-oriented concept. Neither inference is correct. Maneuver warfare has applications across the spectrum of war from air to surface, from tactics to strategy, from operations to logistics. I'm afraid that the title had generated some semantic confusion and excessive debate over definition of terms.'[78]

In sum, two key product champions of maneuver warfare, Gray and Krulak, both understood the importance of misdirection in achieving an innovation. Gray worked diligently to convince the organizational culture that maneuver warfare was grounded in the Corps' storied past. It appears that he attempted to act as disruptive theory predicted, attempting to persuade others that the shift essentially was a peripheral intellectual change that threatened no core interest. In fact, Gray stated that maneuver warfare does not 'conflict with amphibious doctrine.'[79] Essentially, Gray attempted to promote maneuver warfare as a resurgence of Sun Tzu's ancient way of looking at conflict, but indirectly he was promoting the idea as different from attrition warfare. What he was trying to accomplish by such an approach was to prevent the attrition advocates from blocking the shift to maneuver warfare until it was too late.

When Krulak became Commandant, he too was intent on institutionalizing maneuver warfare by the indirect method. To do so, however, he would have to disassociate the movement from Lind. Krulak attempted to do this by encouraging Quantico not to invite Lind to lecture at or observe Marine schools or concept and doctrine centers.

Civilian intellectual process

As the civilian product champion of maneuver warfare for the Marine Corps, Lind did not understand the importance of misdirection and disguising a disruptive innovation as a sustaining change. Consequently, he challenged the Corps directly to accept maneuver warfare. His hubris tended to create enemies rather than advocates, and the result was general unity among senior officers in blocking 'Lind's maneuver warfare crusade'. As Colonel G.I. Wilson notes, 'Lind enjoyed being the maneuver warfare political agent provocateur, and the senior Marine leaders enjoyed attacking Lind and *his* maneuver warfare.'[80] Wilson continues, 'One of the things I am discovering is that many people in the Marine Corps respected the Young Turks stuff, but did not respect Bill Lind. Consequently, the old Marine Corps did not like civilian intervention (as in Bill Lind) asking them to steer a new direction.'[81] Trainor is convinced that 'Gray used Lind as a sounding board and accepted what he considered useful. Lind did not use Gray. Gray used Lind.'[82] The problem was that the rest of the Marine Corps perceived Lind was using Gray, and this perception angered many of the Marine Corps senior leaders. Unfortunately, Gray did nothing to squelch this perception. Trainor notes that 'the other complaint against Gray was he played favorites'. Essentially, it was perceived that if you were not knighted as maneuver warfare 'Bubbas' you were not considered part of the 'Gray club' and not on the fast track for promotion. This perception is not necessarily true, but it was held by a majority of the officers. Apparently, Gray nor Mundy did not try to detach Lind from maneuver warfare during there terms as Commandant and as a result, as Lind became *persona non grata* so did maneuver warfare. Krulak did, however, separate Lind from the Marine Corps effort to institutionalize maneuver warfare, which resulted in a renaissance period, which was continued by General Jones, of Marine Corps officers leading the maneuver warfare innovation.

Political process

A close examination of the process of implementing maneuver warfare in the Marine Corps indicates that the methods product champions followed did resemble those used by product champions for successful disruptive innovations such as amphibious warfare and MAGTF warfare, but with one exception. Perceived civilian intervention by Lind, backed by Gray and Senator Gary Hart, made the maneuver warfare disruptive innovation different. Product champions for amphibious warfare and MAGTF warfare drew officers from the traditional elite into the innovation and created new promotion paths for young officers. General Gray also drew from the traditional elite, but he did not create promotions paths for young 'Gray maneuverists'. Almost all of Gray's 'Young Turks' failed to be promoted. Colonel G.I. Wilson, an original member of the maneuver warfare 'Young Turks', writes that 'the "system" took them ALL out and they have since retired from the Corps a lot more "seasoned" and considerably pessimistic about the Corps and the impact of maneuver warfare'.[83]

Perhaps the most celebrated case of a 'Gray Young Turk' believing his career ended early was Colonel Mike Wyly. A key member of the military reform movement, Franklin Charles Spinny, writes, 'The new thinking reflected in...the Marine Corps (FMF-1, "Warfighting," 1989) was brought about by Defense Department insiders, particularly retired Air Force Colonel John Boyd,...and Marine Colonel Mike Wyly. ...These patriots risked the pain of heresy and the trauma of career ruin by having the courage to challenge our traditional doctrine. In fact, the prime mover of the new thinking in the Marine Corps, Colonel Wyly, is now being forced into retirement 8 months early.'[84]

The forced early retirement of Colonel Wyly, vice president of the Marine Corps University, shocked civilian reformers such as Bill Lind, who stated, 'Wyly deserves a significant portion of the credit for the Marines' stunningly swift maneuvers in Kuwait, accomplished with low casualties that signified 'a switch from the French style of attrition warfare to the German style of maneuver war.'[85] By firing Wyly, Lind said, the Marines are 'taking a principal contributor and slapping him in the face'.[86] A Marine colonel expressed his concern about the signal Wyly's case would send to the officer corps: 'The message is that there is no room in the Marine Corps for mavericks of any kind and intellectual mavericks are the worst kind.'[87]

The above perception by the 'Young Turks' is not entirely correct. Wyly was a brilliant thinker, but he was not a brilliant leader. Generally speaking, it was not his views on maneuver warfare that upset the senior officers, but it was his perceived status as Gray's 'Bubba'. If it was his advocacy of maneuver warfare then others such as Colonels Anthony Zinni and Paul van Riper, outspoken advocates of maneuver warfare, would have endured similar consequences. The difference between Wyly and other senior champions of maneuver warfare was that Zinni and van Riper were both brilliant thinkers and warfighters. Zinni was later to be promoted to a four-star general and held the top position as Central Forces Command and van Riper was promoted to a three-star general and held the top position at the Marine Corps 'Mecca' for training and educating Marines, Marine Corps Combat Development, Quantico, Virginia.

General Gray could not challenge the results of the independent selection board that forced Wyly to retire early without opening himself to charges to cronyism. As David Evans writes, 'It is evident from Wyly's name on the expulsion list that Gray's efforts to inculcate a greater tolerance for divergent views have had only a skin-deep impact on many senior Marine officers.'[88] Others believe that Wyly's early retirement was evidence 'of an old-timers' attempt to punish a renegade'.[89]

Soon after Wyly's retirement, other 'Young Turks' begin to face similar realities. With no senior patrons to advance or protect their careers, some well-know maneuverists, who were active promoting maneuver warfare in the pages of the *Marine Corps Gazette*, were passed over promotion or left active duty early, such as Lieutenant Colonel Scott Moore, Major Tim Jackson, Major Tom Linn, Major Chris Yunker, Captain Bruce Gudmundsson, Colonel Gary Anderson, and Colonel James Lasswell.[90] John Schmitt, author of FMFM-1 *Warfighting*, left the service as a major, and General Myatt, commanding officer of 1st Division (which was praised by Lind as the German maneuver warfare division), failed to select for another star. Retired Marine Lieutenant General Paul van Riper sums up the advancement mechanism, 'The promotion system and Professional Military Education System are held hostage by the antiquated personnel system. This system was built for the Cold War and needs a complete overhaul.'[91] Generally, there were never enough senior maneuver warfare officers on boards to ensure junior maneuver warfare officers would be promoted.

Perhaps product champions failed to protect maneuver warfare junior officers because they also failed to create a system to define a maneuver warfare officer. Maneuver warfare did not carry a warfare specialty and the fitness reports were not changed to identify which officers were qualified. Thus, other than by having a reputation for being a maneuver warfare officer, there was no way for promotion boards to identify one. As Colonel T.X. Hammes notes, 'You can't promote them [maneuverists] until you ID them.'[92] This is an appealing thesis for the Gray 'Bubbas' to advance, but observations of the graduates of the School of Advanced Warfighting (SAW), the elite maneuver warfare officer school, would indicate that their desirability and promotability negates the 'Young Turks' proposition that senior military leaders are not protecting them.

Gray introduced maneuver warfare doctrine with the perception that a civilian had imposed it on the Marine Corps. Consequently, he was unable to bring about any significant social change to accompany the disruptive doctrine such as to restructure the reality or perception of promotability for maneuver warfare officers. In October 2000, Lind would argue incorrectly that,

> Over the past several decades, the Marine Corps has sought to base its claim to expertise in the intervention mission not simply on equipment or techniques, but on its ability to fight a different kind of war – maneuver warfare. ...In

all three MEFs, free-play training, which is the heart and soul of learning maneuver, has all but vanished. The only Marine School that now attempts to teach maneuver warfare is The Basic School. The latest product of the Doctrine Division, the planning manual, is directly contradictory to every maneuver warfare concept of planning – to the (absurd) point of saying that the process is more important than the product.[93]

Thus Lind would argue that the 'true' maneuver warfare officers have all but vanished and the disruptive innovation failed. A better explanation for Gray failing to make the innovation changes he desired lie in his close association with a civilian interventionist that the Marine Corps disliked, Bill Lind. After Krulak caused Lind to vanish from Quantico, maneuver warfare begin to be institutionalized in Marine Corps schools.

Grenada

Lind is correct in claiming that one of Gray's subordinate units, 2nd Battalion, 8th Marines, commanded by Lieutenant Colonel Ray Smith, used maneuver warfare effectively in Grenada. Lind writes:

> Although the Marine Units on Grenada never met much opposition, they did face a number of confusing and urgent situations, in which they did not attempt to follow a rigid plan but rather adapted swiftly to circumstances as they changed. The speed with which the Marines acted and moved was decisive in one interesting case. The Grenadians had about one platoon of troops defending St. George's, which ultimately did not fight. Part of the reason it did not was explained by senior Grenadian officer after his capture. He said the Marines appeared so swiftly where they were not expected that the Grenadian Army's high command in the capital was convinced resistance was hopeless, the best possible outcome in maneuver warfare.[94]

Desert storm (first Iraq)

By the Gulf War, the Marines had trained maneuver warfare for almost four years, although it had not been adopted officially until 1989.[95] At the tactical level, Lind suggests the Marines two divisions fought quite differently, so that we should call one division the 1st 'German' Marine Division and the other the 2nd 'French' Marine Division. Using the maneuver warfare tenets of mission-type orders and thrust vectors instead of phase lines, Lind claims that 1st Marine Division drove the tempo of advance as fast as possible. By contrast, in 2nd Marine Division 'the planning was centralized and rigid, control measures abounded, and the focus was inward, largely on keeping the line even. When units got to the ends of their boundaries, they sometimes stopped until they could draw up new ones'. Lind claims, 'When one battalion commander, responding to new intelligence, quickly pulled his battalion back to avoid an Iraqi fire sack and counterattack, he was condemned for the unpardonable sin of "breaking the line".' He concludes by writing that 2nd Marine Division's approach worked because against this particular enemy anything worked, but in terms of what it said about the Marine Corp's adoption of maneuver warfare doctrine, 'it was dismaying'.[96]

This evidence can be explained differently, however. Within a relatively short time span after the introduction of the constructs of Gray's disruptive innovation, one-half the Marine Corps force fought using maneuver warfare. This fact rivals the speed by which the German Army developed and used the tenets of maneuver warfare in the last two years of World War I.[97]

Operation enduring freedom (Afghanistan)

Lieutenant Colonel Paul (Lester) Kuckuk, an elite maneuver warfare graduate of the School of Advanced Warfighting, writes that 'I Marine Expeditionary Force (MEF) operations during the war in Afghanistan is classic maneuver warfare – a fresh modern example that will be used as a reference for years to come. I MEF avoided surfaces and sought gaps; accepted or refused battle based on their understanding of the purpose of the mission (focus on purpose); delegated decision making authority to the lowest possible level; and communicated Commander's Intent throughout the command.'

Kuckuk argues that the Marine Corps performance in Afghanistan illustrates the point that 'leaders at all levels are imbued with the principles of maneuver warfare-indeed, it has become part of the Marine Corps service culture'. He attributes the success of maneuver warfare as a disruptive innovation because 'the [Marine Corps] service has not tried to make doctrine prescriptive, nor does it make Marines cogs in a machine…it requires that Marines think for themselves'. Kuckuk concludes by stating that 'maneuver warfare has effectively penetrated the operational and tactical thought of Marine leaders at every level. This is not to say that the principles are universally observed, but there are better than even odds that a leader drawn at random has opinions about, and can discuss intelligently, the principles of maneuver warfare. With doctrine, that's as good as it gets.'[98]

Iraqi freedom (second Iraq)

Although the official histories are still being written about the Marine Corps performance during the second Iraq War, early evidence suggests that General Conway, commanding I MEF, used maneuver warfare to fight. One example given by General Trainor is when 'Fedayeen [Iraqi] resistance harried the Allied advance, Army General Wallace, commander of V Corps, wanted to secure his line of communications before driving on to Baghdad. Conway argued successfully to continue the attack with utmost speed. In the end, General Tommy Franks, overall commander of the allied force, and Secretary of Defense Rumsfeld, came down of the side of Conway.'[99]

III Marine expeditionary force

One of Krulak's maneuver warfare champions, Lieutenant General Wallace (Chip) Gregson, is now commanding 3rd MEF in Okinawa, Japan. Gregson writes, 'Maneuver Warfare is the single, indispensable concept, not only in the conventional warfighting sense but also in that area of intertwined political/military activity that is home to what we call "Operations other than War". The current War on Terrorism requires skilful application of maneuver warfare concepts to find the political and military vulnerabilities of our enemies, and effectively attack them. It's no longer enough that our leaders of the future be technically competent, they must also be "fertile in devices", a flexibility of intellectual and practical approach that cannot be accomplished without a thorough degree of comfort with the tenets of Maneuver Warfare'.[100]

Summary of maneuver warfare

In 1989, General Al Gray championed a disruptive doctrine he called maneuver warfare. As evidenced by portions of fighting in Desert Storm in 1991, the evidence supports the claim that at least one of the two Marine Expeditionary Forces used maneuver warfare.[101] During Operation Enduring Freedom (Afghanistan) in 2001, the evidence suggests that the Marine Corps used the tenets of maneuver warfare, as demonstrated by their ability to find and exploit gaps and avoid surfaces in attacking enemy critical vulnerabilities. The early evidence from the Iraqi Freedom in 2003 suggests that I MEF fought using the tenets of maneuver warfare. This evidence combined with III MEF, led by a maneuver warfare General, supports that the Marine Corps is well on its way to achieving a disruptive innovation.

The evidence supports that the different maneuver warfare champions, Gray and Krulak, and Jones in particular followed the patterns of disruptive theory with one exception. The Marine Corps senior leadership perceived Gray's use of Bill Lind to promote maneuver warfare as civilian intervention. This stalled the disruptive innovation and after Mundy relieved Gray most of the Gray 'Young Turks' were vulnerable to attack from the non-maneuver warfare officers. Maneuver warfare could have easily been killed it if it was not for Krulak banning Lind from Quantico when he became Commandant. After banning Lind, Krulak begin placing the best and brightest maneuver warfare officers in key positions and sending them to the School of Advanced Warfighting. One of the Marines Krulak utilized was Colonel Schreiber, a native-born German and graduate of the two-year German *Kriegsakademie* and Vietnam Navy Cross winner. Before his retirement, Schreiber was the head of the Marine Corps doctrine command, and was instrumental in writing and approving a lot of the maneuver warfare doctrine that Krulak would eventually use to sustain maneuver warfare. It seems odd that Gray, who was aware of Schreiber's extensive maneuver warfare background, did not use him instead of Lind who was not respected by the Marine Corps.

A survey conducted for a master's degree thesis at the Marine Corps University in 1999 of 50 randomly-selected Command and Staff College majors and

50 randomly-selected captains attending Amphibious Warfare School revealed that 79 per cent of the respondents believed a cultural shift was required for the Corps to fully integrate maneuver warfare as its warfighting and operating philosophy. Interestingly, 59 per cent believed this cultural shift has yet to occur.

The evidence from the 2001 and 2003 fighting suggest that the cultural shift toward maneuver warfare is much greater today than it was in 1999. The leaders of the Marine Corps apparently are winning the political struggle to transform the officer corps to make it more maneuverists-oriented. They succeeded because they could couch their disruptive innovation as a sustaining change – that is, they couched maneuver warfare as something the Corps already had done many times before. Bill Lind, however, was not content with how long the process might take. He wanted immediate results and thus took every opportunity to point out how different and disruptive maneuver warfare was. It was difficult for Marine maneuverists to separate themselves from Lind and still promote maneuver warfare, but it was easy for opponents to shoot at Lind and anyone standing close to him. Once Krulak eliminated Lind from the maneuver warfare real progress was made in achieving the disruptive innovation.

Admittedly, Gray was successful in creating the intellectual framework within which the maneuver warfare view could be articulated, using definitions common to the Marine culture. What he failed to accomplish, however, was to create the incentive for young officers to accept this alternate view and to create in every Marine the will to suspend old cultural beliefs.[102] A key shortcoming of senior military leaders during the Gray era was their inability to separate Lind from the innovation and thus pursue an indirect or sustaining strategy. Krulak and Jones, however, were able to create incentives for subordinates to think about, propose, and help refine maneuver warfare, which included convincing them that if they joined the effort, their careers would not be blighted when the product champion left.[103] In sum, the evidence supports the case that the Marine Corps did embrace maneuver warfare in its doctrinal publications and it did institute it in the way it organized, trained, or equipped its forces – the real measure of acceptance.[104] It is possible that if Gray had not used Lind to promote maneuver warfare, it may have been instituted in a shorter amount of time.

7 US Marine Corps sustaining innovations and summary of disruptive Marine Corps cases

The pattern of innovation in the two cases that follows is sustaining in nature. The Marine Expeditionary Unit (Special Operations Capable) (MEU(SOC)) is the standard forward-deployed Marine expeditionary organization, which when deployed ashore has sustainment for 15 days.[1] As with all MAGTF organizations, the MEU(SOC) has a standing command element, ground element, air element, and logistics element. Being special operations capable, the MAGTF undergoes intensive predeployment training in specialized demolition operations, clandestine reconnaissance and surveillance, raids, and *in extremis* hostage recovery.[2] The Maritime Prepositioning Force (MPF) consists of three squadrons of four or five multipurpose vessels each, maintained at strategic locations around the globe so as to be within a few days' steaming of any potential hot spot.

Unlike in the previous studies, for the MEU(SOC) and MPF there were civilians who consciously redefined the critical military tasks in response to changes in the strategic environment or changes in technology. Because these two cases are sustaining in nature, new technologies were used to help perform existing missions better, not to change them radically. Thus, disruptive theory predicts the study will not observe product champions angling for control of the officer corps in order to implement the new way of war.[3] Likewise, the fashion in which senior leaders champion a sustaining innovation does not require disguising it. Because these are architectural changes, however, we should see product champions creating and managing small innovation groups to develop new ideas to improve the conduct of existing missions.

Sustaining innovation: MEU(SOC) warfare

Sparked by the foresight of senior military leaders, the Marine Corps was the only service to reorganize around special operations capabilities that could be conducted by the FMF via air or waterborne insertion, at night or in adverse weather, from over-the-horizon, and within six hours notice. Prior to being directed to train for special operations, the Corps possessed the inherent capability in its MAGTFs to conduct certain special operations, but it did not possess dedicated special operations forces.[4]

Engine of change: why and when

The process leading to the development of the Marine Corps Expeditionary Unit (Special Operations Capable) arose from a civilian directive to innovate. Following the aborted Iranian hostage rescue attempt in 1983 and the bombing of the Marine Corps barracks in Lebanon, President Reagan ordered all the services to develop special operations capabilities.[5] In addition, Congress in 1986 directed the creation of a new unified command – the Special Operations Command – as a result of a perceived lack of preparation by the four military services in counterterrorism. The Special Operations Command required each of the services to contribute in different ways. The Marine Corps response was to enhance capabilities extent within the MAGTFs. In other words, senior Marine leaders successfully avoided carving out a special operations force such as Navy SEALs. Instead, all deploying MAGTFs were trained and certified to conduct special operations.

Throttle of change: how

Intellectual process

Reacting to the President's order, General Paul X. Kelly, Commandant of the Marine Corps (1983–87), created a small innovation group to develop a Corps-wide counterterrorism program. The smalll innovation group published OH 7-14/ OH 7-14A, which focused on improving and enhancing the counterterrorism capabilities of selected FMF units. This effort eventually would come under the purview of Marine Corps special operations.

In 1984, Kelly tasked Major General Alfred Gray, Commanding General of Fleet Marine Forces Atlantic, to 'examine potential employment of FMF to conduct maritime orientated special operations and to make recommendations on the formation of a Fleet Marine Force Special Operations Force (SOC) capability'. In his direction, Kelly stated that the special operations capability should remain naval and amphibious in nature and should be organized and employed within the context of the existing MAGTF concept and doctrine.[6]

General Gray initiated the second intellectual process when he in turn established the SOC small innovation group to examine ways to enhance Marine Corps' special operations capabilities. Gray's innovation group recommended developing 'a viable special operations capability in order to provide Fleet Commanders with a "total response" capability'.[7] Next, in 1985 Gray initiated a pilot program to optimize the special operations capabilities inherent in the MAGTF. Focusing on forward-deployed Marine Expeditionary Units, Gray selected a conventional MEU commanded by Colonel Myatt to test and develop a special-operations-capable MEU. Myatt's small innovation group developed a tentative doctrine that consisted of eight missions and 19 capabilities that a MEU(SOC) should be capable of executing. This document was the forerunner to the MEU(SOC) Operations Playbook, published in 1989, which became the Marine Corps doctrine for special operations. The Playbook established 18 capabilities and stressed close integration with SEAL detachments forward deployed on amphibious ships.

Political process

General Kelly had anticipated the 1986 Congressional decision to increase military readiness for special operations in low-intensity conflict. He already had introduced an acceptable solution, which involved training and manning deployed Marine amphibious units as special operations capable. Kelly made it clear that the SOC program did not usurp the main mission of the Marine Corps – amphibious warfare.[8] In a statement to the Joint Chiefs of Staff, he stated, 'what the Corps is undertaking is the enhancement of an existing capability within its current unique and proven organization for combat – Marine Air-Ground Task Force. Our intention is to offer a complementary contribution to existing special operations capabilities that is centered on our established maritime roles and missions'.[9]

A close examination of the process of implementing the SOC program reveals that product champions did not need to create a new promotion path for junior officers because the SOC missions did not constitute new ways of war. General Wilhelm, in his report to Marine Corps Commandant General Carl Mundy entitled *Special Operations Capabilities in Fleet Marine Forces*, makes it quite clear that in the case of the SOC innovation, new methods and technologies were used to help perform existing missions better and not to change them radically. In Wilhelm's words, 'The Marine Corps does not seek an expansion of roles or missions to fulfill the maritime special operations commitments. These capabilities are deeply rooted in our past, consistent with the legislated will of Congress, and part of our 24 hour-a-day capabilities.'[10] He goes on to say, 'The Marines have *always* had an inherent maritime special operations capability…MEU(SOC) program focuses and enhances it.'[11]

To support his assessment that the MEU(SOC) was a sustaining innovation along an accepted trajectory traditionally valued by the Corps, Wilhelm cites many historical precedents, grouped into four broad categories: noncombatant evacuation operations, civil-military operations, security operations, and limited objective attacks and raids. For noncombatant evacuation operations, they include China, Boxer Rebellion, 1900; Egypt, Suez Canal Crisis, 1956; Liberia, Operation Sharp Edge, 1990; and Somalia, Operation Eastern Exit, 1991. For civil-military operations, there were Marquesa Islands, peace negotiations, 1814; Caribbean and Central America, Banana Wars, 1903–33; Haiti, Lake Miragoane flood relief, 1960; and Bangladesh, Joint Task Force Sea Angel flood relief, 1991. Historical precedents for security operations to counter significant threats, protect lives and property, and protect treaty rights and provisions include Priboloff Islands, treaty enforcement, 1891; Philippines, Mount Pinatubo eruption response, 1990; and Haiti, *coup d'état* response, 1992. For limited objective attacks and raids, they include Bahamas, raid on New Providence Island, 1776; Dominican Republic, seizure of the *Sandwich*, 1800; Sumatra, capture of Quallah Battoo, 1832; Cambodia, recovery of the *Mayaguez*, 1975; and Persian Gulf, raid on Maradim Island, 1990.

From the Marine perspective, the Corps always has had an inherent capability to conduct a wide range of maritime special operations. Nevertheless, the presidential SOC directive prompted senior Marine leaders to reexamine this capability and look for potential improvements. That effort resulted in enhancements in equipment and training that became the heart of the MEU(SOC) program.[12]

The MEU(SOC) is a clear example of civilian intervention causing a sustaining innovation. The proposition that civilian intervention can cause a sustaining innovation is consistent with disruptive theory. The civilian order to innovate is not ambiguous because what is being ordered is a familiar, well-defined task that has been done before. In a sustaining innovation, those being ordered to innovate usually have control over everything needed to carry out the order, particularly if what is needed is merely an incremental improvement of existing tasks. Furthermore, senior military leadership would consider an order to refine the way they fight as legitimate. In this case, not only did Kelly considered the SOC order legitimate because it was something the Marine Corps traditionally had performed, but he was also able to anticipate Congress's SOC legislation and move his service in that direction before the other services. As Rosen notes, 'A civilian will have most impact if he can devise a strategy that reinforces the actions of senior officers who already have "legitimate" power in the military.'[13]

Disruptive theory predicts that product champions of sustaining innovations may just as often be mid-grade officers as senior officers. In the MEU(SOC) case, the product champions were General Kelly and his immediate subordinate, General Gray, who directed the SOC small innovation group. For the sake of argument, Colonel Myatt, a mid-grade officer, commanded the MEU unit that developed, experimented, and wrote the doctrine for the SOC. It is apparent, however, that this was top-down innovation being driven by the Commandant of the Marine Corps himself.

Summary of MEU(SOC) warfare

The history of the MEU(SOC) began in the aftermath of the terrorist incidents of the early 1980s. In response to President Reagan's directive to create capabilities to conduct special operations, General Kelly created the Gray architectural innovation group to develop a Marine Corps terrorism counteraction program. Under Gray, the focus of MEU(SOC) essentially was on the FMF becoming better prepared for maritime-based raids and hostage rescue in the Middle East. These SOC tasks were traditional missions that the Corps had performed repeatedly throughout its history.

Sustaining innovation: military prepositioning forces

Traditionally, the Marine Corps has performed nonprepositioned missions. This means the Corps must be prepared to insert itself forcibly into, to fight in, and to win in areas where the United States does not have permanently-stationed forces. In contrast, historically the Army is responsible for preparing to fight and win in those places where the United States has forces prepositioned in peacetime – currently, Europe and Korea.

The Marine Corps' forcible insertion capability was fulfilled by the amphibious assault tactic. But insertion was only the first requirement. The second, equally necessary, was the ability to defeat the opponent. Although the Corps had sufficient armor and air (helicopter) assets to accomplish its insertion mission, it lacked assets to field a mechanized force to match potential adversaries in a land engagement. The solution devised by the President and championed by Secretary of Defense Brown was to procure additional amphibious shipping.

Engine of change: why and when

The MPF program has its roots in the turbulent international events of the late 1970s and early 1980s, most notably the Iranian hostage crisis, the Soviet Union's invasion of Afghanistan, the second oil crisis, and the Iran–Iraq war. All contributed to the requirement for an MPF force as they dramatized US limitations in overseas bases and in strategic airlift and sealift capabilities.[14] As a result, President Reagan directed the Secretary of Defense to devise a program so the United States would not be humiliated again in the eyes of the world. One of the main goals was to cut transit time to the Persian Gulf and save on airlift while fielding major power projection forces. Afloat prepositioning was the solution, with the US Marine Corps taking the lead. The interim program, called Near Term Prepositioning Ships (NTPS), was deployed to Diego Garcia in the Indian Ocean in 1981. The NTPS squadron included two roll-on/roll-off ships and three break-bulk cargo ships, with the 7th Marine Expeditionary Brigade (MEB) designated as its Marine Air-Ground Task Force (MAGTF).

Throttle of change: how

Intellectual process

The 1979 Iranian hostage crises demonstrated severe shortcomings in the US ability to deploy combat forces rapidly. Following presidential guidance to rectify to problem, Secretary of Defense Memorandum for Secretary of Navy of August 1979 directed the Navy Department to program for enhanced mobility for Marine Corps forces.[15] In this directive, the naval services immediately were to procure commercially-available ships to supplement the one Marine Expeditionary Force lift program with equipment and supplies.

The Commandant of the Marine Corps requested the Center for Naval Analyses (CNA) to establish a small innovation group to assess the possibilities for meeting the President's directive for strategic mobility. The CNA group proposed using aircraft to bring combat troops to an airfield in the area of concern and using dedicated, forward-based ships to bring in equipment and supplies to a nearby port. The Commandant and Joint Chiefs of Staff concurred with the innovation group's assessment and recommended to the Secretary of Defense that the

Marine Corps be assigned the mission of providing forces for prepositioned sealift assets and supplies in the Persian Gulf. In March 1980, the President approved the Marine Corps plan.[16]

In February 1981, the Commandant stated that all prepositioning programs were to be considered strategic mobility initiatives, not substitutes for traditional amphibious assault, and not a change in the Marine Corps mission.[17] Also in 1981, the Secretary of Defense directed that the Marine Corps prepositioning initiatives, 'in addition to SWA [Southwest Asia] contingencies, ... will provide a capability to respond to threats on a global basis'.[18] Further guidance from Secretary of Defense would require one MPS and one amphibious MAB by one week to Southwest Asia.

The Commandant directed a second small innovation group that would consist of a series of amphibious conferences between the Navy and Marine Corps. During its fourth meeting, the group identified the need to develop initial doctrine for the conduct of MPF operations. The Commandant directed the head of Marine Corps Schools in Quantico to establish an architectural innovation group at Quantico to develop MPF doctrine and instructed that progress reports be submitted on the first day of the second, fourth, and eighth months following the creation of the small innovation group.[19]

The doctrine developed by the Quantico small group and adopted by both the Navy and Marine Corps paralleled as much as possible established doctrine for amphibious operations. But the sustaining innovation clearly established that an MPF operation was not an amphibious operation because it did not involve forcible entry from the sea. Instead, MPF operations require a preexisting secure area for the offload.

Political process

The MPF concept was developed though the small innovation group process. In addition, lessons learned during exercises such as Bright Star 85, Agile Sword, and Freedom Banner, featuring MEB deployments and MPS offloads, were folded into the concept.

The Commandant of the Marine Corps viewed the MPF operation and amphibious operations as complementary capabilities. Essentially, amphibious operations provided the means for forcible entry, and the MPF operation permitted rapid deployment to areas where force introduction was unopposed and expected to remain so through the arrival and assembly of the deploying force. The salient requirement for an MPF operation is a secure area that will allow for the unopposed arrival and offload of the prepositioning squadron and the assembly of MEB personnel and material. Put simply, an MPF operation is a specific, discrete operation aimed at positioning a MEB for further operations. Understood in this way, the MPF operation is a reinforcement operation of a Marine Expeditionary Unit. Thus, the MPF concept has variety of uses, including to support or reinforce an amphibious operation, occupy or reinforce an advanced naval base, and preemptively occupy and defend key points along sea lines of communication (SLOCs).

Just like the MEU(SOC), this is a clear example of civilian intervention causing a sustaining innovation. The order to innovate supported a traditional way of war fighting. Because the MPS program was not a new way of war fighting, there was no need for a political struggle to gain control of the promotion mechanism to protect younger officers.

Summary of military prepositioning forces

The seeds of the MPF are found in the aftermath of the Iranian hostage crisis. In response to the President's directive for creating a strategic sealift capability for support of the rapid deployment of forces, the Commandant created several innovation groups to study the problem and develop doctrine for MPF operations. MPF tasks were traditional missions that the Corps had performed repeatedly throughout its history and thus did not require a political struggle led by product champions to protect junior officers in a new way of war.

Summary of Marine Corps cases

Overall, the Marine Corps cases show that disruptive innovations can be produced largely within the military itself. Each of the disruptive cases examined suggest that military planners were driven to consider a new way of fighting based on changes in the security environment that resulted in an increased level of interservice and intraservice competition. Certainly the evidence of interservice and intraservice competition predicted by disruptive theory was sustained in the face of various signals from the international system.

Intraservice rivalry

The inchoate disruptive innovation case, maneuver warfare, shows a strong correlation to intraservice competition. A major factor accounting for the stalled disruptive innovation during the Gray and Mundy Commandant era was intraservice rivalry. One of the most vocal opponents of maneuver warfare was the air arm. As noted by General Gray and Colonel Schreiber, Lind had a firm grasp of maneuver warfare ground campaigns, but he had little understanding of how to employ air power. Lind's 1985 *Maneuver Warfare Handbook* was silent on this issue. The Marine Corps would have to wait until Martin van Creveld's 1994 *Air Power and Maneuver Warfare* (written for the Air Force) before the use of air power would be seriously addressed. Feeling that they would always be a supporting effort for the mobile infantry, the air arm did not back Lind's maneuver warfare, as did the infantry. Likewise, the logistics arm was a source of intraservice competition. Correctly, they argued that without logistics, maneuver warfare would be impossible. Once again, Lind and his Marine Corps maneuver warfare advocates failed to address this significant issue. Under these circumstances, enough resentment was generated from two of the three warfighting arms during the Gray and Mundy years to cause the impediment of institutionalizing the maneuver warfare innovation.

Civilian intervention

There is no evidence of civilian involvement in the processes that led to the Marine Corps' adoption of amphibious warfare, air mobility warfare, and MAGTF warfare. In each case, it appears that champions within the service monopolized both expertise and interest; the debate took place almost entirely between Marine Corps product champions and Marine Corps opponents. A 'military maverick' appears only in the amphibious warfare case, but the Commandant of the Marine Corps, General John Lejeune, was directing this maverick, Major Pete Ellis. The evidence did not support the presence of Posen's military maverick, who intervenes between civilian leaders and the military bureaucracy.

Civilian intervention during the General Gray era was present and did have a negative effect on developing maneuver warfare. Civilian intervention was evident in lower-level civilians – especially William S. Lind – who were key players in challenging the Marine Corps' 'outmoded' attrition doctrine. Lind's involvement with Senator Taft and the publication of *A Modern Military Strategy for the United States*, a 1978 white paper on defense, and as a member of Senator Gary Hart's staff gave him considerable political pull on the Hill – enough so that Major James Jones, Senate liaison officer and future Commandant of the Marine Corps, gave Lind's input top priority. In addition, Lind wrote the *Maneuver Warfare Handbook* specifically for the Corps and was closely associated with the maneuver warfare 'Young Turk's' movement. When General Gray was Commandant, Lind was provided advance copies of, and helped to either write, critique, or edit all maneuver warfare doctrine publications, including the much-heralded capstone document *Warfighting* and the Maneuver Warfare Correspondence courses. Both Captain John Schmitt and Captain Tim Jackson, the authors of these publications, had the Commandant's (General Al Gray) permission to consult Lind directly.

Lind visited several of the Marine Corps training sites and provided feedback to Senator Hart, who was monitoring the progress of Marine Corps maneuver warfare. He also was very critical of the Marine Corps in national newspapers and military journals – and several Marines, sensitive to Lind's attacks, responded in print.

On balance, the impact of lower-level civilian intervention and, in particular, of Lind, is difficult to measure. Influence in Congress on military matters remained with the Armed Services Committees, whose members generally were ambivalent to the Congressional Reform Caucus, of which Lind was president.[20] Moreover, separating the impact of Lind while he was an aide for Senator Hart and after Hart left the Senate to run unsuccessfully for the Democratic presidential bid is not easy. The important finding, however, is that Lind was strongly linked to maneuver warfare (a term he coined), and there is little question that most Marines considered him the Commandant's 'Grey Eminence' – the power behind the throne.

The hostile reaction to Lind was not as much to the substance of his argument as to his right to participate in the doctrine development of the Marine Corps (which was the case for General C. Krulak, Commandant of the Marine Corps, who later banned Lind from Quantico). There is strong evidence in the maneuver warfare case that lower-level civilians were linked closely to efforts to develop and push the adoption of maneuver warfare. One of the important findings of this inchoate innovation case is that management failure during the Gray and Mundy era had more to with this lower-level civilian intervention (in the form of William Lind) than with any other factor. Also, there was no external crisis (interservice rivalry) that warranted the wholesale revision of doctrine Lind advocated. Indeed, the United States' main adversary was about to collapse.

Krulak ostracizing Lind is not surprising. Krulak was a strong supporter of maneuver warfare and he wanted the disruptive innovation to succeed. Eliminating William Lind supports the evidence from this study that shows that disruptive innovation is produced largely from within the military itself.

Small innovation groups

The importance of small innovation groups has been apparent throughout all the Marine Corps cases. Because innovation opponents can threaten overtly reform efforts, each product champion in every case formed small groups. The initiative for the modern amphibious warfare concept rested with General Lejeune. As discussed in Chapter 4, Lejeune, aware of Major Ellis' genius as a military planner, established a small innovation group in 1920 of which Ellis was a member. After several months' work, Ellis produced both the defensive and offensive advance base plans that, in turn, served as the intellectual breakthrough to redefine Marine Corps tasks and missions. In 1931, General Fuller created the next small innovation group at Quantico, tasking three mid-grade Marines and a Navy lieutenant to begin translating Ellis' vision to operational tasking. In 1933, General Russell created the Fleet Marine Force, but it is not clear that he created a small innovation group to develop the organizational mechanism that would transform the Marine Corps. On becoming Commandant in 1934, Russell did, however, create a small innovation group at Quantico to finish the task that Fuller's innovation group had begun.

Again, there is evidence of innovation groups being used by product champions in the air mobility warfare innovation. Following the 1946 Bikini atomic test, General Vandegrift created a small innovation group led by Major General Shepherd, tasked to create a new way of conducting amphibious warfare. The group proposed using helicopter mobility to disperse the amphibious assault. Reminiscent of Russell, who created small innovation groups at Quantico, Vandegrift used the Marine Corps schools to develop doctrine. Eventually, this small innovation group effort led to vertical envelopment doctrine.

In 1956, Commandant of the Marine Corps General Randolph Pate created a small innovation group, headed by Major General Robert Hogaboom to determine whether ship-to-shore movement in the future would become the 'all helicopter concept' or should be a combination of helicopter movement and 'traditional' crossing-the-beach operations. The board recommended a mixed technique.

Small innovation groups were used in the creation of Marine combined arms warfare – the MAGTF. The product champion of MAGTF warfare was General

McCutcheon. As a mid-grade officer, he served on two small innovation groups after his return from the Korean War. In 1952, the new Commandant, General Lemuel Shepherd, created a small innovation group, which achieved the intellectual breakthrough for redefining Marine Corps tasks around the MAGTF concept. In 1956, Commandant of the Marine Corps General Randolph Pate established the General Hogaboom small innovation group, which further developed the MAGTF concept to use the helicopter both as a gunship and an infantry transport.

As observed in maneuver warfare case, General Gray, as commander of 2nd Marine Division, established in 1980 the first of many small innovations groups. Called the maneuver warfare board, it was led by a young Marine captain named G.I. Wilson, one of the most famous maneuver warfare 'Young Turks' in the Marine Corps. The group's mission was to begin the process of translating theory into tasks. During the early maneuver warfare exercises at Fort Pickett, Gray established his second innovation group, led by German-born and German Staff College-educated US Marine Corps Lieutenant Colonel K.D. Schreiber. It was Schreiber's innovation group during the Fort Pickett exercises that translated the tenets of maneuver warfare into operational tasks. On becoming Commandant, General Gray established two more small innovation groups. Working directly for Gray, two young captains, John Schmidt and Tim Jackson, wrote the maneuver warfare cornerstone manuals, *Warfighting* and *Warfighting Skills Program* (a maneuver warfare correspondence course).

When General C.C. Krulak became the 31st Commandant he established the Colonel Gary Anderson small innovation group, which would function as the looking glass into potential maneuver warfare operating methods.

Disguising process

Each of the four successful disruptive cases examined strongly suggests that product champions artfully employed misdirection and disguised their disruptive innovations as sustaining. In amphibious warfare, Russell was careful to pass off the formation of the FMF as a mere name change. At the time of the creation of the FMF, Russell was not yet Commandant, and there was no guarantee he would succeed General Fuller. Thus, he was not as free to maneuver as he would be after he became Commandant the following year. As the evidence shows, many senior Marine officers, who were veterans of World War I, thought a shift to offensive advanced operations was a move in the wrong direction. They believed the Corps' primary mission was fighting alongside the Army.

A second reason Russell disguised the FMF was because isolationism was at high tide, and Congress saw the FMF as provocative and interventionist.[21] Advertising that the FMF would be the organization to shift the Marine Corps from defensive to offensive advanced base operations would not sit well. The third reason Russell disguised the FMF was that he needed time to solve the Corps' high-year tenure problem. As the evidence shows, many of Russell's strongest opponents also were the most senior Marines. Russell knew he could build support indirectly for his new way of war by forcing out his senior opponents through forced retirement boards and replacing them with younger officers who were proponents of offensive operations.

An important finding is that Russell's successor, General Holcomb, gave the FMF highest priority and repulsed challenges to the amphibious warfare mission from inside and outside the Corps. Holcomb and Chief of Naval Operations Admiral Leahy agreed that it was politically wiser to downplay the offensive role of the FMF and to stress the FMF's base defense capability, for both understood that Congress would not fund interventionist forces. The evidence therefore strongly suggests that both Russell and Holcomb disguised the FMF's offensive role behind its traditional defensive role.

In the Marine Corps air mobility and MAGTF disruptive innovations, the evidence shows that General McCutcheon played a key role in disguising both. McCutcheon did an excellent job disguising the arming of helicopters as non-threatening to Marine artillery units and fixed-wing aviators. He merely stated that he needed to provide covering fire for the troop helicopters once the helicopter landed. Since artillery and fixed-wing aircraft could not provide the accuracy of fire he needed, McCutcheon needed to arm his helicopters. Once he started arming helicopters he did not stop until the Marine Corps was building the Cobra attack helicopter gunship, which did, in fact, put the artillery gunners out of the close support business.

The disguising concept in the MAGTF innovation case is less important because the process of developing and implementing the doctrine was long and evolutionary. Yet, there appeared to be disguising happening at different points in the case study. McCutcheon initiated a disguising effort to protect the nascent doctrine of Marine air, infantry, and logistics being one weapon system when he opposed the Air Force move to place Marine helicopter transport units under Air Force control. While appearing to be stubbornly defending a traditional Marine Corps issue, McCutcheon really was disguising his true intent – to promote the MAGTF concept – because he believed that if the Air Force won this battle, the MAGTF concept would end.

As predicted, General Gray used disguising. He couched maneuver warfare as an approach that the Marine Corps had successfully used on numerous occasions. The problem General Gray faced is that he also used a civilian 'maverick', William Lind, a self-proclaimed German maneuver warfare expert, to proselytize the new of fighting as a German innovation. Apparently, the Marine Corps was more convinced that maneuver warfare was a German innovation instead of something the Marine Corps had done on occasion in the past. When Lind's crusade to sell maneuver warfare failed, General Krulak, Commandant of the Marine Corps and the second product champion of maneuver warfare, banned Lind from Quantico. Krulak then successfully disguised maneuver as something the Marine Corps always had done.

New junior officer career paths

In the maneuver warfare case, General Gray was initially unsuccessful in creating new career paths to flag rank. The evidence shows that he did not institute a mechanism to identify young maneuver warfare officers. Admittedly, he did protect the young maneuver warfare advocates who followed him from his early flag tours through his tenure as Commandant, but while Gray headed the Marine Corps several of the maneuver warfare 'Young Turks' were either retired early or passed over for promotion. The most controversial event was a Marine Corps board's selection of Colonel Wyly for early retirement, just a half-year before his mandatory 30-year retirement. Wyly's firing sent shockwaves through the Corps. The obvious message was that being a 'Gray' maneuver warfare advocate might not lead to flag rank. In fact, the message was even stronger: being a 'Gray Bubba' could get you taken out. After the Wyly event two of Gray's most brilliant maneuver warfare thinkers were passed over for promotion – Lieutenant Scott Moore, honored as the Marine Corps' most distinguished author for his maneuver warfare publications, and Major Tim Jackson, who authored one of the key capstone maneuver warfare publications.

8   US Navy sustaining innovation
Continuous aim gunfire

Continuous aim gunfire in the US Navy was a sustaining innovation in which new technologies helped perform existing missions better, and not change them radically. Although the dependent variable of this thesis is disruptive innovation, this case is examined from among the Navy studies for three reasons. First, it presents the earliest case assessing how technology produced fundamental changes in capabilities and tactics in naval warfare. Second, this case provides a 'natural experiment' of a sustaining case that can test the proposition that product champions of disruptive innovation manage the process differently than they would if were they promoting a sustaining innovation. Third, the case offers an excellent opportunity to assess the value of civilian intervention and the so-called 'maverick' and 'zealot' product champion in the innovation process.

Continuous aim gunfire comprised an innovation in weaponry. It postulated a new way to aim and fire naval guns. As background, the governing naval gunnery tenet of the late nineteenth century centered on the fact that naval guns were mounted on unstable platforms, rolling and pitching ships.[1] The sequence for aiming and firing required a gunner to estimate the range of the target – normally 1,600 yards in the 1890s – factor in the pitch and roll of the ship, determine the best time to fire based on the ship's motion, and then fire.[2] Naval gunnery proved to be as much an art as a science, dependent on the individual skills of gunnery officers and gun crews. Because a naval gun shoots a ballistic shell, it follows a trajectory similar to that of a baseball outfielder throwing a ball to the catcher at homeplate. Rather than throwing on a straight trajectory, the outfielder elevates the throw to compensate for the force of gravity that makes the ball fall toward the ground while in flight. The gunner elevated the gun barrel based on mathematics and his own experience to ensure the shell would reach its target. He accomplished this by turning a small wheel on the gun mount that operated the elevating gears. At this point, the guns were fixed for range. But guns bolted to the deck of the ship could only remain on target for a single instant in the ship's role. The ship's constant motion prevented a steady aim. The gunner, therefore, waited to fire until that precise moment when the gun was on target, a process based as much on guesswork and luck as scientific precision. Naturally, this method caused naval gunfire to be notoriously inaccurate. As an illustration, in an 1898 study the Navy's Bureau of Ordnance found that during the Battle of

Santiago in the Spanish-American War, only 121 shots hit the mark out of 9,500 shots fired.[3] Continuous aim gunfire innovation resulted in a hit rate increase of more than 3,000 percent.[4]

Engine of change: why and when

Civilian intervention, technological change, and international rivalry were the causes of change for continuous aim gunfire. Lieutenant William Sims convinced President Theodore Roosevelt that the US Navy lagged behind the Royal Navy in naval gunfire accuracy. Upon learning this, Roosevelt intervened and directed the Navy to embrace continuous aim gunfire.

Throttle of change: how

Intellectual process

The first intellectual process occurred with British Admiral Percy Scott – the inventor of continuous aim gunfire. Scott noticed a gunners mate on his ship whose gun was far more accurate than the others. This gunner used the roll of the ship to help his aim. Scott noticed that the gunner unconsciously worked his elevating gear back and forth in a partially successful effort to compensate for the roll of the ship. Observing this, Scott theorized that perhaps what one gunner could do partially and unconsciously might be developed and taught to all gunners as a systematic process. Scott accomplished this task by altering the gear ratio in the elevating gear, previously used only to set the gun in fixed position for range, so as to permit the gunner to compensate easily for the roll of the vessel by rapidly elevating and depressing the gun. Now that guns could continuously aim, Scott added a telescopic sight. These two improvements in elevating gear and sighting eliminated the major uncertainties in gunfire at sea and greatly increased the possibilities of both accurate and rapid fire. In short, this was the continuous aim gun innovation.

The next intellectual step took place when, during a port visit to Hong Kong, American Lieutenant Sims approached Admiral Scott to learn more about reports of Scott's ships being able to fire with 'unbelievable' accuracy. With Scott's assistance, Sims modified the gear on his ship and after a few months of practice by gunners, his experimental gun batteries began making remarkable records at target practice.[5] Over a period of two years, Sims documented the case for continuous aim gunfire in 13 official reports. Sims sent these reports to the Navy's Bureau of Ordnance and Bureau of Navigation. The initial reports disappeared into files, largely ignored.

In the next intellectual step, Sims, not satisfied with merely forwarding his findings through channels, sent his reports to other officers of the Pacific Fleet. During this period he acquired a Fleet-wide reputation of being an innovator, a label he enjoyed. However, senior navy leaders in Washington displayed far less enthusiasm for Sims' claims. Over time, Sims' passionate crusade for innovation turned him into a zealot. He became convinced European naval rivals were aware of US shortcomings in gunnery and felt it was his duty to gain political support to continue his cause.

Despite Sims' frustrations, evidence suggests the Bureau of Navigation was not quite as stagnant as Sims and his supporters believed. While true that the Washington admirals failed to immediately attempt to improve gunfire accuracy by adopting Sims' innovation, the record shows that two full years after Sims documented the remarkable record of HMS *Terrible* in 1900, the British Admiralty itself had not yet fully accepted the methods of Admiral Scott.[6]

Thus, the next intellectual step occurred when the Navy Chief of the Bureau Ordnance conducted an experiment to test Sims' claims. Instead of conducting the experiment at sea, however, it was carried out on dry land in the Washington Navy Yard. Unfortunately, this land-based test did not account for Newton's first law of motion, which naturally operated at sea to assist the gunner in elevating or depressing a gun mounted on a moving ship. These Navy Yard experiments therefore appeared to demonstrate that continuous aim gunfire was not possible, since the effort required to continuously move the gun proved too much for the stationary gunners. Admiral O'Neil, Bureau Chief concluded his letter to Sims detailing the results by stating the 'service was indebted to Lieutenant Sims for his highly commendable zeal and that intelligent criticisms should always be invited rather than shunned', but 'the critic should realize that his opinions' merited only the same consideration as those from any other competent source'.[7]

Under these circumstances, Sims felt he was left with no other recourse but to write to Theodore Roosevelt. Clearly, all naval officers understood that writing a letter of complaint to the President was an act of insubordination condemned alike by naval custom and regulations. By this letter, Sims in effect placed the Navy leadership in Washington on report. Nevertheless, in a remarkable turn of events, Roosevelt began corresponding with Sims and eventually brought Sims back from China, appointing him as the Navy's Inspector of Target Practice.

The next intellectual step occurred as a result of Roosevelt's action. In his role as the Inspector of Target Practice, Sims introduced continuous aim gunfire into the Navy. By 1905, one gunner achieved the remarkable feat of making 15 hits in one minute on a relatively small target at 1,600 yards. Even more incredible, the gunner hit the 50-inch square bulls-eye with half of the rounds.[8]

Disguising process

Neither Sims nor President Roosevelt made any attempts to disguise the continuous aim gunfire innovation. Sims' bombastic letters to senior naval leaders and appeals directly to Roosevelt were clearly overt maneuvers to promote his cause. Likewise, Roosevelt's correspondence to a Lieutenant and his reassigning of Sims from China duty to the head of gunnery tactics constituted overt steps in support of the innovation. An absence of the disguising factor conforms to disruptive theory where one would not expect product champions to disguise their efforts to promote innovations to help perform missions better. In fact, the theory predicts that for sustaining innovations such as continuous aim gunfire we would expect the product champion Sims to be a zealous advocate, which clearly he was.

Political process

The case evidence does not support any attempts by senior navy leaders to create a new promotion path for junior officers or a senior leadership struggle for political power based upon on the innovation.

Summary of continuous aim gunfire

This case examines Posen's argument that civilian intervention produces military innovation in peacetime, either directly or indirectly, through officers he calls military 'mavericks', who provide civilians with the military expertise they lack. In fact this case demonstrates that in sustaining innovations civilian intervention using passionate zealots can cause innovation. In contrast, in the subsequent Navy disruptive innovation cases, civilian intervention and military mavericks do not cause innovations. Since continuous aim gunfire is one of the few cases in the study where civilian intervention is a crucial factor, an analysis of why this is so relevant.

In an detailed analyzes of civilian intervention explaining military innovation, Rosen paraphrased Richard Neustadt's list of five conditions that must prevail if a President's order is to be readily obeyed by his bureaucratic subordinates.[9] First, the order is given by the President himself and expresses a definite decision by him personally. Second, the order is clear and concise and no doubt can be inferred as to precisely what are the objectives. Third, the order is well publicized. Fourth, subordinates possess the necessary means to carry out the order. Fifth, subordinates have no apparent doubt of the President's authority to issue the order.[10]

How well were these conditions satisfied when President Theodore Roosevelt issued an order to the Navy to embrace continuous aim gunfire? At this point, Sims' running battle with senior naval leaders in Washington was well known throughout the Fleet. Thus, Roosevelt recalling Sims from his China duty and appointing him as the Inspector of Target Practice left little doubt the President had become personally involved in the innovation. Sims knew exactly what he needed to do to innovate; his guidance from the President was unambiguous. Continuous aim gunfire comprised a well-defined task Sims fully understood. As a result of Sims' innovations, the President wanted all warships refitted with the new gears and telescopic sights and he wanted gunners to train according to the methods devised by Sims. By making Sims the Inspector of Target Practice, Roosevelt gave him control over everything he needed to carry out the order. Finally, the Navy harbored no doubts as to Roosevelt's authority to issue the order. According to Neustadt, the ready execution of Roosevelt's order to innovate occurred because of the combination of all five factors.[11]

Disruptive innovation theory tends to support Neustadt's model. Neustadt's model explains why a civilian command to carry out a disruptive innovation can be, by its nature, extremely difficult to enforce. By definition, disruptive innovation imposes a new way of thinking and often strikes at the very heart of predominant paradigms. More than likely, a President's order for a disruptive innovation will not be achieved because of the absence of Neustadt's favorable factors.[12] On the other hand, a civilian order for a sustaining innovation will more than likely be achieved because of the presence of Neustadt's favorable factors, because a sustaining innovation such as continuous aim gunfire is merely improving the performance of an existing mission or task. It is much easier for a civilian to satisfy Neustadt's five factors when the improvement gained by civilian intervention is along a performance trajectory that the traditional culture has historically valued.

Although the senior naval leaders in Washington resisted Sims' innovation, they had less of a problem with Sims being a zealot, than they had with the innovation itself. In fact, senior Fleet officers – commanders of squadrons and ships in the Pacific – supported continuous aim gunfire even before Sims emerged as its most visible advocate, and they supported Sims against the Washington admirals despite, and not because of, his belligerent, castigating attacks against those who did not react fast enough to his innovation. However, once Sims convinced Roosevelt of the merits of his innovation, he no longer had to prove it to the Washington Admirals. He just had to teach it to the Fleet.

9   US Navy disruptive innovation
Carrier warfare

American carrier warfare was a disruptive innovation in which naval aviation's product champions successfully transitioned the Navy from the dreadnought, which bristled with giant guns that could shoot shells 20-plus miles, to the large flush-deck carrier, which could launch air strikes from ten times that distance. This shift, Steve Rosen argues, emerged from World War I with the Navy's extensive experiences in sea-based aviation. He notes, 'One might assume that the Japanese attack on Pearl Harbor forced the United States Navy to use carriers by destroying the battleship alternative, but this begs the question of how the carriers, along with a doctrine for their use, came into the fleet *before* Pearl Harbor.'[1]

This case supports Rosen's observations that 'the US ships that defeated the Japanese carrier fleet in 1942 and 1943 were ships laid down in the 1920s and early 1930s – long before Pearl Harbor', and that 'effective doctrine for their use had begun to develop in that same period'.[2] This case also supports Elizabeth Kier and Williamson Murray's assertion that an organization culture receptive to experimenting with new concepts and ideas in annual maneuvers, exercises, and war games is essential for achieving disruptive innovations. The development of American carrier warfare supports the propositions of disruptive theory, in which product champions:

- Create small groups to incubate a new warfighting vision.
- Disguise the disruptive innovation.
- Fight a political battle over the virtues of their ideas and their preferred weapons.
- Protect junior officers and create new promotion pathways for them.

Engine of change: why and when

Carrier warfare arose primarily from the US Navy's efforts to deal with changes in the strategic security environment in the Pacific following World War I. During the interwar period, the Navy found itself hard put to support national interests in the Far East and Western Pacific. Its ability to defend the Philippines – acquired after the Spanish-American War – and to sustain the State Department's Open Door Policy in China became increasingly strained after the 1922 Washington

Treaty for the Limitation of Naval Armaments, which imposed a ten-year moratorium on building new battleships. Construction also was halted on two battle cruisers, but the United States was allowed to convert the unfinished ships into the aircraft carriers *Lexington* and *Saratoga*.

By 1922, the product champions of naval aviation believed that the US strategic position in the Pacific had eroded to the point where it could not be solved solely by fighting battleships from the battle line.[3] Building on the extensive experience in fleet aviation they had gained since World War I, this small group of farsighted and influential aviation advocates saw a solution in sea-based aircraft used in a new way of fighting called carrier warfare.

Four other factors also were driving aviation advocates to develop carrier naval warfare:

- *International rivalry.* At the end of World War I, the Royal Navy was the undisputed leader in carrier aviation. Although Britain was not an enemy of the United States, the resultant rivalry between the two nations helped drive early US Navy carrier developments.
- *Interservice rivalry.* Army Air Force General William Mitchell was advocating the virtues of air power and arguing that the nascent naval arm should come under his control. His push for a separate air force would stimulate naval aviation development.[4]
- *Intraservice rivalry.* The struggle between the Navy's battleship and aviation arms eventually redefined naval warfare, from combat among battleships to strikes from mobile air bases at sea. Some scholars argue that a key element in the maturing of the carrier concept was the intense public debate over the airplane vs. the battleship that pressured aviation advocates, such as Admirals William Sims and Joseph Reeves, to explore the capabilities of aviation at both the Naval War College and at sea on USS Langley.[5]
- *Spotting requirements for the battleship.* Great technological advances in battleship gunnery ranges now allowed accurate shooting over the horizon. With the invention of the airplane in 1903, the potential of aircraft to serve as spotters to support such shooting spurred further development of sea-based aviation.[6]

Throttle of change: how

Intellectual process

The roots of carrier warfare success lay with three key individuals: Admiral William Sims, Admiral Moffett, and Admiral Reeves. Following World War I, Sims returned to the Naval War College where he became a strong advocate of carrier aviation and the new role it might play in naval warfare. Building on the Naval War College's intellectual environment that was ignited by Admirals Alfred Thayer Mahan and Stephen Luce in the early 1900s, in 1919 Sims initiated a process whereby the potential of naval aviation could be determined systematically through tactical and strategic simulations. In determining how aviation should be used, Sims refined the wargames to reflect using carriers in supporting fleet engagements.[7] One of Sims' key contributions was revising the rules of the tactical simulations to accurately reflect the effects of aircraft and naval gunfire. By 1922, Sims had championed the board game 'Maneuver Rules' that contained detailed operating specifications of different aircraft. To his credit, Sims stresses the importance of connecting wargaming rules with actual data and conditions. He notes in letters to the Commander, Fleet Air Squadrons, 'Air tactics are of utmost concern to the [Naval] college, and only from actual work done in the field can we hope to formulate definite and sound ideas concerning them.'[8] Sims continues, 'In operating aircraft in chart maneuvers and game board exercises, various rules are applied which must of necessity be in close agreement with actual conditions if the true value of aircraft to the Fleet is to be appreciated.'[9] The effect of the 'Maneuver Rules' made the tactical games a more reliable indicator of the future contribution of carrier aviation.[10] It is apparent that Sims' tactical games 'contributed substantially to the development of ideas about how to employ the aircraft carrier'.[11] As Barry Watts and Williamson Murray write, 'Ultimately, Newport's war gaming became a key element in the institutional process by which the US Navy worked out answers to fundamental issues that confronted all navies in developing carrier aviation beyond the Royal Navy's achievement in WWI.'[12]

A critical element in the development of naval aviation was the relationship between Sims and Moffett. As discussed in the next section, Moffett was the first chief of the Bureau of Aeronautics. Soon after he assumed this position, Sims contacted Moffett stressing the importance of close cooperation between the Naval War College and the Bureau of Aeronautics. This cooperation, as one scholar notes, 'will turn out to be an important element in developing naval aviation'.[13]

The architects of carrier warfare were Admiral Moffett and Admiral Reeves. Moffett, who served two terms as the Bureau Chief of Aeronautics, provided the political shielding that allowed Reeves to focus on bringing about the carrier warfare innovation.[14] Although Moffett was senior to Reeves, they were allies and they mutually supported each other (although they did on occasion have professional differences on how aviation should progress).

As one of the product champions of naval aviation, Moffett redefined the Navy's critical military tasks in response to changes in the security environment, and he was able to gain control of the officer corps to implement the new way of war.[15] However, he was not a 'maverick' Navy officer but a well-respected battleship commander who had commanded the USS *Mississippi*.

After completing a highly successful battleship tour, Moffett received orders to report for duty to the Navy Department in Washington D.C. Prior to reporting for duty, he paid a visit to prominent members of the Republican party and requested they support his bid to lead naval aviation. They agreed and were successful in convincing President Warren Harding to urge the Navy to appoint Moffett as director of naval aviation. Coincidentally, the senior leaders of the navy also felt Moffett was the best person to head naval aviation and Captain Moffett was appointed in 1921.[16] Whether or not Moffett's political connections were useful in securing the top aviation position, there is documentary evidence supporting that he did use these political connections to receive two additional four-year terms as head of the Bureau of Aviation. This extraordinary long tour was a critical element in the success of carrier warfare.[17] The organization he inherited was in disarray. Without centralized leadership, the department had grown on a largely *ad hoc* basis, with divided lines of responsibility. As William Trimble writes, 'aviation had been grafted onto the existing bureau system rather than integrated into it'. Purposely, the battleship admirals had created an aviation department that had little power and thus would be of little threat to them.[18]

The first intellectual step Moffett took was to increase the power of his department. In April 1921, he testified before the House Naval Affairs Committee about the conditions his predecessor, Captain Craven, had faced in trying to manage the Navy's aviation department. Moffett spoke candidly of 'the great difficulties' under which Captain Craven had been working. 'He was acting practically...as a chief of a bureau and was trying to perform the work of the chief of a bureau without having any executive authority whatever.'[19] William Trimble notes, 'Moffett's arguments for a separate bureau focused on the popular themes of economy and efficiency and the requirement to bring order out of the present "chaotic conditions".'[20] Moffett's testimony led to the unanimous agreement of the House and Senate naval affairs committees to forward a bill creating the Bureau of Aeronautics. President Warren G. Harding signed the bill into law on 13 July 1921 and a week later nominated Moffett as the bureau's first chief.[21] Moffett had achieved a major coup – he now had power equal to the senior leaders of the battleship union, and with that power could begin to establish naval aviation as a fully integrated arm of the fleet.

Between 1920 and 1925 the Naval War College simulations derived three tentative conclusions regarding carrier aviation. First, the battleship effectiveness model of steady stream gun firing did not apply to the pulses of power delivered by carrier aviation. Second, the key measure of effectiveness for carrier strikes was numbers of aircraft in the air. Third, carrier aviation suffered from many weaknesses including the short range of bombing aircraft that forced friendly carriers to get quite close to an enemy formation to launch and recover planes.[22]

As one of the other product champions of naval aviation, in 1925 Reeves began to play a major role in the development of carrier warfare. The next intellectual step occurred when Moffett reassigned Captain Reeves, then head of the Naval War College Tactics Department, as commanding officer of the experimental carrier USS *Langley* to test his ideas.[23] Moffett encouraged Reeves to make *Langley* an operational carrier that could perform several different missions.[24] Reeves recently had supervised the Naval War College games of 1924–25, and in this capacity had headed a small innovation group to develop carrier warfare. Moffett was fighting the 'political' battle with Mitchell and his supporters in Congress, who wanted to form a separate air service, and he needed evidence that naval aviation was making progress. Reeves provided it.[25] With Moffett's support, Reeves begins focusing on developing 'strategy and tactics of the air in its relation to the Fleet'.[26]

The next intellectual step was perhaps the most significant in the development of carrier aviation. Following Moffett's direction to create a new way to fight carriers, Reeves mined the experience he had gained at the Naval War College. While observing several War College carrier simulations, he concluded that the key measure of strike effectiveness was the number of aircraft in the air. Simply put, the greater number of aircraft in the air, the greater chance of succeeding in the strike mission. Believing this was way carrier warfare should evolve, Reeves used the *Langley* to begin testing his concept.

While it is tempting to see Reeves' efforts to get more striking power through more aircraft as the explanation that would bring about a revolution in carrier warfare, there are interservice and intraservice rivalry factors at play that must be considered. Admiral William Pratt, President of the War College, was preparing to testify at the court martial of General Mitchell who had stated that the Naval College War games had shown that aviation had made battleships obsolete. The former Naval College President, Sims, supported Mitchell's claims. Pratt, an advocate of carrier aviation, faced a dilemma. He knew the games conducted when Sims was President did not show that aviation was superior to battleships. He also knew that the games had shown several weaknesses of carrier warfare including the requirement to stay close to their target due to the short range of carrier bombers.[27] Pratt believed that eventually new aircraft with greater ranges would solve this problem. Thus, his challenge during his testimony would be to discuss the weaknesses of carrier aviation without killing the program. In response to these concerns, Pratt believed the answer to his dilemma could be found in the simulations. The simulation rules required the players stick rigidly to the times established by Reeves on *Langley* for various phases of airplane operations from carriers. If Reeves could demonstrate that *Langley* could launch and recover more aircraft, Pratt could use this data to refine the wargame rules.[28]

Focusing his attention on launching and landing, Reeves was able to achieve success in late 1925 when *Langley* launched ten aircraft in less than two minutes. Pratt used this data to counter Mitchell's claim that the navy would and could not develop carrier aviation.[29] Reeves continued to experiment with different methods of increasing the number of aircraft *Langley* could launch and recover. He noted that after landing, a plane would be moved below deck to avoid the possibility that the next plane to land would crash into it. He therefore came up with a barrier to protect the planes on deck, thereby obviating the need to use the aircraft elevator each time to lower a plane into the hangar. This technical solution allowed Reeves to go from embarking only 8–14 planes to 48 planes.[30]

Small group proposition

The next intellectual step occurred in 1927, when Moffett convinced the Secretary of Navy to create a small innovation group.[31] The carriers *Lexington* and *Saratoga* would soon be finished, and as they were quite a bit larger than the *Langley*, Moffett wanted a board 'to explore all aspects of aviation'.[32] Moffett had another reason as well for pushing the innovation group; he wanted to reevaluate the vision of carrier warfare Reeves had been promoting. Launching greater numbers of aircraft for offensive strike missions required a reevaluation of the existing technology, its uses, and its limitations and of the possibilities offered by new technology and concepts. Moffett generally supported the offensive vision of strike warfare, but he wanted to sort out competing versions, which in this case study are represented by aviators who saw the large-deck carrier as the wave of the future.[33] Although Moffett encourage spirited debate within the aviation community over new tactics and technology, he could not afford to have the carrier size debate spill outside the aviation union. The more powerful big gun club – the battleship union – would quickly exploit a crack in the aviation union.

Moffett himself participated in the innovation group, along with Captain Reeves. The board generated three important conclusions:

1   The Navy needed to continue to build aircraft carriers. After examining the forces of Britain, Japan, and the United States, Moffett concluded that only the United States remained below the total tonnage allotted by the Washington Treaty, and that 'definite steps [must] be taken to obtain authorization of the construction of additional aircraft carriers'.[34] After accounting for the *Lexington* and *Saratoga*, each at 33,000 tons, the Navy still had 69,000 tons remaining. Moffett's alternatives were three 23,000-ton carriers, four 17,250-ton carriers, or five 13,800-ton carriers. After considerable discussion, he decided the aviation community should build five 13,800-ton ships, and the innovation board and the Navy's general board endorsed his recommendation.

Based on Reeves's vision, Moffett determined that the size of the carrier did not matter, but that the advantage lay with the side that could launch the most planes for a strike mission. Moffett stated, 'I think... you don't lose much by that displacement [13,800-ton carrier], and it gives you another carrier under the treaty tonnage, so that your whole air force would be more mobile. You could have the air force in a greater number of places.... The more carriers you have the greater the number of planes you can have and the more protection you get.'[35]

As it turned out, Congress authorized only one 13,800-ton carrier, USS *Ranger*, and it turned out to be too small.[36] Following *Ranger*, Moffett decided that 20,000 tons would be about right, and Congress agreed. As William Trimble writes, 'Moffett insisted *Ranger* was not a mistake, but he had to admit, however, reluctantly, that the small aircraft carrier was more attractive in theory than in practice.'[37] While all this hints at a certain lack of orderliness in carrier warfare innovation, the key point is that Moffett had a new vision of warfare and attempted to match that vision to the technology. In the end, his willingness to build carriers that supported Reeves's vision led to carrier warfare.

2   An offensive role for the carrier existed beyond supporting battleships. The board found carriers necessary for 'service of the battle line to furnish fighting airplanes for its protection and a landing place for reservicing

its airplanes; thus leaving other carriers free for scouting and offensive operations at a distance from the battle line too great to adequately serve it'.[38]

3   A mix of aircraft – in priority order, fighters, spotters, scouts, and dive-
    bombers – could defend the battleship task group as well as conduct strike
    missions.[39]

The next intellectual step occurred when Reeves began to address the problem
of operating multiple carriers together. By 1927, Reeves had convinced Moffett
and the other senior naval aviators that it was necessary to deliver a knock-out
blow against enemy air power in the opening minutes of any confrontation
between opposing carrier forces.[40] In Fleet Problem IX, *Saratoga*, under Reeves's
command, detached from the main force of battleships and made a dive-bombing
raid on the Panama Canal. Archival records note that surface ship escorts were
left behind because they did not have enough fuel to keep up. Reeves had demonstrated that the carrier could operate as an independent strike force.[41]

After Reeves's success, Moffett went to Congress in 1928 to begin to educate
them on how senior aviators viewed naval aviation. Appearing before the House
Naval Affairs Committee, he laid out the Navy's plans for the maximum number
of aircraft as determined by the small innovation group. The carrier, he said, was
a 'floating flying field', essential for air operations with the fleet, particularly in
'far-distant waters where shore-based support facilities were nonexistent'.[42]

The next intellectual step occurred in 1931, when Admiral J.J. Clark tasked
Rear Admiral Harry E. Yarnell to develop fleet carrier doctrine. Yarnell created a
small innovation group headed by Captain Jack Towers and including Lieutenant
Commanders Arthur Radford (later a Chief of Naval Operations), Forrest
Sherman, and Ralph Davison.[43]

The final intellectual step occurred in June 1933 with the publication of PAC-10
as the basic carrier doctrine for the Pacific. This created the multi-carrier task
force doctrine. Instead of operating carriers as ships that had to disperse to avoid
air attack, the multi-carrier doctrine created a screening force of cruisers and
destroyers around the carrier to protect it.[44]

Disguising proposition: Moffett and Reeves

There is no question that both the battleship and aviation communities felt that
naval aviation would play a critical role in the next war. The question was what
that role would be: scout for the battleships, spotting for naval gunfire, or operating as an independent strike force?[45] As disruptive theory predicts, Moffett's chief
contribution to carrier aviation was disguising the potential of an independent
carrier strike force, so as not to alarm the battleship admirals. Having commanded
a battleship, Moffett was well aware that the vessel held a near-sacred position in
the Navy. His trick in promoting carrier aviation was to present it as a supporting
asset that would accompany battleships and provide scouting planes for the
dreadnoughts.[46]

Moffett's first problem was to deal with advocates of the seaplane, an alternative to carrier aviation that could be carried on board battleships. A few senior leaders of the battleship union felt that the Navy should focus on developing this concept instead of building carriers. Moffett, who had operated seaplanes with the battle fleet while commanding *Mississippi*, was able to counter the seaplane concept by aligning himself with battleship admirals who wanted more carriers to accompany them and provide spotting and scouting support.[47] Here, Moffett's status as a well-respected battleship officer, rather than a maverick, helped him to sway other battleship officers. Because of his credibility with the battleship union, he was able to convince them that aircraft operating from a carrier would provide better spotting and scouting than seaplanes operating from battleships.

Moffett played a major role in securing future commands for naval aviators during his battle with Brigadier General Billy Mitchell. As part of his campaign to keep Navy aviation in the public eye, Moffett approved several publicity events in September 1925 including risky non-stop seaplane flights between California and Hawaii and visits by the airship *Shenandoah* across the country. Both events ended in fatal accidents that resulted in Mitchell charging Moffett with negligence and incompetence. In response, President Coolidge set up the Morrow Board to consider Army and Navy aviation and the government's role in it. During the Morrow Board hearings, Moffett took the opportunity to promote the idea that his Bureau of Aviation should have the final say on the assignment of aviation personnel, which was a clear reversal of the traditional lines of assignment authority held by the Bureau of Navigation.[48] In a major coup for naval aviation, the Morrow Board recommended that only naval aviators be given command of aircraft carriers and naval airfields.[49] Moffett effectively disguised this major milestone for aviators because *Langley* was the only carrier in commission and its main function was to support battleships. Because battleship command was the pinnacle of achievement for a Navy captain, Moffett avoided spurring a divisive intraservice rivalry by having his 'second-class' aviators command the *Langley*.

Moffett's aviator subordinates also actively disguised the innovation. Reeves promoted placing great numbers of aircraft in the air for strike missions, but he understood this concept was at the heart of carrier warfare and thus the importance of keeping this critical insight from rivals. In a letter to Moffett dated 4 October 1928, for example, he explained how he had hidden *Langley*'s true air strength from visiting Vice Admiral Fuller of the Royal Navy: 'Of course I did not tell Admiral Fuller that we operated not 24, but 36 and could operate 42 and possibly 48 airplanes from the *Langley*.'[50]

While Reeves was focusing on developing strategy and tactics of naval aviation, he was careful in couching his efforts as supporting the battleship. As he noted in his Naval War College tactics thesis, the airplane might be vital for victory in preparing the way for the battleship guns, but 'battles have always been, and under conditions that will probably exist for the next ten years will continue to be decided by the primary weapon [the battleship]'.[51] In taking this tack, Reeves was able to disguise a disruptive tactical innovation as a sustaining innovation for battleship spotting that resulted in the *Langley* gaining two fighter squadrons. Since 1921, the policy had been to assign two fighting squadrons and one observation squadron to the fleet's battleships. When assigned to the battleships, the squadrons' aircraft were fitted with floats. Battleship spotting doctrine, however, had recently been changed and called for a second spotting aircraft from each battleship be ready in case the primary spotting aircraft was lost. Reeves cleverly proposed that they arm the reserve spotters and these should remain attached to the *Langley*. He argued that the armed reserve spotters could protect the spotting planes, while flying in reserve. If a spotting plane was lost, then the reserve plane could replace it. Reeves' recommendation freed up two squadrons for use on *Langley* and effectively doubled the strike power available without causing an intraservice squabble with the battleship officers.[52]

The next disguising effort centered on the result of the 1929 *Saratoga* and *Lexington* fleet exercise off of Panama. After Reeves detached *Saratoga* from the battleships it supported and conducted an independent strike on the Panama Canal, the battleship union demanded that Moffett clarify the carrier's primary mission.[53] Was it an independent striking force, or was its principal task to support the battle line? Moffett knew that if its focus was strike, the carrier would give up much of its ability to maintain air superiority over the battleship forces. He seemed to be caught in a 'Catch-22'. If he stated his true belief that carriers would evolve into the fleet's primary strike force, he risked the battleship admirals sinking his concept of operating carriers independently before it could mature. If he sided with the big gun club, he risked open rebellion among the young aviators he was nurturing in his new way of war.

Moffett's task of selling naval aviation was made more difficult by the outspokenness of younger officers such as Reeves, who proclaimed the obsolescence of the battleship.[54] Reeves supported carriers as independent strike platforms, and he referred to the surface officers who opposed him as 'old coots'.[55] Nevertheless, Moffett was able to disguise his true feelings, and he reassured the battleship community that the carrier remained in a support role. Behind the scenes, however, 'he quietly courted presidents and members of Congress for appropriations'.[56] He also quietly reassured Reeves and the other young aviators that he supported the new vision of carrier warfare, but that he needed time to develop fully the technology to support the vision.

A great amount of credit must go to Moffett's political skill, to his ability to thrust and parry. 'Moffett's tactics, described in detail by several writers, brought to bear what one called his "formidable political muscle and powers of compromise".' Navy historians argue, 'His role is significant in that he was very effective in shielding Reeves from interference, so that the latter could bring about a revolution in carrier warfare'.[57]

In 1931 Moffett finally described his new theory of warfare in a memo to the Secretary of Navy: 'The function of a large carrier should be the same as that of a battleship,...to deal destructive blows to enemy vessels. Its offensive value is too great to permit it to be ordinarily devoted to scouting.'[58] Moffett's new vision of naval warfare had the carrier operating independently, well out of range of enemy battleship guns. Of note, his concept of carrier warfare differed from that of the British, who built and operated their carriers as flat-top dreadnoughts, to survive severe attacks from land-based aircraft. Moffett, in contrast, saw the carrier as a mobile airbase, and his intellectual task was to figure out how many aircraft he could embark on a carrier.

While disguising his true intent for carrier warfare from the battleship admirals, Moffett effectively shaped the public battlefield by targeting prominent political figures and explaining the virtues of naval aviation. He also appealed to the public through the press and through supporting popular films such as *Helldivers* (1931).

Political process

Just as important as defining new tasks for a new way of war is the struggle that ensues when product champions attempt to capture political power within the Navy. Disruptive theory predicts that new career paths are created from within, by senior officers currently holding power, rather than being forced on the service from outside.

Moffett continuously honed his plans to make aviation a viable strike arm. To do so, he needed to cultivate young officers into aviation. Moffett wrote, 'This [Aviation] Bureau has steadfastly and consistently maintained that the young line officers of the Regular Service, graduates on the Naval Academy, make by far the best aviators.'[59] A major breakthrough to support this aim occurred in the mid-1920s when Moffett secured the authority 'to draw the best graduates of the Naval Academy into aviation'.[60]

Although Moffett was extremely successful in disguising both his new form of warfare and his claims on resources, he was unable to avoid potentially divisive personnel issues. Rosen describes the problem: 'The balance of power between aviators and non-aviators within the Navy would be affected by the projected requirements of combat – pilots would be killed more quickly than sailors. A solution to the attrition problem was to expand the number of aviators, who were officers. But if this personnel policy continued, the Navy would wind up with far too many aviator officers.'[61]

Moffett's partial solution was to intervene in the promotion process and advocate a policy whereby only aviators would be selected to command aircraft carriers and air stations. Naturally, nonaviators bitterly opposed such a policy, but they did not have a good answer to Moffett's charge to the Navy's General Board that 'the older established order of things cannot always be applied to the new art of Aeronautics'.[62] What Moffett accomplished was the creation of a viable career path to keep aviator officers employed after their relatively few years of flying were over.

By 1943, aviators had begun to take senior positions on the staff of the Admiral in charge of the Pacific. The transformation was completed in 1942, when Admiral Ernest King became the first aviator Chief of Naval Operations.

Summary of carrier warfare

The start of naval aviation can be traced to the invention of the airplane as a radical technology and battleship gunnery range as an order of magnitude sustaining innovation that allowed accurate naval gunfire beyond the horizon. Early champions of naval aviation saw the potential utility of aircraft as spotters for battleships. A key person in the carrier warfare story is Sims who encouraged a spirit of intellectual curiosity at the Naval War College. The wargames and simulations at the College provided the cauldron where the future disruptive champions of naval aviation could experiment with aviation. Both Moffett and Reeves, the product champions of carrier warfare, had strong ties to the Naval War College and were able to test their visions using the simulations.

During naval aviation's first few years, Moffett and his supporters agreed that aviation's primary role was reconnaissance. Just as in World War I, naval aircraft would be the 'eyes of the fleet'. With the commissioning of the experimental aircraft carrier *Langley* in 1922, however, Moffett began to develop the air arm into an offensive force. He created small innovation groups – which he managed and sometimes participated in – to focus on developing different aspects of carrier striking power. While the nascent doctrine of carrier warfare was percolating, Moffett was able to disguise the effort as a sustaining innovation that would merely increase the effectiveness of the reconnaissance mission.

Reeves was Moffett's tactical aviation champion who understood that to get more striking power from the air arm required more aircraft. In achieving this goal, Reeves cleverly disguised complying with the new sustaining spotting doctrine, which now called for a second spotting aircraft to take over immediately if the first was lost, by a disruptive reserve spotting innovation that allowed two fighting squadrons assigned to the battleships to being assigned to the *Langley*. By promising the *Langley*'s fighters would perform duel duties of fighter and reserve spotting, Reeves could now operate two 18-plane fighting squadrons – a feat where the *Langley* almost doubled its air combat power.[63]

Moffett's political struggle to create a new promotional pathway for aviators was not as easy disguising carrier warfare as sustaining innovation. The battleship union stood in his way at nearly every turn. Although Moffett was the head of the newly established Bureau of Aviation, the Bureau of Navigation chief controlled all personnel assignments. Through astute political maneuvering and after intense infighting, Moffett managed to wrest away control of aviator assignments. In fact, he was able to convince Congress that only aviators should command carriers and naval air stations.

In 1933, Moffett was killed in a flying accident. With his death, Admiral Ernest King replaced him as the aviation bureau chief. In 1942, Admiral King became the first aviator to be selected as Chief of Naval Operations. With King's selection, the paradigm shift was complete; not only did the aviation union control the Navy, but the carrier had replaced the battleship as the Navy's capital ship. The disruptive innovation of carrier warfare was complete.

10 Disruptive innovation

Japanese carrier warfare

Japan's forging of a naval air arm, in which aircraft carriers had an independent and decisive role, within its battleship-dominated Imperial Navy was one of the most noteworthy disruptive innovations of the interwar years. Although the British had demonstrated the future role of naval aviation on more than one occasion during World War I, the Imperial Japanese Navy emerged from the conflict with little experience in aviation. Amazingly, however, by late autumn 1941, Japanese carrier air was the most potent offensive air force of any navy. Within the first five months of the Pacific war, the Japanese Fleet Air Arm had not only decimated enemy forces at Pearl Harbor, Port Darwin, Trincomalee, and Colombo with aerial attacks, but also had sent HMS *Repulse* and HMS *Prince of Wales* to watery graves, the first time in naval warfare that dreadnoughts under way were sunk by air attack.[1]

One might assume that the Japanese Navy had merely copied its British counterpart in building carriers, and borrowed US doctrine as to their use. But this begs the question of how champions of Japanese naval aviation were able to implement carrier warfare when the Japanese Fleet Battle Instructions left no doubt that the battleship divisions were the main weapon and their task was to fight the decisive battle against the main naval force of the enemy.[2] As we shall see, Yamamoto Isoroku, a pivotal champion of naval aviation, envisioned carrier warfare as a new way of war fighting and was able to implement it by disguising the aviation arm as a sustaining innovation of the existing 'big ship, big gun' doctrine.

A full examination of the disruptive doctrine for Japanese carrier warfare must begin just before World War I. At that time, three significant changes in the international security environment triggered Japanese carrier warfare: (1) the advent of aviation; (2) the emergence of Japan as a dominant military power in the Pacific; and (3) the concurrent rise of the United States as a Pacific military power.

The advent of aviation was a major technological revolution outside the control of the Japanese military. In 1909, six years after the Wright brothers' historic flight at Kitty Hawk, the Japanese Navy decided to develop aviation.[3] Initially, the Japanese Navy established a joint committee with the Army to study the military use of aircraft. The Army sent several officers to Europe for flight instruction, and they returned in 1910 with two aircraft. Dissatisfied with the Army's lack of interest in transforming those aircraft for waterborne use, however, and unhappy about Army control of the aviation budget, Navy leaders broke away to form their own aviation committee and officer exchange program.[4] Almost immediately, the intense interservice rivalry between the Japanese Army and Navy fueled the fires of development, and the stage was set for the first successful Japanese maritime air operation. In 1914, the Navy employed a seaplane carrier whose embarked seaplanes reconnoitered and bombed German forces at Tsingtao.[5]

Although Japan gained this initial aviation experience ahead of the Western naval powers, Japanese naval aviation languished for the next several years, as Japan watched from afar the remainder of the European conflict. By 1918, the navies of Great Britain and the United States were experimenting actively with aviation in reconnaissance roles and for antisubmarine patrol.[6] The aircraft's obvious advantage was the increased range at which it could detect and observe an enemy fleet, and the Western naval powers developed this capability significantly. By the close of World War I, there was a new military reality, driven by technological advancements – the British conducted the first attack against a land target from the prototype of the modern aircraft carrier, HMS *Furious*.[7]

Realizing it had fallen considerably behind the Western naval powers in the employment of naval aviation, after the end of the war the Japanese Navy undertook an assessment of US and British approaches and emphases. At the time, the United States relied mostly on seaplanes and land-based aircraft for its naval air. The modern Japanese Navy, which had modeled itself after the British Navy from the beginning, turned again to its mentor in forging its aviation arm. It was a logical choice; Britain had come, by the end of the war, to emphasize carriers as the basic naval aviation force.[8]

Interservice rivalry: aircraft and carrier development

Seeing a role for aircraft in naval power projection beyond the range of shipboard weapons, the more enlightened Japanese battleship leaders supported building carriers. It was clear, however, that this would be merely a supporting role to conventional surface warfare, as was torpedo warfare. The 'Big Gun Club' leaders wanted the Navy's focus to remain on battleships.[9]

Having the same status as torpedo warfare meant that aviation would not be a separate bureau or department. Remarkably then, with no aviation department, no doctrine for carrier employment, and no aircraft suitable for flight operations from a carrier battleship leaders approved in 1919 the construction of Japan's first aircraft carrier. With extensive British technical assistance, in 1921 the Japanese launched the *Hosho*, one of the world's first postwar carriers.

While the *Hosho* was under construction, the Navy once again felt threatened by Army progress in aviation – a French air mission invited to Japan in 1919 at the request of the Army was enabling rapid advances in military aviation. Again, interservice rivalry spurred development. Japanese naval leaders sought the assistance of the British Navy in improving the proficiency of its own air arm. In 1921, the British government sent an unofficial civil aviation mission to Japan headed by Sir William Sempill, a former Royal Air Force pilot.[10] The one-year mission consisted of well more than 100 aircraft and thirty handpicked pilots and engineers. In early 1923, a British pilot, William Jordan, made the first take-off from and landing on Japan's new carrier. By the time the Sempill mission returned home, Japanese naval aviation had closed the gap with the United States and Britain.

During the same period, another change in the international environment pushed Japan toward carrier construction. The 1922 Washington Navy Arms Limitation Treaty, which imposed construction limits on heavy cruisers, had the unintended effect of triggering carrier construction worldwide. Like the US Navy and Royal Navy, the Imperial Japanese Navy circumvented the cruiser limits by converting heavy cruisers to carriers.[11] Consequently, the Japanese battle cruiser *Akagi* and the battleship *Kaga*, under construction at the time of the Washington Naval Treaty, were completed as aircraft carriers.

A significant administrative reorganization in 1927 would drive the development of naval aviation technology as well.[12] Spurred by the success of the Japanese Army's air organization, the Imperial Navy instituted sweeping bureaucratic changes.[13] Control and coordination of naval aviation administration, training, and technical research – previously divided among several bureaucratic departments – moved to a single organization, the Naval Aviation Department, reporting directly to the Navy minister.[14] The repeated reappointment of key military officers to head the department, most notably Yamamoto and Inoue Shigeyoshi, a first-rate naval theorist and advocate of air supremacy, was crucial in protecting the innovation from its enemies within the Navy.

With the establishment of the Naval Aviation Department, Japanese naval aviation technology accelerated. Previously, Japanese aircraft, although manufactured domestically, were dependent on foreign designs. Rear Admiral Yamamoto, chief of the Aviation Technical Bureau from 1930 through 1933, championed a plan to make his country self-sufficient in both aircraft design and manufacture. Big Gun Club admirals supported Yamamoto's building initiatives as they recognized the growing importance of having an air arm that could perform reconnaissance, the *raison d'être* for naval aviation, as well control the air space above the battle fleet so scout planes could spot the fall of shot from the battle line.[15] Besides developing a technology infrastructure to build aircraft, Yamamoto carefully campaigned for a variety of specialized aircraft that could perform not only fleet reconnaissance, but also future offensive roles such as attack (against both maritime and land targets) and fighter protection of friendly forces.

Yamamoto's self-sufficiency and diversity initiatives at the Naval Aviation Department started a process that would 'result in the design, development, and production of some of the finest aircraft in the world'.[16] The new B5N Type 97 torpedo bomber was given double responsibilities as an attack and reconnaissance aircraft when it outperformed two reconnaissance prototype models. By autumn 1936, Japan was producing the best carrier fighter in the world, Mitsubishi's A5M4, as well as the Mitsubishi G3M medium bomber.[17] Based on input from front-line fighter pilots in the 1937 China War, the Aviation Department adduced the requirement for a fighter with the endurance to escort the G3M bombers, maneuverability to engage enemy fighters, and firepower to destroy enemy bombers. In March 1939, Mitsubishi unveiled one of history's most ingenious planes, the A6M2 carrier fighter, soon to be known worldwide as the Mitsubishi Zero, with a range of more than 1,500 nautical miles and a top speed of 332 mph.[18] During the early debates about the role of carrier aviation, Yamamoto did not argue that carrier aviation would eclipse the battleship in importance, but his understanding of the critical new tasks carrier aircraft would perform is evident in his actions to build a diverse offensive air arm. Unbeknownst to the battleship admirals, Yamamoto's campaign for aviation diversity would ensure carriers would not play a subordinate role forever.

The establishment of the Naval Aviation Department also had a profound impact on carrier development. Led by Yamamoto, the department engaged the battleship community over building bigger carriers with less armament. The *Akagi* and *Kaga* had been commissioned in March 1927 and March 1928, respectively, each capable of carrying 60 aircraft. Battleship leaders insisted that both be armed with several 8-inch heavy surface weapons for use in possible gun action with enemy warships. Fleet maneuvers and War College wargames, however, underscored the offensive potential of carrier aviation[19] and highlighted rapidly improving attack and fighter aircraft capabilities. The *Akagi* and *Kaga* could no longer accommodate the considerable increase in size and power of naval aircraft, which required longer flight decks for take-off.[20] Aviation proponents began to view the size of the carrier's air group as more important than the carrier's surface armament. With this in mind, Yamamoto urged removal of the *Akagi* and *Kaga*'s main batteries. A compromise was struck between Yamamoto and the gunnery-minded officers, and they agreed to remove the forward guns to make room for a larger flight deck. At issue was the function of the carrier and the obvious offensive potential of carrier aviation versus the doctrine of decisive battleship engagement.[21]

Until the mid-1930s, interservice rivalry fueled the development of aviation and carrier technology in a sustaining way, to support battleship warfare. Yamamoto, among others, contributed to carrier aviation development that helped the Japanese Navy better perform an old mission, scouting for the battleships.[22] Aircraft also could observe and direct battleship gunfire against enemy targets, improving accuracy. This use of carrier aviation was a sustaining innovation that helped the battleship fight better and did not challenge existing roles and missions within the Imperial Japanese Navy. The question is: What caused aviation proponents to advocate an offensive role for the new air arm and how did they succeed in overcoming the primacy of big-gun battleship doctrine? A second perceived shift in the security environment would ignite an internal Navy rivalry that slowly would transform carrier aviation into a disruptive innovation – carrier warfare.

Intraservice rivalry: carrier warfare disruptive innovation

Two significant changes in the international security environment led senior Imperial Navy officers to champion the carrier warfare innovation: (1) Japan's desire to expand its empire throughout Asia, and (2) the United States' emergence as a Pacific military power after the acquisition of the Philippine Islands. These two competing factors created military challenges that appear to have been at the heart of Japan's intellectual conception of carrier warfare. By the end of World War I, Japan's industrial growth had brought with it a dependence on foodstuffs and raw materials from Southeast Asia. As the Great Depression of the 1930s began to take hold, Japan adopted a policy of southern expansion to seize the resource-rich colonial territories of Britain, France, and the Netherlands, areas it considered legitimate for taking.[23] This desire for economic expansion created a new strategic requirement for dealing with the 'real or imagined' encirclement threat posed by US forces in the Philippines and the US fleet operating from Pearl Harbor.[24] The Japanese Navy's role would be to prevent the US fleet from operating in southern waters.

Navy intraservice rivalry fueled the debate over how to translate this requirement. Japanese battleship admirals asserted that battleships could keep the US fleet at bay by following the defensive doctrine of the decisive big gun battle. According to this strategy, war would unfold with the US Navy conducting a westward offensive using its battle fleet and would culminate in a decisive fleet engagement on Japan's terms somewhere in the western Pacific.[25] The defensive doctrine's key assumption was that Japan would be able to lure the US Navy into battle at a time and place of its choosing.[26] Yamamoto found this wait-and-react strategy a recipe for ultimate defeat.[27] He believed the US Navy would fight on its own terms, which meant it would use its overwhelmingly superior naval strength to simply attrite Japan's navy in a long war.[28]

Dissatisfied, Yamamoto instead adopted a two-part intellectual and political strategy to create a new theory of victory. At the intellectual level he advocated an offensive doctrine of preemptive mass aerial strike to prevent the US Fleet from disrupting the southern operation.[29] To achieve this offensive capability, he would have to champion a disruptive innovation – the aircraft carriers' critical task would change from being 'the eyes of the battleship' to being an independent strike force that would replace the battleship as the dominant naval weapon.

Intellectual strategy

The first part of Yamamoto's strategy was to effect a redefinition of the tasks the Japanese Navy would have to perform in the next war. In Yamamoto's mind, the unmet military challenge threatening Japan's southern operation was the dominant US fleet at Pearl Harbor. He argued to the Naval General Staff that the 'US fleet in the Hawaiian Islands, strategically speaking, is tantamount to a dagger being pointed at our throat. Should war be declared under these circumstances, the length and breadth of our Southern Operations would immediately be exposed to a serious threat on its flank. In short, the Hawaii Operation is absolutely indispensable for successful accomplishment of the Southern Operations.'[30] Yamamoto considered the defensive doctrine centered on the battleship obsolescent. Although he did not originate the concept of a preemptive carrier strike at Pearl

Harbor, he believed an independent naval air arm could best execute an offensive strategy against the US Fleet stationed there.

As early as 1909, the Japanese Navy had identified the US Navy as its sole enemy, and since 1927 the Japanese Naval Staff College had conducted tabletop wargames using two carriers to simulate an attack on Pearl Harbor and US units there.[31] Despite the wargaming, however, the use of aircraft during fleet exercises was not innovative and did not challenge the Gun Club's fixation on the decisive battleship encounter. Carrier doctrine still lagged behind aviation technology as little thought was given to the use of carrier aircraft as important components in offensive operations.[32] Although carrier aviation was growing in offensive strike potential, big gun advocates saw its main function as sustaining surface fighting by increasing the effectiveness of friendly battleships over their enemy counterparts. Aircraft accomplished this defensive tactic by securing 'command of the air' over the surface battle space.[33]

Apart from securing command of the air and providing an alternative to surface weapons, carrier aviation delivered offensive striking power several times farther than long-range naval gunfire. Because the basic doctrine of the Japanese Navy was to strike the enemy at a distance from which he could not retaliate, Yamamoto focused on developing attack aircraft that could 'outrange' those of the US Navy.[34] As his 'outranging' attack aircraft joined the fleet, air advocates began promoting the concept of preemptive mass aerial attack, which would require that the fleet carriers be concentrated rather than dispersed.

The idea of massing carriers caused a great naval debate. Battleship admirals favored the doctrine of carrier dispersal, so that a numerically-superior foe could not wipe out all the Japanese carriers at once.[35] The Gun Club also theorized that scattering the carriers would extend defensive air cover for the other fleet units that delivered the main offensive thrust. The revised battle instructions of 1934 were clear that the battleship divisions were the main weapons in a fleet battle and their task was to engage the main force of the enemy. In support of the fleet battle, dispersed naval air units were to establish air superiority before the outset of action.[36] But this meant that the Japanese would have considerable difficulty gathering and organizing their planes for a simultaneous attack in great force on a given objective.[37]

Under the doctrine of dispersal, the Navy's two carrier divisions – with two carriers each – operated separately during the fleet maneuvers of 1937.[38] But the first years of the war in China demonstrated the importance of massing attack aircraft, both for bombing impact and for defense against enemy fighters. This wartime experience, combined with the lessons learned from employing carrier forces *en masse* during the table-top wargames, led to the conclusion that carrier forces must be concentrated.[39]

The conceptual breakthrough that would render obsolete the dispersal concept came in 1940 from Commander Genda Minoru. While watching a newsreel in London of US carrier operations, Genda solved the problem of concentration versus dispersion: by operating in a single box formation, carriers could mass attack aircraft and still be able to defend themselves by launching a concentrated formation of fighters for combat patrol. His box-formation concept induced Yamamoto as commander of the Combined Fleet to undertake a number of operational experiments. One of the lessons learned was that they would need an overall carrier commander to implement standardized training and doctrine.[40] Until this point, the Navy's carriers were assigned to a carrier division generally comprising two or three carriers. The fleet commanders, most of them in the battleship tradition, and even carrier division commanders were content with this arrangement, which left carriers to function as strictly adjunct components of the battle line.[41]

First disruptive innovation group: In December 1940, Yamamoto directed his chief of staff, Fukudome, to form a secret disruptive innovation group to develop the concept for a preemptive air strike on Pearl Harbor.[42] Yamamoto wanted as part of the group a flier whose past experience had not influenced him in conventional operations. He selected Rear Admiral Onishi Takijiro a career airman who was highly respected throughout the Naval Air Corps as a man of intelligence and foresight.[43] To assist him in his study, Onishi temporarily transferred Minoru Genda from his carrier duties to the disruptive innovation group. Genda was a brilliant staff officer of the First Air Fleet who shared Onishi's belief in the key role of naval air power.

With Admiral Onishi providing oversight, Genda made an exhaustive study to determine whether Yamamoto's Hawaiian attack could be executed. By January 1941, he concluded that it could, if all six of the fleet's large carriers were assigned to the operation, and his report convinced Yamamoto that the idea of a carrier-borne air assault was sound.[44] Genda's draft concept of operations on how to attack Pearl Harbor also buoyed the idea that battleships would support the combined air arm attack, which consisted of both torpedo planes and dive bombers.[45]

Second disruptive innovation group: While Fukudome, Onishi, and Genda were working on their secret concept of operations, in January 1941 Yamamoto formed a second innovation group that consisted of key members of his staff. This disruptive innovation group was lead by Captain Kuroshima, Commander Watanabe, and Commander Akita Sasaki, Yamamoto's air officer. The group's task was to evaluate several courses of action proposed by Yamamoto.[46]

In late March, Yamamoto combined the two disruptive innovation groups and tasked them with solving the torpedo challenge of Pearl Harbor.[47] By the end of April, most of the operations officers of the Combined Staff were participating.[48] Yamamoto's disruptive innovation groups produced an important dividend for aviation proponents: they effectively worked out the offensive strategy and concepts of operations for carrier warfare. With these tools, Yamamoto supported the formation of the First Air Fleet, the naval arm that would execute his innovation. In other words, his disruptive innovation group was established and producing before First Air Fleet activated.

First Air Fleet: Apparently, Yamamoto's support for the formation of the First Air Fleet originated in the new warfighting paradigm generated from his disruptive innovation groups. A key theme emerging from the groups was the need for a separate air arm to execute preemptive air strikes. In April 1941, Yamamoto approved creation of the First Air Fleet, headed by Vice Admiral Nagumo Chuichi and composed of the First, Second, and Third carrier divisions.[49] Although the US Navy was experimenting with carriers working together and eventually would form a permanent air arm in 1942, the Japanese First Air Fleet was the first permanent administrative and operational carrier force in the world. By December 1941, the First Air Fleet was the most powerful offensive naval air force of any of the three major navies.[50]

Political strategy

Yamamoto's intellectual redefinition of naval warfighting was accompanied by a political strategy to implement his innovation. In the political process, Yamamoto disguised carrier aviation as sustaining battleship warfare while also leading a political struggle to create a new, stable career path for younger officers committed to the new way of war.[51]

Yamamoto recognized early that aircraft and carrier technology could enable a new military capability. Stepping beyond the Navy's myopic focus on supporting old capabilities, he was able to generate a new doctrine to suit new technology. In doing so, however, he was careful to air his opposition to battleship orthodoxy only to members of the Air Club. Addressing young pilots, Yamamoto remarked, 'Even though the thick-headed [Gun Club leaders] have modified their outlook somewhat, they still don't grasp the realities of air power, so you young men will have to renew your efforts in training and study [to convince them].'[52] Later he commented, 'You young airmen shouldn't insist on the abolition of the battleship, but rather you should think of it as a decoration for our [navy's] living room.'[53]

Timing, however, posed a challenge to Yamamoto's championing of his disruptive technology. Carrier aviation gradually was progressing – from spotting for the battle line, to providing long-range reconnaissance, to attacking enemy carriers prior to the decisive battleship engagement, and eventually to striking enemy fleets and land bases[54] – but the mainstream leadership, the Navy General Staff, was willing to continue experimentation only as long as each evolutionary stage validated the decisive battleship engagement concept. Quite simply, battleship leaders saw new weapons such as aircraft and carriers as adjuncts to their classical strategy. Rather than irritating the Navy brass by vehemently promoting carrier aviation during its development as the most important feature in naval warfare, Yamamoto took a long-term view and disguised his disruptive innovation as sustaining battleship doctrine.[55]

Yamamoto's championing of naval air was risky, particularly in a time of political instability for Japan, when the 'Young Turks' frequently incited military coups and attempted assassinations of senior leaders, a popular method of eliminating opposing military and government officials.[56] Following his appointment to the senior political post of Navy vice-minister in late 1935, Yamamoto risked assassination by stubbornly opposing a war against the United States, resisting the tyranny of the Army and defying the right wing of the Navy.[57] His biographer writes that after only a few days in office, 'a "super-patriotic" political society declared [Yamamoto] a primary target for assassination [and he began] receiving a steady stream of hate mail'.[58] In summer 1939, a group of young naval officers began talking about eliminating Yamamoto. In fact, on 14 July the 'Young Turks' sent a threatening letter asking Yamamoto to resign, noting that they had placed a price of 100,000 yen on his head.[59] The Navy responded by stationing a 24-hour guard around Yamamoto's house.[60] Nevertheless, Yamamoto apparently resigned himself to the possibility of sudden death. He emptied his office of his personal effects,[61] and shortly after receiving the letter he was appointed as commander-in-chief of the Combined Fleet, because the Navy Minister was sure if Yamamato remained ashore he would be assassinated.[62]

Champions of disruptive innovations in Yamamoto's time quite literally flirted with dismissal and even death. Faced with these threats, Yamamoto aimed to assuage his opponents by disguising the First Air Fleet's ultimate purpose as an independent strike force. He accomplished this by supporting the Gun Club's concept that the decisive surface engagement would be preceded by a battle for control of the airspace over the contending fleets, and that this was the sole mission for carrier air. By taking this tack, Yamamoto signaled to the Gun Club that he agreed with the battleship doctrine that viewed 'air superiority not as a means of developing an independent air strike mission, but as a means of support for the Japanese battle line'.[63]

Dealing with the Gun Club was only half of Yamamoto's challenge. He also had to contend with a bombastic Air Club who favored bold innovation in doctrinal and organizational concepts that made air warfare superior to surface warfare.

Admiral Ozawa and the air fleet concept

Although the Imperial Navy entered the China War in 1937 favoring dispersal of its carriers, the first years of the conflict demonstrated the importance of massing attack aircraft, both for bombing impact and for defense against enemy fighters. Air Club advocates eventually began to argue that the massing of aircraft required carriers to be concentrated as well. Extending this concept, Rear Admiral Ozawa Jisaburo, commander of the First Carrier Division, in early 1940 urged Yamamoto to authorize the formation of an 'Air Fleet' within his Combined Fleet. Under such a unified air command, all air units – both carrier and land based – would come together to train and fight.[64] 'Yamamoto, while undoubtedly recognizing the profound strategic implications of such a tactical concentration, twice deflected Ozawa's recommendation, realizing, one supposes, that time would be needed to overcome traditionalists within the Combined Fleet itself.'[65] In June 1940, Ozawa boldly bypassed Yamamoto, writing directly to the Navy Minister with an outline of his ideas for an Air Fleet.

Ozawa's Air Fleet concept provoked heated discussion among senior Gun Club members. Although Yamamoto easily could have killed off the concept, he supported Ozawa's proposal quietly. Overt support was too great a risk for two reasons. First, the timing of Ozawa's proposal did not fit with Yamamoto's campaign plan for developing an air arm. In fact, Yamamoto was upset that Ozawa had made the debate public before he had had time to shape the battlefield properly with the Navy General Staff for his Pearl Harbor air attack concept. If the Gun Club had defeated the Air Fleet proposal, it would have killed Yamamoto's means for conducting an air attack on Pearl Harbor, which he surreptitiously had started developing, at least conceptually, in March 1940.

Second, open support of the Air Fleet concept would have put Yamamoto under direct criticism from the Gun Club, which could have resulted in his dismal as commander of the Combined Fleet. Although nettled by Ozawa's poorly-timed proposal, Yamamoto was politically astute in managing Ozaswa's efforts – supporting him quietly so as not to receive too much criticism and thus being able to remain effective as an *éminence grise*. In his capacity as a behind-the-scenes power broker, Yamamoto arranged an air demonstration for the highest government officials to promote the importance of the air arm concept and his aviation build-up to the Navy. Intuitively, Yamamoto sought to create a tipping point for the Air Fleet concept. On 11 October 1940, he shrewdly staged a review of the fleet off Yokohama, for the inspection of the emperor, who liked and supported Yamamoto. In addition to assembling scores of battleships, cruisers, and destroyers to strengthen the appearance that he supported the Big Gun faction, he also included a half-dozen new aircraft carriers and a 500-plane flypast to display Japan's massive naval air strength.[66] Following the fleet review, Yamamoto took a delegation of high-ranking naval officers to Osaka, the big industrial city, to meet with a conference of Japanese bankers. Over three days, he outlined his plan for building naval aviation, eventually persuading the bankers to support building more aircraft.[67]

When he felt the danger of repercussions had subsided and the Gun Club was convinced the Air Fleet would be supporting the surface fleet, in December 1940 Yamamoto authorized its implementation. Essentially, he had disguised the disruptive air arm concept as a sustaining innovation of the old battleship doctrine.[68] By design, the Air Fleet concept created a single command responsible for training and fighting the carriers together. The Navy's carriers, however, still were assigned to carrier divisions, which, in turn, were assigned to support different fleets of battleships.[69] Simply put, this arrangement still left the carriers as adjunct components of the battleline.[70]

The real concentration of naval power came five months later. In April 1941, Yamamoto championed the First Air Fleet sustaining innovation that moved the carrier divisions from the battleship fleets to the First Air Fleet. How did Yamamoto convince the Gun Club to move the carriers? Again, it was a matter of timing. Because the decisive air battle gradually was becoming part of battleship doctrine, the Gun Club could understand why Yamamoto thought it prudent to promote the First Air Fleet to concentrate the carriers for training to support the battleships.[71]

At this point, the Gun Club saw the First Air Fleet, in its massing of carriers, as a evolutionary tactical concept. Yamamoto, however, saw it as a revolutionary strategic concept. Given this difference in warfighting perspectives, Yamamoto was able to form the First Air Fleet by disguising it as tactical innovation whose tasks still fit within the traditional mission of supporting the battlefleet. He argued that the massing of the carriers in the First Air Fleet was a critical component of the battlefleet concept, as the carrier force would shape the battleship fight by delivering a preemptive strike against enemy carriers.[72] In making his case, Yamamoto wisely did not claim the First Air Fleet was an independent tactical formation capable of undertaking the decisive battle on its own. The main contestants of the decisive engagement remained battleships.

Yamamoto disguised the First Air Fleet by convincing the Gun Club that with the swift improvement in aircraft capabilities, the massing of carriers was a perquisite to conducting the decisive air battle, which would precede the final big-gun all-out battle. The Gun Club bought this logic. Of course, in Yamamoto's vision of the upcoming war with the United States, the big gun battle would never happen.[73] His assumption was that Japan would fight a short and decisive war.[74] Knowing his was the weaker naval force, Yamamoto hoped to knock out the US Fleet in one bold stroke prior to Japan's *Blitzkrieg* of the vast southern regions. Then, before the United States could rebuild, Japan would consolidate its conquered positions and negotiate a peace.[75] Such a plan did not require decisive surface battles, as carrier aviation had the potential to be the decisive new weapon of sea power.

In reality, the Japanese battlefleet did not take part in the first several months of World War II.[76] Instead, without raising the alarm of the Gun Club, Yamamoto deployed only his air arm in a deadly array to wreak havoc in the Pacific. Interestingly, the Japanese Navy waited until March 1944 to rewrite its battle doctrine to make the battleships subordinate to carrier task forces.[77] As Mark Peattie writes, 'For all its striking power, however, the First Air Fleet [from 1941–44] was still not regarded by the Japanese naval leadership as the main element of the Combined Fleet. According to Japanese naval orthodoxy, that role was still reserved for "big ships and big guns".'[78] Yamamoto could not overcome this orthodoxy so he disguised the disruptive innovation as a sustaining one and used the air arm as a strategic weapon.[79]

Vice Admiral Inoue

One of Yamamoto's principal allies, Inoue is an instructive example of running afoul of the entrenched naval orthodoxy by using a direct or nondisguised engagement strategy. In January 1941, Inoue wrote a memorandum entitled *A New Theory on the Armament Plan* and submitted it to the Navy Minister. Essentially, he argued that Japan's victory over the United States depended on naval air power, not naval surface power, and he fervidly expressed his views that decisive fleet engagements involving battleships was outdated thinking. With control of the air being the important feature in warfare, the Imperial Navy should stop building *Yamato* class battleships and focus on building air power.[80] This direct attack did not work well, as it had the unavoidable consequence of poisoning relations with the Gun Club. The result was disastrous. In August 1941, irritated Big Gun leaders, tired of Inoue's caustic criticism of the Navy's shipbuilding plans and of his air power theories, fired him as chief of the Naval Aviation Department and ostracized him to the far edge of the empire at Truk.[81]

It would have been a mistake for Yamamoto to underestimate the Gun Club's ability to fire him as well. So to guard himself against the fate suffered by Inoue, Yamamoto promoted carrier warfare as a supporting element to battleship warfare. He could not save his friend from being fired, but the experience undoubtedly taught him that a headlong attack of the Gun Club's belief in battleships was futile. Consequently, Yamamoto had to disguise his disruptive innovation.

Developing junior officers

Disruptive innovators create and sponsor new career paths through which younger believers can be promoted. Yamamoto was no exception. Throughout his career, he endeavored to create a young officer corps of like-minded officers.[82]

Following a sea tour as commanding officer of a cruiser, Yamamoto, then a captain, requested assignment to the Navy's flight school. Although there was considerable antagonism within the Kasumigaura Aviation Corps toward someone with so little connection to aircraft muscling his way in, Yamamoto emerged to become executive officer of the Navy's aviation school.[83] He insisted that the young aviators conform to battleship standards. Besides learning how to fly, they also had to be able to communicate their ideas and impart to others in writing what they knew. A future disciple of Yamamoto who was stationed with him at Kasumigaura was Onishi Takijiro, who later would champion the Navy's kamikaze effort at the end of the Pacific War.[84]

Encouraging young men to join the air arm, Yamamoto ceaselessly and enthusiastically promoted naval aviation. In one incentive plan, 'he paid 6 yen, up to a 30-yen maximum, for pilots who took off from and landed on a carrier, even in daytime, or for those who took off from a land base and landed on a carrier, or vice versa. The extra pay was given regardless of a pilot's rank. Enlisted men received half the sum paid to officers under the incentive plans. Yamamoto's system continued through the war and was responsible, to a degree, for the rapid rise in the efficiency of pilots and observers'.[85] Yamamoto also decided to promote night flying. Night flights were very dangerous, and air officers were not keen on making them. Notes one report, 'He paid 6 yen to pilots, observers, and flight engineers for each night flight, increasing this sum to 30 for the fifth and subsequent night flights. You had to be airborne at least five minutes each time.'[86]

When Yamamoto approved the First Air Fleet, he could not overcome the seniority of the Navy promotion system and prevent Nagumo from taking command of it. He could, however, influence who would be the first air officer, and he appointed Genda as air operations officer.[87] Where he could, Yamamoto handpicked officers for billets in the First Air Fleet. For instance, he selected Lieutenant Commander Fuchida, who had come to Yamamato's attention with his successful 'air attack' on the fleet in the 1940 maneuvers, to be flight commander of the carrier *Akagi*, and subsequently to lead the attack on Pearl Harbor.[88]

Summary

Yamamoto coupled his visionary view of naval aviation with his political acumen in military bureaucracy to champion carrier warfare. It appears that civilian intervention did not provide the initiative for Japanese carrier warfare. Instead, the available evidence supports the proposition that Yamamoto, holding traditional military credentials and not a military maverick, recognized that structural changes in the strategic environment could be met by a disruptive innovation – a naval air arm. In the beginning of Japan's naval aviation development, interservice rivalry was the engine, as the Navy did not want to fall behind the Army. As naval aviation evolved, however, the engine of change shifted to intraservice rivalry, with the burgeoning Air Club in competition with the established Gun Club.

Yamamoto believed the Navy's battleship strategic orthodoxy was a recipe for ultimate defeat. The Japanese Combined Fleet would be attrited in a long war in which the United States could bring its vastly superior industrial might to bear.[89] Air power presented a solution. If, however, Yamamoto's offensive air strike strategy was to eclipse the Gun Club's defensive doctrine, he would have to disguise the fleet air arm's primary task as supporting the battleships. This is what he did.

Disguising occurred when Yamamoto was able to mass carriers within the fleet air arm for the purported purpose of eliminating the opponent's carriers prior to the decisive battleship battle. In supporting the creation of the First Air Fleet, the battleship admirals did not realize they were promoting the birth of carrier warfare, in which carriers eventually would have an independent and decisive role.

Military victory begets political strength. Armed with a succession of air arm victories during the first few months of World War II, Yamamoto reached the tipping point he had hoped the Japanese Navy would achieve – the moment when carrier warfare surpassed battleship warfare as the decisive way to fight.

11 US Navy disruptive innovation

CWC – naval combined arms warfare

Admiral James L. Holloway's carrier battle group concept was an important organizational change in the fleet's approach to fighting battles. When used to their best advantage, the combined resources of the carrier battle group provided a robust counter to the three-dimensional and simultaneous threats of modern warfare.[1] Creating a naval combined arms force, however, required effective command and control that could take advantage of the complementary characteristics of different platforms. Command and control refers both to the process and to the system by which the commander decides what must be done and ensures that his decisions are carried out.[2]

The product champion of naval combined arms warfare was Admiral Thomas Hayward, and his command and control doctrine was called Composite Warfare Commander (CWC). The CWC concept provided the command and control not only for naval forces operating in a carrier battle group, but also for carrier battle groups operating together. It divided missions according to the environment in which they occur (air, surface, subsurface, etc.), giving individual warfare commanders authority for defensive and offensive operations in their areas. The unique feature of CWC is that functional warlords, such as the antisubmarine warfare (ASW) and antiair warfare (AAW) commanders, carry out specified tasks on their own initiative without having to consult the overall commander for detailed instructions.[3] They have the authority to make tactical decisions and control the forces assigned to them, keeping the task group commander fully informed. If the warfare commanders task the same asset in conflicting missions, such as asking the aircraft carrier to launch different types of aircraft at the same time, the task group commander would weigh in and give direction and assign priorities. Thus, in the CWC concept, control by the task group commander is maintained by negation.

Engine of change: why and when

The creation of CWC was driven by two needs. The first was the need to control the multiple weapon systems of each platform within the battle group. As the Navy began to move away from its Vietnam-era focus on strike warfare, warfighters began seeking a better balance between power projection and sea control as the means to accomplish the Navy's missions.[4] In addition, the acquisition of advanced technologies increased the requirement for coordination; for example, the use of long-range tactical missiles such as Harpoon and Tomahawk had to be coordinated with tactical aviation. Unfortunately, while weapon technologies advanced, command and control facilities and means remained limited. Consequently, the battle group fell short of a robust force that could conduct an array of simultaneous missions. Battle group commanders desperately needed a command and control system that allowed them to direct the whole rather than just the parts.

The second need was to operate several carrier battle groups together. As Commander-in-Chief, Pacific Fleet, Admiral Hayward conceived Sea Strike, a war plan that proposed taking three carriers to attack the western Soviet Union.[5] Executing multicarrier operations in the Pacific would not be easy, however. The command and control metric was based on World War II experience and technology, and it would prove inadequate for modern three-dimensional naval warfare.

Throttle of change: how

Intellectual process

Vice Admiral James Doyle Jr. was the initial product champion of CWC. As Commander, Third Fleet, in the mid-1970s, he was responsible for the training and work-ups for the ships and aircraft in Third Fleet that would deploy to the Seventh Fleet. Essentially, this meant exercising the various components of the deploying carrier battle group – carriers, cruisers, destroyers, submarines, amphibious ships, and air wings – as a whole. One of Doyle's key concerns was that the battle group lacked a sufficient command and control structure to provide the framework for fighting in all warfare areas simultaneously. He noted, 'It was obvious that the traditional approach whereby the task group commander aboard the carrier was responsible for the minute-to-minute, hour-to-hour, day-to-day tactical control of strike [power projection ashore], AAW and ASW was not working.' Doyle continued, 'There was too much responsibility for one person and one staff, and the displays, equipment, and instrumentation aboard the carrier were inadequate to perform the mission.'[6]

The first intellectual step Doyle took was to form a small innovation group led by Bernie Schneiderman, a civilian analyst employed by Third Fleet. Schneiderman's task was to develop a command and control system to support Admiral Hayward's objective of integrating carrier task group resources.

For several years strike carriers had used a special functional commander for antiair warfare. This AAW coordinator was usually the senior surface warfare officer commanding a cruiser. Schneiderman reasoned that if the Navy was willing to put all antiair resources under a single coordinator, then the same approach might work for other warfare areas. Thus, his group created a tactical concept called Composite Warfare Commander,[7] with an ASW coordinator and antisurface warfare (ASUW) coordinator equal in status to the AAW coordinator.[8] This would allow the aviation battle group admiral to concentrate on the primary mission of the force, usually strike warfare.

The challenge faced by Schneiderman's small innovation group was to determine how the various warfare areas would be commanded, now that the Navy had designated big deck carriers to perform both strike and sea control missions. Years before CWC, a rear admiral embarked on an ASW carrier searched the seas for enemy submarines as part of the sea control mission. Supporting the ASW carrier was a screen of destroyers commanded by a destroyer squadron commander aboard a destroyer. The ASW carrier and the destroyers were called a Hunter Killer Group, and their mission was to locate Soviet submarines.[9]

In the CWC concept, the ASWC and ASUWC were coequal to the AAWC, but the AAWC function was often performed on board the cruiser, which possessed adequate command, control, and communications facilities. Similar facilities did not exist on board destroyers. This meant that the ASW commander, who by seniority and experience would be the destroyer squadron commander, would be at disadvantage if he were located on a destroyer.

The next intellectual step occurred when Captain Stu Landersman joined Schneiderman's small innovation group. As commander of Destroyer Squadron 23, Landersman volunteered to test the ASW part of the experimental tactical memo written by Schneiderman on the Composite Warfare Command concept.[10] The CWC doctrine did not mandate that Landersman locate on the carrier, but such a move made sense given the limited communications assets on the destroyers. The ASW coordinator on board the carrier, along with the submarine element coordinator (a submarine commander who directed the movements of the submarine), would have a small command center in a corner of the carrier's Combat Information Center from which to coordinate the search for and prosecution of submarines. Landersman met considerable resistance from some members of the surface warfare community who disapproved of him not being on a destroyer when commanding destroyers.

On board the carrier, the aviation staff welcomed Landersman, but they told him that he would be a part of their staff when he issued orders. Landersman objected, pointing out that his role as ASW coordinator was a separate function. After considerable wrangling and Landersman threatening not to play unless he was allowed to be a separate but equal subordinate commander as provided in the novel CWC doctrine, the aviators permitted him to participate on his own terms.[11] By being obdurate, Landersman ensured the new doctrine linkages among existing components (destroyers, submarines, and carriers and their staffs) would be given a chance to be tested.

Over the next two years, Landersman and his Destroyer Squadron 23 staff would participate in 12 major fleet exercises, playing most of the coordinator and commander roles of the CWC concept. The next intellectual step for Landersman as ASW commander was to use all the different battle group resources in coordinating ASW operations. This included coordinating direct-support submarines, maritime patrol aircraft, carrier ASW aircraft and helicopters, ships with towed arrays, and ships with hull-mounted sonar. Landersman also included SOSUS (underwater cables on the ocean floor) and tactical fighter and attack aircraft to find submarines.

The next intellectual step involved Admiral Hayward, the future Chief of Naval Operations. While Commander-in-Chief, Pacific Fleet, Hayward had learned of Landersman's success in attacking exercise submarines. At the beginning of his last deployment as squadron commander, Landersman paid a call on Hayward at the admiral's request and was tasked to start a school where he would teach battle group tactics to senior commanders.[12] This was important for two reasons. First, Hayward became the new product champion of CWC. Second, since battle group tactics had never been taught before, Landersman decided to teach CWC command and control.[13] The only guidance Hayward gave was that Landersman was to pattern the school after the Royal Navy Maritime Tactical School at HMS *Dryad*.[14]

Disguising process

Landersman considered relationships between Navy communities an important part of the CWC concept; to be successful, any coordinated effort required the full support of every community represented in the battle group. He also believed that the best antisubmarine system was another submarine, and getting submarines to play in the CWC concept was a major factor in its success.

The CWC concept provided for a submarine to operate directly for the admiral of the battle group. Traditionally, aviators on board carriers were not allowed to coordinate the movements of submarines – submarine admirals retained operational control – but under CWC, specially-certified submarine officers were to ride the carrier and direct the employment of the submarine, allowing aviators and destroyermen to coordinate their efforts with the submarine. Selling this arrangement could have been difficult, but Landersman knew the submariners were trying to gain support for their new SSN-688 program. If he could prove submarines could operate in integrated direct support of carrier battle groups, the submarine community would have an additional mission with which to justify its program.

Captain Jerry Holland, an innovative submarine squadron commander, was designated by the submarine operating authority to be the submarine element coordinator on board the carrier with Landersman. Not only did the Landersman-Holland team prove that submarines could provide direct support for carriers, but Holland would eventually write the doctrine for how the submarine would link into the CWC concept.[15] Landersman's goal of getting submarines to play in CWC was a major success, and he had gained a strong ally – the submarine community – as well.[16]

To counter opponents of the CWC concept, Landersman began to see the overall ASW commander effort as more offensive than defensive. With this shift in perception, Landersman and Holland used long-range P-3 air assets to coordinate with distant-stationed submarines to detect, localize, and attack enemy exercise submarines long before they could attack the carrier. The result in some exercises was an unheard-of 50 or so constructive submarine kills.

The next intellectual step for Landersman occurred when he was invited to convince the Atlantic Fleet to adopt CWC. Vice Admiral Tom Bigley, who was serving as Commander, Second Fleet, had been Admiral Hayward's deputy commander in the Pacific and had been closely associated with the origins of CWC. Admiral Bigley 'secretly' flew to San Diego, where he enrolled as a student in Landersman's school. Afterward, he became the biggest proponent of CWC on the East Coast and brought Landersman to the Atlantic Fleet.

Eighteen months after the Pacific Fleet formally began teaching CWC to its senior battle group commanders, the Atlantic Fleet established a battle group tactics course. Concerned that the schoolhouse did not have 'commander' in its title, the new Atlantic commander requested that Landersman change the name of his school from Office of Executive Director for Pacific Fleet Tactical Training to Tactical Training Group Pacific, and that Landersman be called commanding officer, Tactical Training Group Pacific. Landersman agreed – but only if the commanding officer, Tactical Training Group Atlantic, would teach the CWC curriculum that Landersman had been instrumental in developing. The Atlantic commander agreed, and Landersman accomplished the operational goal of spreading the CWC innovation. This is a good example of disruptive theory whereby a product champion used deception (accepting a name change) to promote his innovation (CWC).

A tough challenge Landersman faced was that cruiser-destroyer group commanders were not allowed to command carrier battle groups in the Pacific, or even to be on carriers. As a result, cruiser-group commanders had less CWC because they were denied access to carrier battle groups. Opposition to CWC was overt, however, from surface warfare officers serving on the staff of the admiral who owned all the surface ships. (This admiral did not own carriers.) Landersman recalls that opponents argued CWC would cut out the surface warfare community from valuable naval operations. They also said that professional naval officers did not have to be taught group tactics; they learned them in the progression of increasingly responsible jobs at sea leading up to ship command. Landersman explains his opponents' position: 'Anyone who had completed a successful ship command could step into the AAWC, ASWC, and ASUWC roles. It did not take special training, and CWC was nothing new, it was one of the options in ATP-1 [navy tactics manual] for command and control of Navy task groups.'[17]

Landersman saw the situation much differently. He continued to promote CWC to the aviation and submarine communities by demonstrating how effective the innovation was in hunting enemy submarines, and these two communities accepted his argument and the CWC doctrine. But in private discussions with disgruntled surface warfare officers, Landersman pointed out that most of the time the three principal warfare area coordinators (AAWC, ASUWC, and ASUWC) were surface warfare officers, each operating under a command-by-negation policy that allowed the coordinator more freedom to act than any other command arrangement. Landersman states, 'So, I reasoned that surface warfare officers were running the battle group.'[18] Again, this supports the disruptive theory prediction that the product champion will disguise his innovation to mislead its potential opponents.

Political process

As the long-term product champion of CWC, Hayward protected those subordinates practising the new way of war. For example, after having failed three times to select for promotion, at the end of his tour teaching CWC Landersman requested and received approval to retire. A few weeks before Landersman's retirement, Hayward called and requested that he instead become a member of Hayward's newly-created Strategic Studies Group at the Naval War College. Landersman accepted and after the one-year tour was finished planned again to retire. Once again Hayward called, offering him a Chair of Naval Tactics at the Naval War College. Although Landersman had reached mandatory retirement age, Hayward offered to let him stay in.

Attending CWC school was an important part of younger officers' training in the new way of war. Opposition to CWC by the Pacific surface community, however, meant that surface warfare quotas went unfilled and that the more senior students failed to show up or were permitted in their orders to leave before the course ended. To counter the opposition movement and to protect Landersman, Admiral Hayward paid a visit to San Diego – and the only person he went to see was Landersman. Of course, the senior Navy leaders in San Diego learned quickly that the Chief of Naval Operations had been in town, but not to visit them. When Landersman asked one of Hayward's aides why the CNO had visited only him, the aide stated that the admiral had thought it best to demonstrate protection and give Landersman some clout.[19]

The next political step was to ensure that junior officers learned CWC and to tie such knowledge to promotion opportunities. At Destroyer School, later renamed Surface Warfare Department Head School, the surface Navy created a new billet – tactical action officer. This officer received rigorous training in CWC tactics, which required memorizing considerable facts about both the US and Soviet navies and how they fought. Eventually, department heads were not allowed to graduate until they had passed the tactical action officer course. This meant all surface warfare junior officers had to be proficient in CWC if they wanted to advance.

Summary of CWC

The CWC disruptive innovation created a new way to fight. For the first time, the Navy had a command and control system that permitted it to take advantage of all its weapon systems and allowed it to operate carrier battle groups together.

The key to this innovation was the performance of two product champions, Doyle and Hayward. Throughout the long innovation process, they created small innovation groups and protected the members of these groups who were attempting to institutionalize the concept. Two key members of those groups were Schneiderman, who created CWC from existing doctrine, and Landersman, who skillfully promoted CWC to the aviators and submariners as something that would enhance their ability to perform the missions in which they were most interested. Simultaneously, Landersman promoted the innovation to the surface community by noting it offered them the perfect opportunity to command the entire battle group by negation, a considerable feat when surface warfare staffs were not even allowed on the carrier.

CWC was fully institutionalized when it became Navy doctrine, and it is now part of the curriculum at Navy schools such as Surface Warfare Officers School. There, graduating department heads receive a strong working knowledge of CWC, an understanding they take with them as they rise to positions of higher responsibility, including, perhaps, command of carrier battle groups.

By disguising the CWC innovation to the aviation community, Landersman accomplished what Zumwalt had failed to do – regain control for the surface warfare community of fleet operations.

12 US Navy sustaining innovation
Carrier battle group concept

Before the mid-1970s, the three principal combat missions of the US Navy were strike operations against enemy forces ashore, antisubmarine warfare (ASW), and amphibious assault. Generally, however, Navy amphibious ships and craft in conjunction with naval infantry – the Marine Corps – performed the amphibious assault function outside the carrier battle group concept.[1] To carry out these missions, from the late 1950s on, the US Navy deployed separate specialized carrier task forces for strike operations against the shore and for ASW. The centerpiece of strike operations was specialized attack carriers and their specialized attack aircraft.[2] The centerpiece of ASW operations was specialized ASW carriers and their specialized ASW aircraft.[3] A secondary mission for both types of carriers was defending against enemy aircraft and surface ships, and each carrier was specially armed to perform this function.

By the mid-1970s, most of the strike and ASW carriers, which had been built immediately following World War II, were reaching the ends of their useful service lives. In addition, the force needed to be cut back, as the ASW threat had diminished now that hostile German U-boats and Japanese submarines no longer roamed the seas.

Engine of change: why and when

A changing security environment, reduced budgets, and intraservice rivalry were the primary drivers behind the carrier battle group concept. Admiral James Holloway recognized the need to reverse the slide in ship numbers to meet the rising Soviet naval threat. In addition, he had to deal with an administration that wanted to focus on NATO reinforcement and not on building new carriers, and with the legacy of Admiral Hyman Rickover, who had convinced Congress to pass legislation that required all new major combatant ships be nuclear-powered.[4]

Throttle of change: how

Intellectual process

The product champion of the carrier battle group (CVBG) was Admiral Holloway. In 1976, Secretary of Defense Donald Rumsfeld created a small innovation group to study different-sized carriers and carrier operations. Holloway, a member of the group, led discussions on a range of size options, including a small 20,000-ton vertical take-off and landing aircraft carrier.[5]

Based on that group's findings, Holloway opted for a few high-performance nuclear-powered carriers, combined with a large number of conventional surface warships. He defended his choice of supercarriers, stating, 'The most telling disadvantage of the small carrier is that it is less cost-effective than the big deck. Although the CVLX-45 [a mid-size carrier of 44,000 tons] cost more than half as much as the CVN-72 ($2.08 billion compared to $3.0 billion) it has less than half as many aircraft.'[6]

The next intellectual process for Holloway was to create a multicapable carrier, that is, a carrier that could perform both strike and ASW. Given that technology advances were allowing more diverse combat capabilities to be packed into a given ship or aircraft, Holloway redesignated all the specialized attack and ASW carriers as CVs or CVNs, and all carriers were reconfigured to include new mixes of strike, fighter, ASW and other combat aircraft.

The next step for Holloway was to create the carrier battle group concept. This sustaining innovation improved the performance of naval assets along a trajectory the traditional Navy had valued for more nearly 40 years. Before World War II, the US Navy had created the specialized task force operational concept, combining ships of different types in operational formations to carry out specific tasks. This concept had evolved from previous concepts wherein all battleships and heavy cruisers deployed together as a battle force, and all the light cruisers deployed together as a scouting force.[7]

The centerpiece of the carrier battle group was the newly redesignated CV or CVN. Other units included a multipurpose air wing, cruisers, destroyers, attack submarines, and a multipurpose combat logistics ship.[8] In the 1980s the multiple capabilities and overall power of the CVBGs were strengthened by the introduction of new warfighting weapons and systems, principally the Aegis antiair warfare system and the Tomahawk cruise missile. Navy analyst Peter Swartz notes that these systems not only greatly increased the total combat power of the CVBGs across the spectrum of naval warfare, but also distributed that power more evenly across a variety of surface combatants and submarines, as well as the carrier.[9]

Summary of carrier battle group concept

Changes in the security environment, reduced budgets, and intraservice rivalry appear at the heart of the intellectual development of the carrier battle group concept. Faced with a growing Soviet naval threat and the nuclear propulsion restriction engineered by Rickover, Holloway led a small innovation group to study the problem. His solution was to build a reduced number of large-deck nuclear-powered carriers that could perform both the strike and ASW missions. Around this new multimission carrier, he created the carrier battle group concept. The components of the CVBG were themselves multicapability ships, submarines, and aircraft. Unlike the specialized strike carrier task group and the ASW carrier task group, Holloway's task group could accomplish both missions effectively. The carrier battle group concept was a sustaining innovation because the new technologies were used to help perform existing missions better, and not to change them radically.

13 US Navy disruptive innovation – aborted

Project 60 – defensive sea control warfare

In 1970, Admiral Elmo Zumwalt, Jr attempted an innovation in the way naval forces fought. His new warfighting vision of defensive sea control not only required new equipment and platform components, but also a new architecture to link old and new in a novel way. Instead of focusing on power projection, as in Vietnam (and elsewhere), he wanted to shift to open-ocean defensive sea control to counter the rising Soviet maritime threat.[1] 'Project 60', quite simply, consisted of a promise Zumwalt made during a pre-nomination interview with Secretary of Defense Laird to shift to a new way of fighting and to provide a plan for implementing the innovation within 60 days after becoming CNO – hence the title.[2]

Engine of change: why and when

The origins of Project 60 and defensive sea control warfare can be traced to four separate, but interlocking developments. First, strong signs during the late 1960s indicated that the Soviets were building an array of forces to not only challenge the US Navy's forces in the Mediterranean, but also to rapidly become a global naval power, which could challenge US naval forces on the open ocean. Zumwalt was especially concerned with sweeping the potential Soviet submarine threat from the seas.[3] In a brief to Secretary of Defense Laird in 1970, Zumwalt stated, 'Their [Soviet] submarine activity is four times as intense as ours and covers all the sea lanes of the world…the Soviets have more attack submarines than we do. And they are building at a rate of 10–14 a year; we are building three.'[4] He went on to argue, 'In just two years, the Soviets have produced at least 6 new designs in submarines. Their new attack submarines are $3\frac{1}{2}$ to $5\frac{1}{2}$ knots faster than ours.'[5] In fact, in 1970 Zumwalt stated, 'The Soviets have a two-ocean Navy. If our Naval forces are reduced below the level of end FY 70, we will no longer be able to oppose them simultaneously in the Atlantic and Pacific Oceans.'[6]

Second, Zumwalt faced severe force structure challenges. The block obsolescence of World War II-era ships reaching the end of their active life cycle, which required they be retired, came to pass on Zumwalt's watch. Third, he faced declining budgets. The Navy's total budget in 1970 decreased almost 9 per cent. Finally, Zumwalt inherited a Navy previously led in succession by three aviator CNOs.[7] Consequently, for the previous 12 years the main effort of the Navy revolved around building bigger and better carriers and planes. Zumwalt argued that many of the resources needed to keep the entire force up to date was 'gobbled' up by the insatiable demand of the Navy to deliver air strikes. He later wrote, 'The Navy not only obsolesced, but lost a significant portion of the forces it needed to control the seas, particularly anti-submarine vessels, planes, and weapons of many sorts.'[8] As a surface warfare officer, Zumwalt felt these carriers were being built at the expense of new surface ship construction. These four trends marked the central features of the external environment Zumwalt would have to contend with as the CNO.

Throttle of change: how

Intellectual process

As posited by disruptive theory, the first intellectual step Zumwalt took after learning he would be CNO was to form a small innovation group headed by then-Rear Admiral Worth Bagley, then Commander of Cruiser-Destroyer Flotilla Seven and Group, Seventh Fleet. The key members of the group consisted of Zumwalt himself and Bagley. Zumwalt demanded two traits for the head of this small group and Bagley possessed them both – being an intellectual and a surface warfare officer. Zumwalt called Bagley 'a brainy destroyerman rather than a brainy aviator or a brainy submariner'.[9] As Jeffrey Sands writes, 'he [Zumwalt] needed someone who would be inclined to agree with his intention to raise the visibility of the surface community'.[10]

Bagley headed the small group for one month and then turned it over to then-Rear Admiral Stansfield Turner when Bagley departed Washington D.C. for two months to resume his command duties in Vietnam. Turner, a Naval Academy graduate and Rhodes Scholar (and future head of the CIA) immediately translated Zumwalt's warfighting vision into specific tasks. Word quickly spread through the Navy E-ring Corridors of the Pentagon that the surface union was creating revolutionary warfighting changes and the submarine and aviation barons immediately responded by sending 'volunteers' to assist Turner's innovation group. As predicted by disruptive theory, Turner not only reported directly to Zumwalt but the CNO directly involved himself in the process of the group. For example, during this time Zumwalt spent 'an average of two hours a day on it [Project 60]'.[11] After Bagley resumed as head of the small innovation group, Zumwalt renamed it OP-00H.

The next intellectual step Zumwalt took was to create a second small group at the Center of Naval Analyses under the direction of Dr Phil DePoy and Erv Kapos, the director of CNA's Operations Evaluation Group. Both DePoy and Kapos held dual memberships in both small innovation groups. Reminiscent of Hans von Seeckt after World War I, and as predicted by disruptive theory, Zumwalt created several small innovation groups, which he personally managed to create a new way of fighting. The key questions that Zumwalt and his disruptive innovation groups struggled with were 'What might be done? What can be done? What should be done? What can we do?'[12]

In August 1970, Turner presented an interim version of Project 60 to all Navy flag officers in Washington. Immediately, thereafter, Bagley returned from Vietnam and relieved him. Bagley then circulated the conclusions of Project 60 to all flag officers for acquiescence.[13] Project 60 contained well over 20 initiatives and many of these dealt with personnel problems. In this study, the discussion is limited to the development of new warfighting tasks and components necessary to institutionalize defensive sea control.[14]

Zumwalt considered sea control to be the Navy's most important mission, closely followed by power projection.[15] Placing sea control ahead of power projection as the Navy's most important non-nuclear mission signaled a major shift in priorities. Actually, the mission of sea control languished for several years as aircraft carriers roamed unimpeded off the coast of Vietnam between Dixie Station, the southern operating area, and Yankee Station, the northern operating area in the Gulf of Tonkin.[16] Although the Navy built a brown-water riverine force, which Zumwalt actually commanded prior to becoming CNO, he contended that the Navy's myopic focus on carrier strike warfare had hindered the service from developing other mission areas. Zumwalt blamed the phenomena on Admiral Moorer, the CNO from August 1967 to July 1970, whom he claimed possessed deep-rooted prejudices in favor of carrier air. Naval historian George Baer notes that Zumwalt felt strongly that aviators 'saw the chance to validate the carriers' role in strike support and sacrificed other Navy functions so they could pursue that validation'.[17] Zumwalt went on to write, 'I think the second-rate Navy effort [of coastal control and river patrol] at that point grew of out of decisions deliberately made by Admiral Moorer. Air strikes meant glory for the Navy. He did not want to waste the Navy's resources fighting the war inside Vietnam.'[18]

Baer concurs with Zumwalt, concluding that the Vietnam War threw the Navy off balance by emphasizing strike warfare at the expense of an array of sea-control functions.[19] Uncontested logistic support from the United States and unopposed maneuver off the coast of Vietnam caused the Navy to become complacent about traditional missions such as air and sea control. Baer writes, 'It reinforced carrier-air doctrine instead of encouraging naval officers to think of carrier air strikes as only one of several Navy missions in an age of flexible response.'[20]

None of this is to deny that power projection ashore is important, but because the NATO lift mission could not be performed by air alone, Zumwalt argued the Navy's primary mission should be to provide and protect sealift across the Atlantic.[21] He wrote, 'The Soviet Naval threat, our commitments abroad, and the credibility of our sea-based strategic deterrent demand that the sea control mission be assigned priority of resources at the expense of projection of power ashore.'[22] Consequently, Zumwalt concluded, 'Heavy reliance on sealift is an integral part of the US role as a sea power. It emphasizes the absolute need to be able to control the seas if the nation is to exist.'[23]

The challenge Zumwalt faced, however, derived from air base shortages in Europe. NATO plans called for using all the Navy's carriers in the role of power projection ashore. This meant sea control forces must protect the projection forces rather than protecting sea-lanes of communication. Zumwalt stated, 'There would be little left to provide more than random security to the sea lines of communications. We would then be ceding to the Soviets this linch pin of rapid reinforcement upon which NATO depends to stabilize the conflict on land and reduce the likelihood of escalation.'[24]

A NATO conflict would require, Zumwalt argued, movement of naval forces from the Pacific, resulting in the abandonment of the Pacific area west of Hawaii, and cession of control of those waters to the Soviet Far East Fleet. Likewise a conflict in the Pacific could cause the Navy to lose sea control in the Atlantic and the sea lines that supported NATO.[25] Zumwalt felt he faced not just a problem, but also a classic combined-arms dilemma – a no-win situation. That is, for the Navy to counteract either a NATO or Korea conflict, it must shift carriers from one ocean to another and in doing so make itself more vulnerable to another conflict.

Defensive sea control innovation

From the small innovation groups Zumwalt created and managed came the solution to meeting the Soviet naval threat – defensive sea control. Its aim hinged on shifting the Navy's primary mission from power projection (Korea and Vietnam) to sea control against Soviet submarines and surface fleets. The strategy depended on four new classes of sea-control ships.[26] An extremely austere carrier called the 'Sea-Control Ship' was at the heart of the strategy. Capable of carrying fourteen helicopters and three Harrier VSTOL aircraft, the ship's principal peacetime purpose was to operate in dangerous waters such as the western Mediterranean and the western Pacific. This would allow big-deck carriers, the Navy's most important warships, to withdraw from these areas and deploy out of reach of an enemy first strike. Operating the big-deck carriers in such a way would allow them to respond to a first strike – and therefore deter it.[27]

In a wartime situation the positions of the two carriers would be reversed. The big-deck carriers would fight their way into the most dangerous waters, destroying the enemy fleet beyond cruise missile range using its air wing. As they approached the littoral they would shift air attack from enemy vessels to shore installations and provide air support for the land battle. The Sea-Control Ships, in turn, would operate in mid-ocean, providing protection in the form of escorts for the anticipated 20 or so convoys of merchantmen, troop transports, and naval auxiliaries in need of air protection.[28]

Another feature of the strategy depended on developing and deploying cruise missiles on surface ships.[29] Such a tactic would shift some of the long-range power projection from the aviation union to the surface union. Eventually, Zumwalt wanted long-range cruise missiles deployed on submarines as well. This deployment of cruise missiles would provide offensive firepower for forward-deployed Sea-Control Ships, which would be scurrying to get out of range of Soviet missiles. Another critical feature of placing cruise missiles on surface ships rested on their ability to shape the follow-on campaign or battle to the Navy's advantage in both time and place. For example, surface-ship cruise missiles would allow naval forces to attack fixed inland defenses thus enabling the commander to gain the initiative, preserve momentum, and control the tempo of operations.

Means for accomplishing strategy

The next intellectual step focused on determining what ships should be built to accomplish the new warfighting strategy. Zumwalt's solution again came from his small innovation groups. The problem Zumwalt's innovation groups faced was that an all-High Navy would be so expensive that it would not have enough ships to control the sea. 'High' was short for high-performance ships and weapons systems. Some missions required high-end ships, but other missions did not. These missions, such as sea-control, could be accomplished with what Zumwalt coined as Low-end ships. 'Low' was short for moderate-cost, moderate-performance ships and systems that could be built in relative large numbers. By using a high-low mix the Navy could be in enough places at the same to time to cover both sea-control and power projection.[30]

The solution his small groups proposed came to be called the 'high-low mix'.[31] This innovation sought to increase the Navy's ability to achieve sea control while retaining some forces for projection of power ashore. Such a plan demanded retention of forces at the current 1970 levels. Because the Navy already planned to buy several types of 'high mix' ships – SSN-688, LHA, DD-963, CVANs, and a nuclear-powered guided missile frigate – Zumwalt focused on 'low mix' ships. As he later wrote, 'The innovative part of this program was the Low.'[32]

This plan included a smaller carrier concept called the Sea-Control Ship, which was capable of performing both sea control with ASW aircraft and power projection with strike aircraft. Such a ship was designed to enhance surface ship capability for the sea control mission, in the face of the Soviet anti-ship missile. Thus when faced with opposition at sea, the carriers, now operating both strike and ASW aircraft, could be used to protect the sea lines of communications. If sea control no longer posed a problem, as in Vietnam, both big deck carriers and Sea-Control Ships could operate in an air attack role.[33] His original plan would have provided two or three Sea-Control Ships for the cost of one CVN ($300 million in FY70 dollars).[34]

Disguising process

Zumwalt's next intellectual effort attempted to achieve consensus for Project 60. He circulated Project 60 to all the senior Navy officers and demanded immediate responses and recommendations. As one might expect, Zumwalt received no significant changes as he was determined to meet his self-imposed 60-day deadline.[35] As disruptive theory predicts, senior leaders succeed in achieving the disruptive innovation if they use misdirection or disguise the innovation as merely a sustaining change. But, Zumwalt did neither of these tactics: instead he told senior Navy leadership that he wanted 'revolutionary' change.

Frustrated with the sluggish response of the Navy's rudder to his orders, near the end of his term Zumwalt circulated Elting Morison's Continuous Aim Gunfire innovation case study to all flag officers.[36] Ironically, as discussed in Chapter 8, Continuous Aim Gunfire Case also exemplifies a sustaining innovation. Although we can commend Zumwalt for his efforts to shift the Navy's course in a new strategic direction, he selected a sustaining innovation as the model for achieving a disruptive architectural innovation. As disruptive theory predicts, different management methods are needed to institutionalize a disruptive innovation. Instead of disguising the defensive sea control innovation as a sustaining change, Zumwalt stressed how different the change was and why the aviation and submarine unions should change to support his initiatives. The result, of course, was that these two unions did everything they could to block Zumwalt's innovation.

Although Zumwalt used top-down leadership and follow-up processes to attempt to break through bureaucratic inertia of the Navy, he also tried to use both vertical and horizontal persuasion to implement his innovation. Vertically, he first attempted to persuade the Secretary of Navy and Secretary of Defense. Once he felt he had their support, he sought support from executive branch members including the National Security Council where he frequently met with Henry Kissinger, OMB's director George Schultz, and the Assistant Director for International Affairs and Defense, James Schlesinger.[37] In addition, Zumwalt took considerable effort to lobby Congress.[38] Horizontally, he attempted to establish two-way communications with the fleet, but with any senior military leader who espouses such beliefs, subordinates see such attempts as one-way communication – from senior to junior.

Another intellectual process employed by Zumwalt rested in his reorganization of OPNAV. One of his motives for such a decision was that he hoped to give the surface community more visibility on the Navy's Pentagon staff. As Jeffrey Sands writes, 'Admiral Zumwalt believed that it was necessary to increase the visibility and influence of the surface community to a level more equal with the aviation and nuclear submarine communities.'[39] After significant opposition from senior Navy leaders, Zumwalt compromised on his initial reorganization plan and settled for a structure whereby there would be three warfare Deputy CNOs (or so-called 'barons') consisting of undersea warfare, surface warfare, and air warfare.[40]

Political process

Disruptive theory predicts that product champions in positions of power will use their position to ensure officers favoring the new way of war succeed them in position. Zumwalt clearly attempted to position as many surface warfare officers in key positions as he could. He also took care to protect those senior surface officers closest to him. In fact, he carefully placed these officers into critical positions traditionally held by the aviation union. As an illustration, the Sixth and Seventh Fleets, in the Mediterranean and the Western Pacific respectively, were considered the plum jobs afloat and, traditionally, command of these positions rotated between aviators and surface officers. After Arleigh Burke's eight-year tour as CNO, the next three leaders of Navy were aviators who assigned these posts exclusively to other aviators.[41] Zumwalt wrote, 'this had provoked some hard feelings among surface officers, including me'.[42] Subsequently, Zumwalt considered it a major accomplishment when he helped sponsor then-Vice Admiral Isaac Kidd's shift from First (subsequently called Third) to Sixth Fleet. Throughout his tour as CNO Zumwalt continued to promote non-aviators as carrier battle group commanders and fleet commanders.[43] He failed, however, to ensure the success of the political process that would ensure that changes would transcend his watch.

Speculative evidence supports the claim of a post-tour 'purge'. As Admiral Turner recalls, 'Almost as soon as Zumwalt left, though, the Navy began to turn back the clock. Zumwalt's men were pushed aside, and in 1975, although I was given the fourth star of full Admiral, I was "sent away" to Naples.'[44] Perhaps the most notorious anti-Zumwalt purge case became that of Captain 'Dirk' Pringle, an aviator and the only CNO Executive Assistant never to make Rear Admiral in the history of the Navy. Zumwalt described Pringle as a 'brilliant flyer'.[45] His lack of promotion resulted, in all likelihood, from his association with Admiral Zumwalt. Sands writes, 'Admiral Zumwalt does acknowledge having encountered difficulty in convincing many non-surface officers in the senior naval leadership of the wisdom of his strategic agenda. He had quite deliberately turned to the surface community, and particularly the destroyer community, to build his agenda and staff the key decision nodes within the OPNAV organizational structure.'[46]

Disruptive theory also predicts that product champions take the necessary course of action to ensure the long-term success of the innovation by creating new, stable career paths for younger officers who are committed to the new way of war. Zumwalt consciously decided not to perform this task. In fact, he made conscious efforts to distance himself from the younger officers serving him so they would not be negatively effected by the association.[47] His aide in Vietnam, Lieutenant W. Lewis Glenn, Jr believes he survived the purges and made rear admiral largely because Zumwalt did not want him 'tarnished by his brush. The people he kept around him were the commanders and the captains that he could make flag officers out of...I was a lieutenant, and I think he felt that in my case it would be counterproductive to me to get too close. And I think there's probably some truth in that. There's still a backlash in the Navy'.[48]

Summary of Project 60 and defensive sea control warfare

Perceptions of structural changes in the security environment, theoretical developments involving new military capabilities to meet threats, and intraservice rivalry appear to be at the heart of the intellectual development of Defensive Sea Control Warfare. As a young Navy Captain, Zumwalt, the product champion of the innovation began promoting in the pages of *Proceedings* the concept of a 'high-low' mix.[49] This was his technological response to the increasing Soviet naval threat and the aging surface ship fleet. By itself, however, the high-low mix was not a warfighting innovation, but merely the components of a new warfighting architecture. How Zumwalt planned to link these components together for the new mission of defensive sea control encompassed the disruptive innovation.

He intended to build a small, less-expensive aircraft carrier called the Sea-Control Ship to operate in areas dominated by the Soviets (western Mediterranean and western Pacific). Big-deck carriers would operate outside Soviet cruise missile range and provide second-strike capabilities if the Soviets initiated a first strike. Once the shooting started, the Sea-Control Ships would retreat to the open ocean and perform sea control missions such as convoy duty, while the big-deck carriers would fight their way back into the areas thus vacated.

The concept of defensive sea control warfare died soon after Holloway, an aviator, relieved Zumwalt as CNO. According to disruptive theory, Zumwalt did some things right, but he did many things wrong. Starting with the positive, many characterized Zumwalt as being 'convinced of the need for change, intolerant of impediments to change, and entrepreneurial [i.e., adopting and adapting proposals from others] in his approach to the substance of change'.[50] As this study argues, two types of entrepreneurship exist – sustaining and disruptive. Although Zumwalt attempted to achieve a disruptive innovation, he applied the sustaining model. Consequently, his disruptive innovation did not succeed as he hoped. Precisely because Zumwalt's basic blueprint for institutional change derived from the continuous aim gunfire model of sustaining innovation, he sowed the seeds of eventual failure. Zumwalt, quite simply, failed to realize that sustaining innovations such as continuous aim gunfire can be revolutionary in character, but also remain sustaining because they merely improve the performance of an established way of warfighting. In other words, revolutionary advancements do not always imply disruptive innovation.

His major success lay in managing the intellectual process of the Navy. A Center for Naval Analysis report noted, 'As an effort to lay out and publicize in advance what Admiral Zumwalt planned to do during his term of office, Project 60 is an effort without precedent in US Naval history.'[51] He established the necessary small innovation groups and he participated in them directly. Although performing superbly in managing the small group intellectual effort, he failed in managing the disguising process and political process. Zumwalt clearly did not attempt to disguise his disruptive innovation as a sustaining change. Instead, he purposely coupled his defensive sea control innovation with an overt agenda aimed at increasing the power and prestige of the surface warfare community. By doing so, he unwittingly sank his innovation effort by fueling the ire of the aviation and submarine unions who associated the new way of fighting with the promotion of surface officers who, in turn, gained power at the their expense.

Two notable failures occurred in this example of the political process of disruptive theory. First, Zumwalt failed to create a new promotion path for junior officers who believed in his innovation. Second, he failed to shape the political battlefield remaining after he left the Navy. Innovation, quite simply, must transcend the tenure of individuals who start them.[52] Instead, believers in Zumwalt's reforms faced purges. As Admiral Worth Bagley noted, 'many believe that what Admiral Zumwalt set into motion was, for the most part, not given support after he left'.[53] The defensive sea control architecture that Zumwalt promoted ended after he left office. Likewise, a subsequent aviator CNO, Admiral Hayward, killed the surface effect ship innovation. Furthermore, according to Zumwalt, the Sea-Control Ship (the small carrier) was killed by Admiral Rickover's intervention with Congress, while Bagely believed the Sea-Control Ship failure had as much to do with aviators supporting big-deck carriers as it did with Rickover attempting to kill low-mix, non-nuclear ships.

While it is true that the Defense Sea Control innovation failed, many of the new technical developments that Zumwalt began to build eventually came to fruition in the disruptive Offensive Sea Control innovation. Rather than using them architecturally as Zumwalt envisioned, the components were used in a sustaining architectural way which would be eventually described as 'The Maritime Strategy' warfare. As disruptive theory predicts, the causality for these sustaining successes resulted from a sense of consensus among senior aviator and submarine leaders who perceived these initiatives as merely improving the performance of power projection ashore – an established way of fighting that the aviation and submarine unions have historically valued.

14 US Navy disruptive innovation

Maritime action groups – surface land attack warfare

The surface land attack disruptive innovation is a study in how two sustaining technical innovations – Aegis radar and the conventional Tomahawk land attack cruise missiles – triggered a novel linkage of platforms and a new way of warfighting.[1] Surface land attack was a major shift away from carrier battle group operations to small task groups operating independent of the carrier to conduct strike warfare against land targets. Previously, strike warfare was the exclusive mission of the aviation community. With the capabilities of Aegis radar and Tomahawk missiles, however, small groups of ships, maritime action groups, could perform strike warfare in a much different way.[2] In contrast to the Cold War years when the inviolability of the carrier battle group structure prevailed as the Navy's operating doctrine, the post-Desert Storm years saw Aegis and Tomahawk surface ships departing ahead of and separate from their associated carrier.[3] This new way of war fighting was disruptive in nature not only because surface ships did not have to remain with the carrier and could operate in packs, but also because surface ships could now hit targets with greater precision and at longer ranges than could strike aircraft launched from carriers.[4]

Clearly, the Gulf War set the stage for a new vision of how naval forces could be used. As Bradd Hayes phrases the argument, 'Tomahawk and Aegis were central in the discussions leading to the Naval Service's white paper, . . . *From the Sea*, and were indeed "catalysts for doctrinal innovation."'[5] . . . *From the Sea*, a new strategic concept, codified the new strike capabilities of surface ships against land targets.[6] The paradigm shift was complete when Admiral Mike Boorda, an Aegis and Tomahawk-trained surface warfare officer, became Chief of Naval Operations. To illustrate this shift: the Navy was hesitant to use Tomahawk missiles in retaliation for Libyan terrorist acts in the 1980s, and instead used naval strike aircraft. The reasons for not using Tomahawk were the result of intraservice rivalry rather than technical considerations. As Robert J. Art and Stephen E. Ockenden argue, 'The Navy – specifically, the carrier admirals – did not want the Tomahawk antiship missile (TASM) because it represented a clear and present danger to the mission of the carrier-based aircraft.'[7] With Admiral Boorda as CNO, however, the Navy showed no reservation in using Tomahawk when it retaliated against Iraqi terrorist threats to former President Bush following Desert Storm.[8] Admiral William Owens, Vice Chairman of the Joint Chiefs, noted a year after Admiral Boorda became CNO, 'Today's surface combatants do not spend all their time at sea in company with or in proximity to aircraft carriers, even when assigned to carrier battle groups. This operational separation almost certainly will increase.... Overall, then, they will move toward greater operational independence *vis-à-vis* the carriers.'[9]

What makes the case unique is that disruptive theory predicts that product champions disguise disruptive innovations, but the evidence from this case shows two different product champions disguising the potential disruptive effects of sustaining innovations. Interestingly, the study found little disguising by the product champion who linked these new technologies in a new way of warfighting. Strong interservice rivalry may explain this non-disguising phenomenon.

Engine of change: why and when

Surface land attack warfare resulted from the naval service's efforts to deal with the changes in the strategic security environment wrought by the end of the Cold War and intense interservice rivalry.[10] Two key events determined the new security environment. First, the fall of the Soviet Union caused a fundamental shift in naval thinking away from open-ocean confrontation with the Soviet Navy and toward a much more flexible use of naval force in littoral environments. Traditionally, the carrier battle group was based on the premise that navies fight navies and so, absent a blue-water Soviet threat, the Navy's *raison d'être* vanished. Other nations might harass the United States at sea in a regional contingency, but any country's attempt to build a serious blue-water challenge would take years and enormous expense.[11] Now that the US Navy held unchallenged presumptive sea control, it needed another mission.

The second key event, Desert Storm, provided this mission – littoral warfare. In Desert Storm, the Navy learned it was not 'joint' enough, and this shortcoming limited its contributions in the overall campaign. This led to a higher level of interservice rivalry between the Navy and Air Force and caused the Navy to reevaluate how it would enhance its effectiveness in the joint arena. Ultimately, this interservice rivalry spurred the Navy to shift its focus to fighting in the littoral, as documented by...*From the Sea*, and specifically to the disruptive innovation surface land attack warfare.

As Captain Edward Smith, an original member of the small innovation group that produced the innovation noted, 'from beginning to end, it [...*From the Sea*] remained an internal Navy Department reaction to those changes [resulting from Desert Storm] and not one directed or pressured from outside'.[12]

The two key technology innovations that led to surface land attack warfare were Aegis phased array radar and the Tomahawk cruise missile. Both were developed to compete with evolving Soviet technologies.[13] The Soviets' rapidly expanding nuclear missile arsenal and subsequent building of naval ships to launch them created a requirement for the US Navy to be able to find and sink these ships before they could strike. Since surface land attack warfare builds on Aegis and Tomahawk innovations, this case first will develop these events as mini-cases and then analyze the larger disruptive innovation of surface land attack warfare.

Throttle of change: how

Aegis intellectual process

The challenge facing carrier battle groups was to devise a system to protect it from saturation missile attacks staged by Soviet aircraft and cruise missiles.[14] A myriad of Soviet naval aviation breakthroughs such as the TU-22M 'Backfire' bomber, which could carry several varieties of high-speed air-to-surface missiles, had made such attacks a likely Soviet tactic. To counter this threat, the carrier had to be capable of tracking, targeting, and engaging large numbers of incoming contacts. This is much easier said than done. The required radar would have to rely on something other than a mechanically aimed antenna.[15] The solution was an electronically aimed, or 'phased array', radar, which could move from one target to another almost instantaneously, to distribute radar beams and defensive missiles among several targets.[16]

The first intellectual step was taken when the Navy appointed Captain Wayne Meyer to manage the Aegis (for the shield of Zeus) phased array radar system, which RCA, the prime contractor, had begun developing. Generally speaking, all three warfare unions agreed that a new generation of Anti-Air Warfare (AAW) surface escort for carriers was needed, but critics of Aegis complained that RCA's phased array concept would cost too much to develop and field in the fleet. A central concern was the radar's weight and power requirements. Zumwalt tasked Meyer with determining whether the Aegis system could be scaled down and procured at a lower cost. Meyer responded by stating that a scaled-down version would be a waste of money and that the most prudent course of action was to continue pursuing the original system.[17]

As discussed in the Project 60 case, Zumwalt was attempting refurbish an obsolescing surface fleet with a high-low mix of new ships, and Rickover was pressuring him to buy expensive nuclear-powered Aegis ships. Zumwalt knew he could not afford to buy sufficient numbers of an Aegis ship, but he could not cancel the program because this would leave the Navy without a medium-range air defense capability – a situation the aviation union would not tolerate. Zumwalt finally agreed to develop the Aegis system, but he delayed deciding what type of ship it would go on. Although Meyer now had the resources, he did not know on what platform Aegis would be carried nor did he have a voice in whether it would be nuclear- or conventionally-powered.

The next intellectual step occurred when Captain Meyer gained control of both the Aegis system and the Aegis platform design effort. This was accomplished when senior naval leaders in favor of the Aegis system assigned him as Chief of the Surface Missile Systems Office in Naval Ordnance. He also retained his position as head of the Aegis Project.[18] The importance of Meyer having this dual responsibility is that he could synchronize the pace of design and production of the platform with the developing of Aegis. In other words, Meyer could slow platform development so it would not constrain his ability to build the Aegis system he wanted.

Disguising process

During this time, Meyer realized that 'the Aegis system was changing from just an AAW sensor/weapon system to one which could direct *all* AAW weapons and sensors for an entire Carrier Battle Group'.[19] He believed this was the right direction, and it was technically feasible. Aegis systems could revolutionize how carrier battle groups fought. Instead of operating as an AAW escort, Aegis ships would become the command centers for CVBG operations. Meyer gained support for an enhanced Aegis radar within the Navy's shipbuilding community, but he effectively disguised the system's true potential from the powerful aviation community.

The key challenge that remained for Meyer was deciding what platform would carry the Aegis system. Rickover's influence in Congress combined with Admiral Holloway's (the new CNO) support of Rickover's nuclear navy were too powerful for the Secretary of Defense, who supported Meyer's conventionally-powered Aegis ship, to override. In the end, Rickover would win, and Aegis would be programmed for installation on USS *Long Beach*, a nuclear cruiser.

Meyer, however, used Rickover's nuclear-power advocacy to hide his own non-nuclear Aegis program. Behind the scenes, he was secretly developing an Aegis destroyer powered by gas turbines as a companion to Rickover's nuclear-powered Aegis cruiser. Eventually, however, this led to a showdown in Congress. Rickover attempted to block Meyer's presentation of the conventionally-powered destroyer, fearing that he would lose a cost-benefit comparison between the two platforms.

The next intellectual step was Meyer's performance before Congress during the confrontation. He proved to be Rickover's intellectual match. Educated as a systems engineer at MIT and the Naval Postgraduate School, Meyer quickly gained the confidence and respect of members of Congress and their staffers. Performing as teacher, an engineer, and a competent manager, Meyer convinced Congress to fund both conventionally-powered and nuclear Aegis ships.

With the focus on what type of ship would employ the Aegis, Meyer was able to move well beyond an improved AAW system without the knowledge of the aviation community and to achieve his ultimate goal of revolutionizing surface battle tactics with the introduction of Aegis command and control systems. As an illustration, Admiral Thomas Hayward, the new CNO, testified to Congress that, without Aegis, existing carrier battle groups would be at great risk from Soviet missile attack.[20] Hayward was supporting a mid-range AAW system, not Meyer's new vision of Aegis, but as Meyer states, 'We delivered a war machine that far exceeds what we were supposed to deliver.'[21]

Summary of Aegis

As disruptive theory predicts, Meyer started off as a mid-grade officer who was a passionate zealot in promoting Aegis as sustaining innovation. He sold Aegis as a new technology that would be used to perform intermediate-range AAW better and by doing so, was able to overcome aviation's opposition to the new capability, which it believed would diminish its traditional mission of battle group protection.

From the very beginning, however, Meyer realized that Aegis had the potential to create a new way of fighting. Aegis, quite simply, could manage not only intermediate range AAW, but also the entire battle. Recognizing this, Meyer strove to develop Aegis to its full potential, and he accomplished this by disguising its true performance characteristics. Supporting this theme, Meyer sold Aegis as a counter to the intermediate-range AAW threat, which served as a brilliant camouflage for his true vision of Aegis as a battle force integration system.[22]

Throttle of change: how

Tomahawk intellectual process

Prior to the development of sea-launched surface-to-surface missiles, the Navy's efforts focused on championing surface-to-air missiles. Since carrier aviation was the centerpiece of naval warfare at sea, the aviation union supported the surface union in developing long-range missiles that could provide an outer air defense for the carrier battle group. The aviation union did not support US surface vessels acquiring long-range surface-to-surface capabilities as long as carrier aviation could provide them.[23] The first intellectual step toward developing the surface-to-surface Tomahawk missile took place when Secretary of the Navy Paul Nitze directed Rear Admiral Elmo Zumwalt to initiate a study on cruise missiles after the 1967 sinking of the Israeli destroyer *Eilat* by an Egyptian Styx cruise missile.[24] Immediately, the aviation union balked at Zumwalt's proposal to develop a long-range missile, so to gain support, the surface union agreed to limit the range of the Harpoon missile.[25]

The future product champion of the Tomahawk missile, Commander Walter Locke, served as the guidance project officer for Harpoon. The next intellectual step took place when Locke began to study the advanced cruise missile (ACM), which would have a range of more than 300 miles. By design, the ACM could be launched from vertical launch tubes, which meant that Locke would gain support from the submarine community because the ACM could be launched from both submarines as well as surface ships. The Navy appointed Locke as the director of the ACM project. He teamed up with Dr John Foster, Director of Defense Research and Engineering, to promote the ACM.

In early 1972, the Secretary of Defense attempted to start strategic cruise missile development, but supplemental funds were never appropriated. Although Foster testified about the technical feasibility of a strategic cruise missile, naval aviation had already convinced Congress any missions that Foster identified could be accomplished with the carrier assets. Thus, Congress had no interest in developing a long-range tactical cruise missile. Nevertheless, Locke was intrigued by the potential value of extending the range of ACM, and on his own broadened the definition of ACM to include strategic cruise missiles and used some ACM funds to develop them.[26]

Unexpectedly, Locke received renewed interest in his program from Congress after the signing of the May 1972 Soviet Union and US SALT I agreement.

Although SALT limited ballistic missile production, it said nothing about the production of cruise missiles.[27] Locke and Foster used the SALT agreement to convince Congress to fund a strategic (nuclear-capable) ACM. The logic the product champions used was that strategic ACMs on submarines 'would add more deterrent per dollar than any other of our schemes'.[28] Congress liked the strategic implications of the submarine-launched cruise missile (SLCM). They were impressed by the fact that the SLCM added a fourth leg to the traditional strategic triad of sea- and land-based intercontinental ballistic missiles and bombers.

The next intellectual step occurred after Congress funded strategic cruise development. Locke was convinced that the tactical version of the cruise missile held untold potential and on his own began to develop it in conjunction with the strategic variant. As Gregory Engel notes, 'Congressional support of the strategic variant of the cruise missile paid for the bulk of the research for the conventional Tomahawk missile.'[29] Locke successfully disguised the development of the tactical ACM by linking it to the strategic version.

By the summer of 1973, Locke was convinced that the success of the strategic cruise missile guidance system proved that with further development, the missile 'could be better than any other nuclear armed weapon'.[30] He believed that the strategic Tomahawk was almost good enough for attacking land targets with a conventional high-explosive warhead. Supporting this theme, Locke noted, 'I saw an opportunity for an expanded cruise missile capability. I believed that the Tomahawk would make a much greater contribution as a conventional, non-nuclear land attack missile.'[31] Locke noted four reasons why: 'First, an extremely accurate missile without a nuclear warhead could hit and destroy a specific target with little damage to the surroundings.' Second, 'accuracy would reduce the number of weapons needed to destroy a target'. Third, ' "pilots" lives and aircraft could be saved by using conventional land attack cruise missiles'. Fourth, 'enemy defenses would find Tomahawk difficult to "see" and even more difficult to shoot down because of its small size and low altitude flight'.[32] Finally, Locke believed that 'delivering a precise blow to the enemy with little risk to the attacker or nearby civilians would be a significant change in war. This was the promise that air power enthusiasts had repeatedly made but so far seldom delivered'.[33]

Disguising process

The next intellectual step was Zumwalt's Project 60 initiatives, which included development of surface-to-surface and subsurface-to-surface missiles. Locke now had a senior product champion for Tomahawk in Zumwalt. He also had gained the support of Admiral Rickover, who would be the product champion for submarine-launched Tomahawk. By carefully building support in Congress and the submarine community and by disguising the Tomahawk mission as ambiguous, so as not to raise undue suspicion in the carrier community, Locke was able to proceed with development of the tactical Tomahawk. Engel paraphrases Locke's argument: 'As long as a strategic cruise missile appeared to be the goal, the tactical anti-ship version could be treated as a fortuitous spin-off. So, although the Navy drafted a requirement for an anti-ship version of the cruise missile in November 1974, it purposely paced its progress behind the strategic version.'[34]

The next step for Locke and Foster was to gain the support of William Clements, the Deputy Secretary of Defense. President Richard Nixon had picked Clements to evaluate all defense projects and cancel those that lacked promise. Again, Locke was able to sell the Tomahawk program as a strategic program that would give the United States leverage. The angle the product champions used was that the SLCM provided a valuable strategic reserve. 'Cruise missiles launched from SSNs would not threaten strategic stability as a "first launch" threat (since their yield would be too small to target ballistic missile silos) but would increase stability by providing an invulnerable reserve. Its mission would be to deter second, or follow-on Soviet strikes should a nuclear exchange occur.'[35] Afterward, Clements became a strong supporter of the Navy's cruise missile program.

The next intellectual step came when Zumwalt decided to seek approval for the placement of strategic Tomahawk onboard surface ships as well as submarines. Using a study that said placing Tomahawks aboard SSNs would divert them from their primary mission of ASW, Zumwalt convinced Deputy Secretary of Defense Clements to support his decision to designate all cruisers as platforms for Tomahawk. Clements, in fact, went one step further, declaring that he would not approve another shipbuilding program unless Tomahawk cruise missiles were included.[36]

The subsequent intellectual step occurred when Locke briefed Dr Albert Wohlstetter, a member of Admiral Holloway's CNO executive panel. Wohlstetter believed that 'such a weapon [Tomahawk] could achieve nuclear effects using conventional warheads thus greatly enhancing strategic stability'.[37] This was the first time that the development of a non-nuclear Tomahawk to support the strategic mission was raised. Holloway liked Wohlstetter's non-nuclear Tomahawk concept and directed Locke to increase the accuracy of the missile. As Engels argues, 'Pursuit of a conventional land-attack variant was a watershed for the Tomahawk.'[38] Over the next few years, Locke was able to create a digital scene-matching area correlation guidance system, and by 1980 he had perfected the conventional Tomahawk, which had unheard-of accuracy. Now, by placing the conventional land-attack Tomahawk on a variety of surface ships, the Navy had increased the Soviet's targeting problem dramatically. While all this hints at a major technological breakthrough, Engel probably was closer to the mark when he said, 'The real doctrinal breakthrough was that surface combatants could now mount land-attack operations independently of the Carrier Battle Group in situations where only a limited air threat existed.'[39] Based on the above, Wohlstetter's support of tactical Tomahawk is what allowed Locke to received directed funding. Clearly, however, if Locke had not disguised his efforts, he would not have been able to convince Wohlstetter of the tactical possibilities of Tomahawk.

The aviation community remained opposed to the tactical Tomahawk and argued that their inability to be recalled after launch was a major concern. Consequently, the missiles were considered but not used in the 1986 Libya strike. Again the aviation community successfully argued that if the conventional

Tomahawk strike was unsuccessful, it could derail the strategic program for which the missile was originally designed and by which it was sold to Congress and the Navy (or at least to the aviation community). Locke makes the argument that Admiral Watkins, the CNO and a surface officer, 'failed to lead in the establishment of doctrine for a new and different capability; therefore, the operational forces did not embrace Tomahawk. The lack of readiness in 1986 resulted from Admiral Watkins' 1982 decision to give the conventional land attack missile the lowest priority'.[40]

Locke sums up the Navy's problem with Tomahawk, stating, 'Unfortunately, the Navy leadership then lacked the vision to grasp the importance TLAM/C [tactical Tomahawk].' He paraphrases Dr Wohlstetter's argument, 'Any large organization tends to have a problem with innovation.' Locke attributed the Navy's tactical Tomahawk stagnation problem directly to Watkins. 'Admiral Watkins especially craved order. When he was the Vice Chief of Naval Operations, he behaved as a critic disturbed by the development process and the complications of a joint program [such as Tomahawk].'[41]

Once Watkins departed as CNO, Locke went to Vice Admiral Joseph Metcalf, by then Tomahawk's sponsor. He argued that the Navy must follow the Air Force example of using the Global Positioning System (GPS) to guide the Tomahawk. Metcalf agreed and directed that tactical Tomahawk be built with GPS.[42]

In 1982, Admiral Locke was relieved of his position as manager of the Tomahawk program because senior naval leaders feared another Rickover, who had created a strong political base within Congress. Engel writes, 'Because Admiral Locke had effectively bypassed naval leadership to overcome numerous problems, he was considered an outsider.'[43]

Summary of Tomahawk

Initially, Locke sold the Tomahawk as a new technology that would be used to help perform the existing strategic mission better, and not to change it radically. Such an innovation is sustaining in character because it sought to improve the performance of the strategic triad along the dimensions of an established performance trajectory that the Navy and Department of Defense historically valued. Indeed, sea-based ICBMs were one leg of the strategic triad, and strategic Tomahawk appeared to be a derivative of the sea-based leg that would allow a smaller strategic missile to be placed on attack submarines as well as surface ships. The overall result was increased strategic flexibility and a more difficult targeting problem for Soviet planners.

As disruptive theory predicts, a product champion and project manager such as Locke can be a passionate zealot because he is promoting a sustaining innovation. Indeed, Locke was an intellectual mid-grade officer who was a passionate zealot in promoting strategic Tomahawk. Along the way, however, he recognized that a tactical (or conventional) Tomahawk had the potential to create a new theory of victory in naval warfare and realized there would be an ideological struggle over the creation of technologies that would contribute to this new way of fighting.

As disruptive theory also predicts, the product champion of a disruptive innovation will disguise the innovation to protect it as it develops. This is exactly what Locke did. He promoted the strategic Tomahawk as his overall goal, while working on the spin-off tactical missile. In contrast to his public passionate zeal for strategic Tomahawk, he disguised his rhetoric for the tactical version by promoting it privately among his supporters.

Throttle of change: how

Surface land attack warfare: intellectual process

The product champion of surface land attack warfare was Admiral Frank Kelso, the Chief of Naval Operations. The concepts and doctrine for surface land attack warfare were actually part of a larger Navy white paper called...*From the Sea*. The contents of this document, which later was amplified in *Forward...From the Sea*, developed an array of ideas for how the naval services should go about reinventing itself for fighting in the littorals. One of these warfighting concepts that the Navy successfully institutionalized was surface land attack warfare.[44]

The first intellectual step occurred six months after Desert Storm ended when Admiral Kelso, Chief of Naval Operations, and General Mundy, Commandant of the Marine Corps, directed Vice Admiral Leighton 'Snuffy' Smith and Major General H.C. Stackpole, the Navy and Marine Corps staffs' respective directors for operations and planning, to create a small Navy and Marine Corps innovation group. The group was asked to assess 'what had or had not changed in the national security environment; and what implications these changes held for the roles and missions of forces'.[45]

Leighton 'Snuffy' Smith and Stackpole, in turn, created a small innovation group consisting of 18 captains/colonels and commanders/lieutenant colonels drawn from the Navy and Marine Corps staffs. The small innovation group reported twice a week to a six-member flag committee headed by Rear Admiral Ted Baker and Major General Caulfield. When the innovation group began addressing capabilities required to pursue littoral warfare, Baker and Caulfield expanded it to 25 members in order to capture operational expertise from the fleet and theater commanders.

The next intellectual step occurred when Baker and Caulfield divided the small innovation group into four smaller groups. Navy Captain 'Rusty' Petrie, a serving air group commander (CAG), headed one group, which was responsible for strike warfare. Using Admiral Owen's Sixth Fleet maritime surface action group as a model, Petrie proposed that Tomahawk surface combatants using dispersed firepower could perform the mission of strike warfare – a mission previously reserved for aviators.[46] Captain Bradd Hayes, a member of the strike warfare group, considered Petrie's decision to support surface land attack warfare as the defining intellectual moment of the disruptive innovation. For the first time, an aviator in command of a carrier air wing openly supported a surface combatant force operating separately from the carrier battle group performing the strike warfare mission. This event marked the beginning of a paradigm shift in naval warfare.

Six months after the creation of the small innovation group, Baker and Caulfield briefed the new warfighting doctrine for three days at the secluded Quantico Marine Corps Base to what was reported to be the largest convocation of Navy and Marine Corps flag and general officers ever assembled – 'by one reckoning 80 per cent of the three- and four-star officers in both services and some 50 additional flag officers'.[47] At the conclusion of the conference senior leaders felt that a general consensus had been forged regarding the new direction for the naval services. Captain Edward Smith, an original member of the small innovation group, relates the sentiment of one senior flag officer, who stated, 'This ("...*From the Sea*") is now our future. Let's get on with it. Get aboard, or get out.'[48]

The 1991 Gulf War marked the first time Tomahawks were used against an enemy. US Navy surface ships and submarines launched hundreds of cruise missile attacks against Iraqi targets throughout the war. Based on this performance, senior naval leaders had little reason to disguise the fact that Aegis-and-Tomahawk-capable surface ships could operate independent of the CVBG and conduct strike warfare without risking having an aviator shot down and held as a political bargaining chip.

Following the Gulf War, Bradd Hayes noted that 'Presidents facing a crisis now are just as likely to ask "where are the Tomahawks?" as they are "where are the carriers?" Conventional Tomahawks are now considered one of the weapons of choice to make political statements against rogue states.'[49]

Surface land attack warfare political process

Both Tomahawk and Aegis helped usher in new roles for junior officers. As disruptive theory predicts, a new distribution of power within the service must emerge from the ideological struggle, as well new paths to power (i.e., flag rank). As Captain Bradd Hayes notes, command of an Aegis-and-Tomahawk-equipped ship has in fact become the 'holy grail for surface warfare officers and is viewed as one of the surest paths to Flag rank in their community'.[50] On a recent 2000 flag list for surface warfare officers, a majority had commanded or served as executive officers on Aegis and Tomahawk ships.

Disruptive theory also predicts that new career paths are created from within, by senior officers currently holding power, rather than being forced on the service from outside. This is certainly true in the case of Aegis/Tomahawk cruisers and destroyers. Once these ships were introduced into the fleet, assuming command of one was considered the fast track for promotion. Admiral Boorda, a surface-trained Aegis and Tomahawk officer, was the first such officer to reach the rank of CNO. This achievement marked a complete paradigm shift in surface land attack warfare. Although an aviator, Admiral Jay Johnson, assumed the top job after Boorda's death, Admiral Vern Clark, another Aegis and Tomahawk officer, relieved him.

Summary of surface land attack warfare

Surface land attack warfare innovation resulted from a new theory of victory whose product champion was Admiral Kelso. The Navy's move to littoral operations and specifically surface land attack warfare occurred despite the absence of civilian intervention and the lack of an immediate or highly visible strategic vulnerability. Instead, it occurred as a result of a high level of interservice rivalry. The Army and Air Force had outperformed the Navy in Desert Storm and as a result the Navy transformed its core mission to littoral warfare to remain competitive in the joint arena. The naval services reinvented themselves as a 'sea-air-land' team that would provide the joint bridge, and door, into the littoral theater of operations.

The cruise missile and Aegis system were perhaps the most significant weapon systems of the 1970s and early 1980s. Both Locke and Meyer were passionate zealots in promoting what appeared to be sustaining innovations; however, each realized early in the development process that these weapon systems had the potential to lead to a new way of warfighting. As disruptive theory predicts, both disguised their efforts to push Aegis and Tomahawk technology to the limit. By camouflaging their efforts they were able to decrease intraservice opposition from the tactical air communities and in doing so created systems that performed far differently from how they were promoted publicly.

In the Tomahawk case, Locke managed the tactical cruise missile as a shadow program, a mere spin-off from the strategic program. It was this disguising activity that allowed the conventional program to flourish. The aviation community had no desire to support the conventional Tomahawk because it not only threatened the traditional aviation strike mission, but also competed for scarce funding. Ironically, as the importance of the strategic Tomahawk increased (a program the aviation union generally supported), its momentum carried forward the conventional program, which represented a clear and present danger to the mission of carrier-based aircraft.[51]

Cruise missile development did receive support from senior-level intervention, but this fact supports disruptive theory, which posits that as long as the technology innovation is sustaining in nature and enhances the performance of a mission, civilian assistance is most welcome. In an unexpected twist of the theory, Locke used civilian intervention to support strategic Tomahawk, which in the long run indirectly benefited tactical Tomahawk as a spin-off innovation. Arguably, however, if civilian intervention had supported tactical Tomahawk directly, it would have been unsuccessful because the aviation union would have used its powerful Congressional resources and bureaucratic mechanisms to block its development.

Meyer's successful management of Aegis is almost the same story as Locke's Tomahawk. Meyer successfully disguised the full potential of Aegis from the aviation community while gaining its support for what appeared to be a mid-range AAW development. The key difference between the two innovations is that Locke pushed cruise missile technology to its limits without any new doctrine on how it might be used. Meyer also pushed the Aegis technology to its limits but he had an accepted, albeit limiting, doctrine in place on how it would be used. The key similarity, however, is that both product champions were working to a hidden agenda.

The actual disruptive innovation of surface land attack warfare did not come about until both Aegis and Tomahawk were in place onboard surface ships and the Navy faced a strong rival challenge from the Army and Air Force. As predicted by disruptive innovation theory, the product champions – Admiral Kelso and General Mundy – created a small innovation group to create the new way of fighting. Using a model developed by Admiral Owens in Sixth Fleet operations during Desert Storm and after the successful firings of several hundred Tomahawks during the war, Kelso and Mundy did not have to disguise their disruptive innovation. Locke and Meyer already had performed the necessary misdirection during development of the two weapon systems. When faced with this paradigm shift, Captain Petrie, representing naval aviation, accepted the new way of warfighting and did not attempt to block it.

Perhaps more so than in any other case examined, the correlation here between the Navy and Marine Corps in how its senior leaders manage innovation is very strong. The two services approaches to forging...*From the Sea* strongly suggest that they rely on the same causal mechanisms as described by disruptive theory.

15 US Navy disruptive innovation

Tactical collaborative network and a summary of disruptive Navy cases

Inchoate disruptive innovation: the development of network-centric warfare

The Navy's development of network-centric warfare is an inchoate disruptive innovation. It is following a pattern similar to the German development of the *Blitzkrieg* where the doctrinal rethinking took place largely before the technology and the experiments. The doctrine for network-centric warfare is found in *Naval Warfighting* – a 1994 combined Navy and Marine Corps capstone doctrine that emphasizes the tenets of maneuver warfare.[1] Championed by Admiral Frank Kelso and Marine Corps General Mundy, *Naval Warfighting* is essentially a derivative of Marine Corps maneuver warfare, a doctrine that stresses the principles of initiative, exploitation, combined arms, and independent action by commanding officers at all levels.[2]

To bring this doctrine to fulfillment, the technological advances of network-centric warfare focus on linking ships, aircraft, and shore installations into highly integrated networks. In doing so, the Navy's aim is to transform the fleet from 'platform-centric' to 'network-centric' operations, whereby battlegroup ships and aircraft share command-and-control and target data through seamless networks. Using these networks to share information, the aim of network-centric warfare is to turbo-charge 'speed of command' in order to generate a higher tempo of action than the enemy.[3] In an environment where chaos is the rule, not exception, 'speed of command' will help naval forces adapt to rapidly changing situations and exploit fleeting situations at much higher speeds than the adversary. This, in turn, will permit our naval forces to disrupt the enemy's ability to function.

The focus of this case is on an unresponsive acquisition system for developing the components (devices and systems) for achieving network-centric warfare. As the Commander of Pacific Forces, Admiral Dennis Blair complains, 'Transformation means continuous change and requires continuous experimentation, but the current [navy] acquisition system is out of touch...Our acquisition system is fundamentally broken, especially in the area of information technology.'[4] The problem is that the Navy's present acquisition system is designed to minimize risk, which means it lacks the agility to react to disruptive technological innovations in the later stages of the acquisition process just prior to full-scale production.

A sustaining innovation: Cooperative Engagement Capability

One of the legacy technological pillars of navy network-centric warfare is the Cooperative Engagement Capability (CEC). CEC was designed during the Soviet era to link dispersed navy guided missile ships and communications relay aircraft operating in a particular area into a single air-defense network. As Rear Admiral Phil Balisle writes, 'Before CEC, a ship had to wait for an attacker to cross its radar horizon before it could detect, track, and engage, delaying an intercept until the attacker was well within its own sensor range. With CEC, a missile ship now can engage an attacker before it reaches the ship's radar horizon and can provide midcourse guidance to an intercept at the radar horizon of its own control radar.'[5] Although CEC is a breakthrough technology that will bring about dramatic changes in military operations, it is a sustaining innovation because it enhances a core competency – theater air and missile defense – along a performance trajectory valued by the Navy.

Sensor-netting software and special-purpose hardware allows CEC ships to exchange radar detection information in a way that supports guided missile engagements on one Aegis combatant based on radar data provided by another Aegis combatant. Generally speaking, a CEC system installed on a ship or aircraft includes CEC software, computer processor, and a high-powered directional phase-array antenna for receiving and transmitting information. Sensor data is processed and transferred to the data distribution system (DDS), which transmits the data at extremely high rates to other CEC network participants. Meanwhile, the processor aboard each unit fuses the sensor data received from other network players for use by the ship's Aegis air defense weapons system.

In the works since the early 1980s, the Navy has sunk more than $2.5 billion to develop the first network-centric system. After CEC passes the final operational evaluation, the Navy plans on installing the system at an estimated cost of nearly $80 million for each unit.[6]

Several significant problems exist for this 15-year-old technology. The result has been a disagreement within the surface force between the champions of CEC and Aegis-equipped ships. For example, in January 1999, Rear Admiral George Huchting – the senior Aegis manager for nine years – stated that, 'CEC was invasive to the Aegis system.'[7] The central dividing point was that the Aegis community felt it was being forced to incorporate a questionable system. CEC advocates have countered by arguing that integration testing revealed glitches in the Aegis software that limited CEC-Aegis interoperability.

Following the lukewarm CEC-Aegis integration test, Ronald O'Rourke, the lead defense specialist for the Congressional Research Service, Congress's watchdog for defense programs, reported three sets of CEC hurdles to Congress.[8] First, significant interoperability problems exist between the CEC system's software and the software used to run the Navy's Aegis air defense system. Although the Navy is addressing these problems, O'Rourke is not convinced the Navy's solution will fix the challenges of incorporating shared radar data with complex weapon systems. At risk is the Navy's goal of CEC becoming the backbone for a Department of Defense (DOD) initiative to link all air radar sensors called 'Single Integrated Air Picture' or SIAP.[9] In the Navy's SIAP vision, CEC would provide the joint weapons engagement capability. If product managers cannot successfully integrate CEC and Aegis systems, then the DOD will have been defeated in its goal of achieving joint-service networked air defense.[10]

Second, CEC requires an enormous amount of bandwidth. Bandwidth can be compared to the diameter of a water pipe, which can vary from small kitchen size to large tunnel size. The reason for the large size diameters is the CEC user-push approach, which disseminates all detected data, useful or not, on the network. Presently, the Navy does not have a plan to manage the limited data-transmission bandwidth capability that would be severely strained by the CEC system. Using a sports illustration, CEC requires each network user to watch all the Sunday NFL games provided by the radar network despite the user's interest in select teams. Likewise the radar network simultaneously monitors all the games that each participant is providing. In the CEC network, large amounts of bandwidth are necessary for everyone to use the network simultaneously. As the number of units increases, network bandwidth requirements rise. As a result, there are limitations to the number of sensors that can participate on a CEC network.

Third, CEC is a 15-year-old technical construct that does not benefit from modern network concepts and innovations. Therefore, although CEC has validity as a warfighting concept, it is in great need of a 'technology refresh' – a term used by Lockheed Martin officials who were recently hired by the Navy to infuse new ideas into CEC.[11] Ronald O'Rourke ends his CEC report to the 107th Congress by noting that Tactical Component Network (TCN®) could possibly be the needed technology refresh.

In sum, at the heart of the CEC technical and interoperability issue is the fact that it is a single purpose 'device' trying to be a network foundation. That is, CEC is a set of equipment with which all network combatants must integrate. The mission requirements of each combatant become subordinate to the mission requirements of CEC as existing equipment must be reconfigured to support CEC. This results in numerous interoperability issues and restrictive physical and functional coupling between all participants, both technically and programmatically. Extending the point, if the Internet worked liked CEC does, whenever a user changed its web page, it would have to reprogram its computer.[12]

Disruptive innovation: *Tactical Component Network*

One of the disruptive technological pillars of Navy network-centric warfare is the TCN®. TCN® is a disruptive technology innovation that brings about dramatic changes in military operations along a new warfighting trajectory not traditionally valued by the Navy. Its breakthrough is that it transcends the Navy's vision of linking Aegis weapon platforms. The September 11 attack underscores the reality that the United States requires a robust and coordinated homeland defense. The foundation for this defense must be an integrated warning, information, and coordination network that can link DOD resources with Federal, State, and local agencies. The evidence supports that by embracing TCN® as the backbone architecture for network-centric warfare, the Navy has a magnificent opportunity to spawn a truly Joint disruptive innovation in how homeland defense is conducted.

TCN® is a software product created by a small defense company and designed to better enable information exchange and synthesis among military combatants. Based on modern network concepts, TCN® is a collaborative tracking system with Internet-like attributes.[13] Each sensor user has the equivalent of a web page that any other unit can browse. Thus TCN® can be thought of as a 'generic enabler' – as opposed to a gizmo or device – that avoids all of the pitfalls of the old CEC device concept. With TCN®, a user or data source merely plugs into the Web. Element independence is maintained and bandwidth is consumed in miniscule amounts compared to the CEC method.

Engine of change: why and when

The Warren Citrin story

TCN® is the brainchild of Solipsys Corp. engineers, including Warren Citrin and others who were on the original CEC design team as employees of the Johns Hopkins University Applied Physics Laboratory (APL). Frustrated that the acquisition process prevented CEC from benefiting from technology advances, Citrin co-founded Solipsys in 1996 and shortly thereafter began an internally-funded effort to investigate solutions to the myriad CEC problems he had noted over 12 years as lead engineer.

Prior to the 1996 creation of Solipsys Corporation, the maker of TCN®, the founders and many of the company's employees worked for the Johns Hopkins University Applied Physics Laboratory (APL). While at APL, they were the developers of a mid-1980s processing device called CEC. Although quite an advanced processing structure for its time, the CEC was severely limited in its applicability beyond its original intent of allowing an Aegis cruiser to launch Standard missiles using target information from other Aegis cruisers operating in the vicinity. According to Citrin, 'the Navy and APL marketing of the CEC got out far ahead of the system and CEC gained a following and budget well beyond its capability to deliver. The CEC was sold as the solution to every joint service and international network problem'.[14]

By 1996, the CEC (the device), which had not substantially evolved technically, was a large, well-funded program with hundreds of APL, Navy laboratory, civilian and uniformed military and industry contractor careers entwined in its success. CEC was nonetheless widely viewed as a cutting edge technology about to revolutionize military operations. In Citrin's view, CEC was a Commodore-64-era networking concept and technology in a time of Internets and Pentium 4s.[15]

TCN® avoids the CEC problems by invoking the concept of 'collaborative tracking'. In this method, each sensor contributing to the network is provided with a set of software applications that run in commercial off-the-shelf computers. Using a well defined applications program interface (API), each sensor collaborates with all other sensors in the network in a manner that ensures only contributory information is exchanged in the creation of the network track picture and in accordance with the needs of each information user. This results in the minimization of the amount of communications capacity required to support even enormous networks of sensors. Using the sports example, TCN® allows each network user to tune into the NFL games of their choice. Similarly, the radar network only pulls those NFL highlights it desires from each net participant. In the TCN® network, less bandwidth is necessary as selective data routing permits for efficient bandwidth rationing.

TCN® has demonstrated the ability to support thousands of tracks to CECs specification accuracy using low cost radios already in the military inventory such as EPLARS and PRC-117. It also supports DDS linkages for the Aegis combatants that need communications attributes consistent with remote Standard missile engagements. The result is a network of participants that are physically and functionally independent. As the key to a viable network foundation, it is a network concept that supports the mission requirements of the participating combatants rather than the other way around, as with the CEC processing and communications structure.

Throttle of change: how

Intellectual and political process: Warren Citrin story

The founders of Solipsys set out to champion innovations that would meet the needs of the times rather than those of an earlier era. Having gone through a decade of experience developing and testing CEC, the new company would start from first principles and develop, on its own internal funding, a set of products that would do the job intended for CEC but in a much more applicable, cheaper, and extensible way.[16]

Small group proposition

Upon departing APL, Citrin immediately hired several engineers and formed a small innovation group. Citrin provided direct oversight of the innovation group. The intent of this group was to develop the product that would ultimately be called TCN®. The technical principles that guided TCN®'s development can be summarized as follows:

- A product, meant to be used in a wireless network, must not consume additional communications capacity simply because additional clients join the network. Capacity consumption should only grow with the growth in meaningful information to be exchanged.
- The addition of new contributing information sources to the network should not require software accommodations on the part of any other network client.

- Each network client should be able to define the information that they wish to receive and each source of information should have the ability to select those network clients authorized to receive that information.
- The network structure should permit varying levels of security and data authorization with 100 per cent anti-tamper assurance.

Citrin believed that these principles would result in a defense network product that would meet the true technical requirements of a viable information network.[17]

Political process (no disguising)

Having created the 'ideal' network foundation, Solipsys engineers put together a technical presentation that they would use to describe TCN® to the Navy acquisition leaders. By 1998, TCN® was highly developed, both conceptually and technically, and presentations by Solipsys to the Navy were in full swing. Citrin did not disguise his disruptive technological innovation as sustaining CEC. Instead, Solipsys engineers described not only the benefits of TCN®, but also the shortcomings of CEC. Having been the original developers of CEC, they were able to articulate the coming train wreck that CEC represented if the Navy continued to push that concept as the answer to all questions. The presentations were not limited to government people, but were given to laboratory and industry members as well, including APL and the new CEC design agent, Raytheon St. Petersburg. According to Citrin, there was no credible technical dissension to the Solipsys exposition of TCN® or the shortcomings of CEC. The advantages of TCN® over CEC were clear and irrefutable. Citrin writes, 'One would have expected, at least from Navy and government acquisition officials, an expression of gratitude that Solipsys had developed the long-sought-after solution to some extremely vexing technical and operation challenges for the armed forces. And they had done it at no cost or risk to the government, unlike the CEC that had already cost $2.5 billion dollars and was on the ropes technically. Instead, the Navy and the laboratory, APL, which was charged with giving them disinterested technical advice, declared open war on Solipsys and TCN®.'[18]

The events that took place in 1998, when TCN® was originally disclosed to the community, stunned Citrin. Despite its obvious advantages, TCN® fell victim to the archaic Navy acquisition bureaucracy. The acquisition bureaucracy stifles innovation through institutional inertia. This paralyzing inertia comes from many sources. One notable source in the realm of defense information technology is the military laboratory community. Years ago, before the digital commercial revolution, the university and military labs had the defense information technology (IT) field much to themselves. As times changed and industry, both large and small, caught up with and even surpassed the labs in this arena, they became a community with a diminished mission. This has led to an environment in which the laboratories act as industrial product developers, competing with civilian industry. This, in and of itself, would not be a problem. The labs, however, retain their status as trusted government agents and product assessors, even in areas wherein they are competing. The resulting conflicts of interest and impediments to innovation are enormous.

This problem is central in the TCN® case. In this instance, the Johns Hopkins University APL has a large stake in the CEC. Having originated the system, APL continues to be heavily funded for CEC and CEC-related products. Although Raytheon is the industry design agent for the product, APL continues as the developer of new CEC IT upgrades. Having CEC as the focus of future joint and coalition networking efforts is very important to APL's business plan for the next decade or more. Yet APL acts as the Navy's Technical Direction Agent for these matters, including the determination of matters that materially affect the selection of TCN® or CEC as the networking basis. An APL employee acts as the Navy Technical Director in a quasi-governmental position overseeing these very issues. As one might expect, APL has been the most vocal detractor of TCN®. Although APL and the labs are certainly not the sole forces hampering innovation in the defense community, their role in the TCN® and CEC case illustrates an acquisition process that kills innovation and particularly disruptive innovation.

Another obstacle to innovation is the often used 'sunk cost' argument. With $2.5 billion of sunk cost into the CEC, the surface warfare community has decided it is better to ignore TCN®. In fact, it is preparing to sink another $4 to 8 billion into CEC – the cost of outfitting and maintaining 215 Navy units.[19]

As reported by Greg Schneider of the Washington Post, 'It's just that the Navy has invested too much time and money in its current system – the Cooperative Engagement Capability, or CEC, built by Raytheon Co. – to think about changing course now, according to Navy officials.'[20] Rear Admiral Phillip Balisle, who oversees surface warfare, confirmed using the sunk cost argument when he stated, 'CEC – we need it today…I do not have the luxury, no matter how attractive the [TCN®] option is, to simply wipe the slate clean…'[21]

Sadly, CEC product champions are making the classic reasoning error called 'the fallacy of sunk cost' when they are influenced by what has been spent. A fallacy is, quite simply, an illogical way of thinking that is very appealing to almost everyone. Derived from economic decision theory, sunk costs are expenses already incurred which cannot be recovered regardless of future events. One instance of decision-makers not facing up to sunk costs and instead letting them sway subsequent decisions was the argument to continue the Vietnam War so as not to waste those lives lost already. Since sunk cost cannot be regained, the rational action is not to use them to determine the merits of choosing between CEC and TCN®. Thus, for decision-making purposes, sunk costs are strictly irrelevant. What does count, however, is a rational assessment of the expected marginal cost and benefits of committing further resources.

Engine of change: why and when

TCN® amphibious story

One senior warrior has begun to champion the TCN® cause, however. Rear Admiral Paul Schultz, Commander Amphibious Forces Seventh Fleet, began quietly promoting Expeditionary Network-Centric Command and Control. He called his innovation the Modular Command Center (MCC), with TCN® as the backbone architecture.

As a cruiser-destroyer sailor with extensive network-centric experience as the OPNAV CEC program sponsor, Admiral Schultz found deficiencies in the situational awareness available to his Amphibious Ready Group (ARG). Schultz was familiar with TCN® and its potential to drastically enhance situational awareness aboard amphibious ships. He was also aware of the TCN®/CEC debate in the mid 2000s that had reached a fevered pitch and many articles were beginning to appear in the defense press.

Throttle of change: how – small group proposition

As commander of amphibious forces Seventh Fleet, Schultz formed a small group to manage the TCN® disruptive innovation. He designated Ensign John Wallace, a limited duty Aegis combat systems officer, as the lead action officer of the small group. He also worked closely with Mun Fenton, an Office of Naval Research employee who was a TCN® advocate.

Political process: disguising proposition

Looking to provide a situation awareness improvement to the amphibious ready group, Admiral Schultz attempted to work with the surface Navy community to field the MCC and assess the TCN® technology in an operational environment. Fearing a threat to the CEC acquisition program from TCN®, the surface community turned him down.[22] In late 2000, Congress appropriated funds to install TCN® on Seventh Fleet ships in the Pacific. This was done over the objections of the surface Navy community that had, by this time, become very fearful of the TCN®. To circumvent the objections of NAVSEA, Solipsys was given a contract to install TCN® on the ships by the Naval air community, NAVAIR. This unusual arrangement – installing a new technology on ships by the air community – proved to be a turning-point in TCN®'s fortunes.

With support from interested congressional leaders and the Office of Naval Research (ONR), Admiral Schultz and his staff were able to put the MCC/TCN® to sea in Fall 2001.[23] Working diligently but quietly, he was able to meet his objectives without confronting the oppositional CEC community.

Admiral Schultz successfully tested MCC with TCN® during pre-deployment work-ups in the fourth quarter of 2001. As anticipated, the results were superb. By disguising his innovations and efforts to circumvent an entrenched Navy acquisition community, Admiral Schultz used the methods of previous product champions such as Admiral Moffett. Moffett disguised the nascent concept of carrier warfare by telling battleship admirals that carrier air would be used primarily in a spotting role for naval gunfire.

TCN® proved to operate well in the CTF-76 Amphibious Ready Group of the Seventh Fleet and new, powerful supporters began to appear. By late 2001, the new Assistant Secretary of the Navy, John Young, and others in Congress began to understand the potential benefits of the new technology and also began to recognize the familiar resistance in the entrenched surface navy acquisition community.

Another important area was homeland defense. The TCN®/CEC debate transcends the Navy's vision of linking Aegis weapon platforms. Citrin argues that a TCN® innovation would not only facilitate the linking of USW to surface and air network-centric warfare, but also would provide the foundation for a network-centric homeland defense.

Acquisition system supporting sustaining innovations

The most startling aspect of the CEC/TCN® debate has been the unwillingness of the established acquisition community to embrace the potential offered by the TCN® technology. As Warren Citrin discovered, introducing a disruptive technological innovation without disguising it as a sustaining innovation will most likely result in a 'delay and defer' strategy employed by the program managers. Essentially, there are several groups contributing to the 'delay and defer' strategy.

CEC Program Office

This is probably the most defensible of the various parties involved in the 'delay and defer' strategy that was employed to hold off any serious consideration of TCN® over the near term. The key argument was sunk cost. The current system required 17 years to evolve from concept to an evaluated system. Any fundamental change to the architecture would require more time and expense. Mark Trainer argues that as interpreted by the CEC Program Office in the late 1990s and early 2000s, 'improving the existing CEC design was simply not in their job specification. The CEC Program Manager, Dan Busch, exemplified this behavior and was totally focused on achieving a successful OPEVAL with the current system and bringing that system through the Milestone Decision for full-rate production. While one can argue the ultimate warfighting value of bringing an old and non-extensible system through this production gate, there is also a valid counter-argument that Captain Busch executed his charter perfectly and without distraction'.[24]

Trainer notes, 'One must be impressed by this dedicated focus. Equally impressive was the extended network of CEC supporters throughout the Navy administration and within the OSD acquisition community, and the way in which the CEC Program Office was able to call in specific individuals to stifle any TCN® traction or momentum.'[25] Trainer asserts that 'For every positive article, presentation, or demonstrated event that emphasized what TCN® could bring to the world of sensor-netting, there seemed an equal and orchestrated response from the operational Navy that would sing the praises of the existing system, even with a 17-year-old legacy and (frankly) horrific integration track record on every target platform.'[26] As an aside, this single-minded devotion is actually incentivized within the structure of the acquisition community. The acquisition process itself is the culprit, ensuring that innovation cannot be embraced past a certain point in the game, where achieving production status and rolling into the deployment phase define success.

Pentagon OPNAV

Virtually all the OPNAV sponsors have achieved flag rank through supporting the existing CEC program. The prospect of championing another technology in the face of this pedigree is understandably not appealing.

Congress

In an era of budgetary challenges within the Pentagon, a highly-placed Congressional supporter can be very attractive. It is Raytheon, St. Petersburg's good fortune to have the current Chairman of House Appropriations as their Congressional Representative – a chair that Rep. Bill Young has occupied for several years. As Trainer notes, 'During the 1997 and 2003, Congressman Young has delivered up more than $600 million on Congressional plus-up to supplement the OPNAV CEC program budget. With something like 400 constituents employed by the CEC program in St. Petersburg, this aggressive support is clearly justified from his perspective. And during this time, CEC was held by its Navy sponsors to be a technological and tactical wonder, so why wouldn't the Congressman support the capability?'[27]

Again, this may help explain the reluctance of an OPNAV sponsor to endorse a shift in technology – and potentially jeopardize a historically strong relationship with the House Armed Services Committee.

Summary

In the TCN® case, the acquisition process is an adversary that threatens to kill a dramatic technological innovation that surpasses the Navy's CEC. Using advanced information networking concepts and Internet-age technology, Tactical Component Network, or TCN®, promises to disrupt the legacy CEC device by sharing radar-tracking data for millions or billions of dollars less while allowing more participants to function within the entire network.

Although Citrin had championed a disruptive innovation that would surpass CEC, he did not disguise his innovation as sustaining. Instead he attacked the CEC community head on. The CEC champions responded with several arguments. One of these arguments was sunk cost.

Using the fallacious 'sunk cost' argument that the Navy has already invested over $2 billion in CEC over 15 years, CEC product champions dismissed TCN® from consideration for use in the Navy.[28] While making the sunk cost argument, the CEC champions rushed to install the CEC system as soon as a Milestone 3 production decision[29] could be obtained. So, from the view of Citrin and the TCN® champions, the CEC acquisition community is anxious to deliver a 'Commodore 64-era' networking solution to the fleet, ensuring that operational units never receive the 'Pentium 4-era' TCN® system.

TCN®, however, did receive considerable support after Admiral Schultz championed it as the information backbone for amphibious ships. By taking the

TCN® innovation into the amphibious community, he effectively disguised it from the CEC community that was focused on Aegis missile ships.

The network-centric warfare disruptive innovation remains incomplete as the navy develops technologies and systems. One of the promising pillars of network-centric warfare is TCN®, however. TCN®'s adversary is not a particular person, but a process managed by acquisition admirals that stifles free thought and crushes new ideas.[30] A process, quite simply, that rewards bureaucrats and insider business interests and punishes innovation. This adversary poses a serious threat to next generation network-centric innovation.

The stakes in this dispute are considerable. Because this innovation could transform the way the naval services fight, the outcome of the TCN® debate will have a powerful impact on embedded and emerging legacy systems and the naval services ability to absorb network-centric advances. What is at stake goes beyond dramatic changes in naval warfighting, however. The outcome of the TCN®/CEC debate could affect our ability to revolutionize America's defense against catastrophic terrorism.

Summary of naval cases

Disruptive innovation

Civilian intervention

The proposition that civilian intervention causes disruptive innovation receives little support from the Navy cases. Although product champions received support from top-level civilians in funding issues, the five disruptive innovations were the result of American and Japanese navy product champions, and the debate that occurred was among navy officers. The study reveals that civilian intervention, of course, could – and did – affect product champions' innovations by altering the balance of resources to favor one program over another, but there is no evidence that civilians delineated for themselves the new way of warfighting. They were simply resizing resource 'wish lists'. An example is the American carrier warfare case, in which Moffett wanted to buy four 13,600-ton carriers but received authorization for only one – USS *Ranger*. Another example is the Japanese carrier warfare case, in which Yamamoto wanted to buy more carriers beyond his allotment, but instead four super 18-inch battleships were authorized.

Interservice rivalry

American and Japanese carrier warfare and surface land attack warfare all showed a strong causal link between intense interservice competition and disruptive innovation. In the CWC and defensive sea control cases, however, the study did not find such a connection. Instead, the engine of change was intraservice rivalry resulting from changes in the security environment. The study draws mixed theoretical conclusions from these two episodes. Although the external variables covary in the same manner, only CWC was a successful disruptive innovation. The implication from the CWC case study is that intraservice competition has more explaining power than interservice competition. Specifically, the desire for the surface warfare community to regain control of Fleet Task Forces was the causal factor in achieving the innovation. Similarly, intraservice competition explains why the newly-acting aviator Chief of Naval Operations, Admiral Holloway, nixed Zumwalt's disruptive sea control for offensive sea control. The critical external factor was the desire of the aviation community to reassert the prominence of the carrier battle group.

Yet, the lack of interservice competition in the aborted defensive sea control case also supports the argument that interservice cooperation stifles innovation. Thus, a variation of the interservice competition explanation (cooperation stifling innovation) could have been responsible for the aborted disruptive innovation. Even without interservice competition, however, intraservice considerations, as in the CWC case, can overcome stagnation when rivalries among intraservice arms are sufficiently great.

Small innovation groups

The case studies reveal that establishing small innovation groups gives product champions significant leverage over the disruptive innovation process. Admiral Moffett, the product champion for carrier warfare, created two small innovation groups for developing carrier doctrine. Heading the groups was Captain Reeves, the intellectual planner who devised the concept of carrier warfare as practiced during World War II. In the Japanese carrier warfare case, Yamamoto used several small innovation groups to create his disruptive innovation.

Another case in which product champions relied extensively on small groups is Navy combined arms warfare – CWC (Composite Warfare Commander). Initially, Vice Admiral Doyle, Commander, Third Fleet, attempted to integrate the training of deploying battle groups, submarines, and long-range land-based ASW planes. Doyle created a small innovation group headed by civilian Bernie Schneiderman, who devised the CWC concept. Soon joining the group was Captain Stu Landersman, the product champion of CWC, and it was he who devised most of the intellectual breakthroughs integrating submarines and SOSUS (fixed underwater ASW capability).

The origins of defensive sea control innovation came from Admiral Elmo Zumwalt. As soon as he learned he was to become CNO, Zumwalt formed a small innovation group headed by Rear Admiral Bagley. Soon afterwards, Rear Admiral Stansfield Turner joined the group, and together they created a general innovation plan called 'Project 60', which included defensive land attack warfare.

After the Navy's lackluster performance during Desert Storm, Admiral Frank Kelso, the Chief of Naval Operations, and General Carl Mundy, Commandant of the Marine Corps, formed a small innovation group in Newport. This group eventually published the white paper '...*From the Sea*', which created the surface land attack innovation whereby Aegis-and-Tomahawk capable surface ships operated independent of the carrier battle group to conduct strike warfare.

Disguising process

The prediction that product champions disguise disruptive innovations was validated in all the cases except surface land attack warfare. In the carrier aviation case, Admiral Moffett and Admiral Reeves initially were careful not to take issue with any of the concepts that discussed the use of carriers to support battleships. By doing so, they were able to disguise the nascent carrier warfare concept while he built more carriers and experimented with new concepts during fleet exercises. Likewise, Yamamoto was careful to tell the battleship leaders that his carrier warfare operations were the prelude to the main event – battleship warfare.

In the Navy's combined arms case, Captain Landersman ignored the direction from his surface warfare superiors and embarked on the carrier to command the ASW portion of the battle group. The aviators, who always controlled the battle group, welcomed him on board, as he did not pose a threat. As Landersman continued to test and evaluate the CWC concept he brought more surface and submarine commanders to the carrier. He was able to disguise what he had been doing successfully: when the CWC concept finally was accepted and implemented in the fleet, the three key warfare commanders – AAW, ASW, ASUW – were all surface warriors. As Landersman had planned it, the surface Navy was running the battle group. A key point in Landersman's disguising scheme is that he made the qualifications for being a CWC warfare commander so specialized that only surface warfare officers who had received the special training at surface warfare schools would have the required expertise.

In the failed defensive sea control case, the study did not find any evidence that even loosely connected the product champions with disguising or misdirecting their innovations. In fact, just the opposite is true for Admiral Zumwalt, who closely resembled a zealot in promoting his innovation. This finding supports the disguising hypothesis that states that if champions do not disguise their disruptive innovation as sustaining the old way of fighting then the innovation will fail. This finding also provides insight into a possible correlation between intraservice rivalry and the disguising factor. The general lesson may well be that if intraservice rivalry is the dominant external variable then the precedents for the product champion disguising the disruptive innovation as sustaining are evident in the Navy cases studied in this book.

In the surface land attack case, however, the central mechanism contained in the disguising model – that the politics of managing disruptive innovations is eased when their proponents disguise them as sustaining innovations – is largely absent from empirical observation. In fact, in this case, a service consciously adopted a disruptive innovation, knew what it was doing, and basically did not conceal it. Arguably, the absence of disguising in this case casts serious doubt on the disruptive theory original claims. But when considering the external factors, the lack of disguising makes sense. Based on the evidence, a situational conditional would consist of a dramatic change in the external environment that would be sufficient to cause a disruptive innovation without the product champion having to use a dissimulation strategy. Such a claim would be consistent with other portions of disruptive theory, which predicts that an increase in interservice competition will cause disruptive innovation. Specifically, the observable condition supporting this phenomenon would be a sudden spike in interservice competition such as the Navy experienced after Desert Storm. Compared to the Air Force, and to some degree the Army, the naval services performed poorly in the Gulf, especially in providing air power. The Navy's perception after the war was that the Air Force was the country's premier service and that unless the Navy learned to fight in the littorals it would continue to run in second place. This interservice competition caused the head of the Navy, Admiral Kelso, to create a small innovation group and to shift the Navy to a new way of war fighting without disguising it as a sustaining innovation.

There was some technological disguising involved in this innovation that did contribute to its overall success. Surprisingly, the product champions of the radical technologies – Aegis and Tomahawk – had disguised their true warfighting potential. Both technologies had been deployed on ships using the old way of fighting for a considerable time prior to Kelso's disruptive innovation. Thus, these disguising events alone did not provide the impetus to cause a disruptive innovation.

An examination of the Tactical Component Network (TCN®) case reveals that disguising proved to be a key factor in the development of this disruptive technological innovation. The TCN® case provides a natural experiment between the non-disguising efforts of its inventor, Warren Citrin, and the subsequent disguising efforts used by Admiral Schultz to promote TCN® in the fleet. With the dominant external factor being intraservice rivalry in the TCN® case, the evidence shows that the CEC acquisition community shunted Citrin's non-disguising efforts. The evidence also shows Admiral Schultz's disguising efforts to promote TCN® as a sustaining (and non-threatening) amphibious innovation led to the Navy experimenting and evaluating the merits of using TCN® as the architectural backbone for the CEC innovation.

Political process

There was a high degree of correlation in all disruptive cases between the product champion's ability to manage the political process of protecting and promoting junior officers in the new way of war and the subsequent success of the innovation. In all four successful disruptive cases, the product champions fought and won the political struggle, capturing political power within the Navy. Even more convincing is the fact that product champions who did not protect and promote their junior officers failed to innovate.

We observe in the defensive sea control case that loyal Zumwalt surface officers began to be purged after Holloway relieved Zumwalt as CNO. Holloway, an aviator, and the rest of the aviation community never forgave Zumwalt for firing a popular aviator admiral in command of the Pacific Fleet. The study equates this purging behavior with Holloway ending Zumwalt's innovation. Clearly, the easiest way to end a disruptive innovation is to end the careers of those officers associated with the innovation. It is apparent that Zumwalt consciously did not protect his junior officers, fearing that too much protection would lead to their persecution after he departed.

Sustaining innovations

Civilian intervention

The tenet that civilian intervention causes sustaining innovations receives strong support from the Navy cases. A close examination of the Tomahawk case strongly suggests that without the support of high-ranking civilians such as Secretary of Defense Melvin Laird and others, the Tomahawk program never would have begun. Disruptive theory predicts that civilian intervention often is responsible for sustaining innovation, and this is observed in Albert Wohlstetter, a civilian who provided the intellectual argument for building a conventional Tomahawk variant. The relationship between civilian intervention and sustaining innovation also is seen in the development of Aegis radar. In 1958 Richard Hunt, a civilian at Johns Hopkins Applied Physics Laboratory, accurately predicted a threat model posed by future Soviet bombers. Civilians at the Center for Naval Analyses assessed the type of radar the Navy would need to deal with the threat Hunt predicted. As with the Tomahawk case, strong civilian support from the Secretary of Defense and his office pushed the program forward.

Disruptive theory also predicts that outside intervention for sustaining innovations is most effective when supported from the inside by military mavericks. The case evidence supports both Captain Meyer and Captain Locke filling the maverick role for Tomahawk and Aegis. An examination of the development of continuous aim gunfire shows a military maverick assisting a civilian product champion. It is apparent that Lieutenant Sims, a military maverick, appealed directly to the President of the United States, who in turn directed the Navy to innovate to continuous aim gunfire.

Interservice rivalry

An important finding in the Tomahawk case is that the high level of interservice competition between the Navy and the Air Force spurred innovation. The study shows that the Navy's development of a cruise missile that had the potential to play a strategic role was seen by the Air Force as a threat to its own roles and missions, and under these circumstances, interservice rivalry was high. The competition allowed the Navy to garner extra resource dollars, which allowed its product champion to disguise the development of the conventional Tomahawk as a sustaining nuclear variant. In the end, the conventional Tomahawk would be an important component in the surface land attack disruptive innovation.

Intraservice rivalry

The intraservice rivalry hypothesis was validated for all five Navy sustaining cases. As disruptive theory predicted, intraservice competition was enormous within both the Aegis and Tomahawk programs. Historically, the air arm's fixed-wing assets had provided the needed air coverage for the carrier battle group, and naval air believed both programs threatened this traditional role of aviation. Again, the aviation arm's attack on the surface arm was motivated by competition for resources. An important finding of these cases is that Aegis and Tomahawk far exceeded the performance criteria to which they were designed. But the fear of product champions of Aegis and Tomahawk that they might be impinging on traditional aviation roles and therefore stirring up opposition could account for their ambivalence about rapidly pursuing new and better applications for these emerging technologies.

As discussed earlier in the TCN® case, the study did find a connection between intraservice rivalry and the disguising proposition.

Small innovation group

As predicted, all the sustaining innovation cases validate the small innovation group hypothesis. The evidence shows that product champions created and managed small groups for all four sustaining innovations. The zealot product champion proposition, however, is neither validated nor refuted. In three of the four cases, zealots furthered innovation. Yet, a zealot product champion did not promote sustaining innovation in the CVBG case. Instead, it was a direct result of the CNO driving the innovation. Overall, zealot product champions may play a role in promoting sustaining innovation, but they certainly are not required.

16 Conclusion

Disruptive innovation theory captures many of the non-linear complexities of the transformation process and explains them in simple, easily digestible form. Although a number of factors contribute to successful innovation, the evidence presented in this study suggests that many of the relations among doctrine, technology advances, innovative concepts, and new ways of fighting can be clustered under two constructs – sustaining and disruptive. This is an important finding. By clustering both technology and doctrine innovations under the two banners of improved performance along sustaining and disruptive warfighting trajectories, important conclusions can be drawn about the larger issues and patterns of championing new ways of war.

Chapter 1 raised two questions critical to understanding the role senior naval leaders have played in managing disruptive innovations. First, how can senior naval leaders achieve a disruptive innovation when they are heavily engaged around the world and are managing sustaining innovations? Second, what external sources cause disruptive innovation? Chapter 2 attempted to answer those questions by borrowing theoretical constructs and empirical observations from both the military and civilian innovation literature and inductively building a disruptive theory using British and German armored warfare development.

Key elements of the disruptive innovation argument centered on the importance of product champions establishing small innovation groups and using dissimulation. Specifically, the intellectual process of creating a new way of warfighting is strengthened by the product champion establishing and managing small innovation groups that are separate from the bureaucratic organization. In addition, the politics of disruptive innovations are eased when those championing the transformation can disguise it as a sustaining innovation. While neither dissimulation nor small innovation groups guarantee success, both are important factors with implications for product champion management and civilian oversight of military innovation. The efforts in Chapters 3–15 deductively tested the disruptive innovation model by examining successful cases of American Navy and Marine Corps disruptive and sustaining innovations in peacetime (including Japanese carrier warfare development), as well as one abortive disruptive innovation in the Navy and one inchoate innovation in the Marine Corps.

External causes of *disruptive* innovation

Engine of change: why and when

The study identified and examined four competing external sources of disruptive change. Three of the four frameworks indicate that disruptive innovation results from conflict in decisive relationships: from civilian-military competition, in which civilian intervention produces innovation through military mavericks; intraservice competition, in which rivalry between branches of the same service produces disruptive innovation; or interservice competition, in which rivalry between the different services produces disruptive innovation. The fourth framework suggests that a state's choices in military doctrine and innovation are shaped by cultural factors.

Civilian intervention

The general hypothesis that civilian intervention produces disruptive innovation either directly or indirectly through military mavericks lacks historical validity. Instead, what has proved more useful in this study is the argument that civilian intervention is effective only to the extent it can support or protect product champions. In the Navy and Marine Corps disruptive cases examined, with the exception of the inchoate maneuver warfare case, civilians exerted little influence. This is entirely consistent with disruptive theory, which predicts that an order to fight differently will be seen by the military as being outside the legitimate authority of civilian leaders.

In the maneuver warfare case, the study found intervention by civilians had a retarding effect on the disruptive innovation. Many Marines considered William S. Lind, who did not have the political pull to cause a disruptive innovation, as possessing some type of power, and they took him seriously when he visited Marine Corps bases. Perhaps more important, however, many senior Marines resented General Gray telling them they should buy and read Lind's *Maneuver Warfare Handbook*. Whether or not maneuver warfare was in the best interest of the Marine Corps, it was seen as Lind's doctrine, and some Marines rejected it for that reason. It would not be until General Krulak became Commandant and banned Lind from the Marine Corps schools and think-tanks at Quantico that maneuver warfare began to take hold in a serious fashion.

Interservice rivalry

In all the successful disruptive cases the study showed a connection between intense interservice competition and disruptive innovation. One of the most important and surprise findings of this study comes from the surface land attack case. This case proved to be the only example in which a service consciously adopted disruptive innovation, knew it was doing so, and made no attempt to conceal its actions. Interestingly, this case also exhibited the highest level of interservice rivalry. Although it is difficult to measure the extent of such competition and its impact, the surface land attack case stood out in that it corresponded to a sharp spike in interservice rivalry between the Air Force and Navy following the 1991 Gulf War.

As in all of the United States' modern conflicts, interservice problems arose during the Gulf War. Air power played the central role in the conflict, and the Air Force and its generals controlled the air campaign. The Navy's limited contributions to the air campaign demonstrated its shortcomings in joint warfare; after four decades confronting the Soviet threat, the Navy had yet to adjust its Cold War relationship to the joint US command structure.[1] Enactment of the Goldwater-Nichols legislation in 1986 increased the operational control by theater commanders (whatever their service) over naval forces. Navy fleet commanders, however, still operated according to the principles embodied in the Maritime Strategy of the late 1980s, wherein they controlled the aircraft and missiles they dispatched from their ships against the enemy. The Navy, because of its traditionally decentralized handling of air power, did not take easily to the Air Force managing all air assets. In sum, the 1991 Gulf War demonstrated that the Air Force's ability to respond to Saddam's invasion of Kuwait outpaced the Navy's.[2]

The pattern that emerges is that during times of high interservice rivalry a lesser degree of disguising is needed to enable a disruptive innovation. When interservice rivalry is at low to moderate levels, disguising becomes an important factor. The aborted and inchoate cases are consistent with these generalizations; they demonstrated that when disguising is not present during times of low to moderate interservice rivalry, the disruptive innovation stagnates or fails.

It is worth noting that an inverse correlation exists between interservice rivalry and product champion disguising;[3] that is, high levels of interservice competition do not require disguising and low to moderate levels of interservice rivalry require high levels of disguising.

Intraservice rivalry

Another surprising finding of this study concerns the relationship between interservice and intraservice competition. The evidence supports intraservice rivalry to be present in all the disruptive cases, with little variation between the cases. A great deal of variation in the interservice rivalry variable existed, however, with the lowest levels of the factor being present in the aborted and inchoate cases. In these two cases, intraservice rivalry was the key factor that caused the innovations to stagnate or fail.

From this evidence, the study infers that the interservice variable is more dominant than the intraservice variable. As a result, during times of high interservice competition, this variable determines the rate and pace of disruptive change. As interservice rivalry decreases, the intraservice rivalry variable grows in importance and determines the rate and pace of disruptive change. Using a warship analogy, when the ship is sinking, rivalry among the gunnersmate, engineering, and boatswainmate communities disappears as all hands focus on damage control.

As the crisis fades and the ship resumes its regular duties, the intra-ship rivalries reappear as sailors rejoin their divisions. When threatened by the Air Force following Desert Storm, the different warfare branches of the Navy agreed they must work together to counter the common enemy. After the interservice threat diminished, intraservice rivalry returned and each branch promoted its own way of fighting the next war.

In sum, an important finding postulates an inverse relationship between interservice competition and intraservice rivalry. The general lesson for students and advocates of innovation may be that it is wrong to focus on civilian intervention to achieve a disruptive innovation; the better path might be through civilian manipulation of interservice competition.[4] The aim of civilian policymakers would be to increase interservice competition to high levels, as experienced by the Navy after the Gulf War.

Organizational culture

A significant observation is that of the major role organizational culture played in enabling disruptive innovations in the United States Navy, Imperial Japanese Navy, United States Marine Corps, and the interwar German Army. Generally speaking, the senior leadership of these militaries demonstrated open-mindedness in examining novel ways of fighting. Each relied on simulations, wargames, and exercises to test novel ways of linking components. The roots of innovation are cultural. As demonstrated by the interwar British armored warfare case, a military can possess the most visionary and innovative thinkers in the world, and still fail to transform if the organizational culture does not support disruptive transformations.

The most serious error in the organizational culture thesis, however, lies in the claim that an innovative culture alone will enable a disruptive innovation. Clearly, it is necessary, but it is not sufficient to achieve new ways of warfighting. Other external factors are equally important such as interservice and intraservice rivalry plus the internal factors of managing a disruptive innovation.

External causes of *sustaining* innovation

In stark contrast to the disruptive innovation cases, civilian intervention assumed a key role in the sustaining innovations. Beginning in 1898 with Congress directing the Marine Corps to form an advanced base force, to 1967 with Secretary of Navy Paul Nitze directing the Chief of Naval Operations to initiate a study on cruise missiles (following the sinking of a Israeli destroyer by an Egyptian SSN-2 Styx missile), the evidence suggest that sustaining innovations are strongly connected to changing security environments seen by civilians, who, in turn, direct the naval services to examine and counter these new threats.

As predicted, civilian intervention to assist military mavericks is a primary means to produce sustaining innovations. In all six sustaining cases, civilian intervention was observed, and in four of the cases a maverick played a central role. Not surprisingly, when civilians intervene in sustaining innovations they are usually successful. This is not to say that without civilian intervention sustaining innovations would not occur. Civilians do not have to intervene to achieve a sustaining innovation, as senior military leaders are self-motivated to promote innovations that enhance current warfighting performance trajectories.

In regard to the other external variables of sustaining innovation, interservice competition proved dominant over intraservice competition.

Internal causes of *disruptive* innovation

Throttle of change: <u>how</u>

In this section, the study summarizes the analysis of the question: How can senior naval leaders achieve a disruptive innovation when they are heavily engaged around the world and are managing sustaining innovations?

Intellectual process (small groups)

The proposition that the process of translating a 'new theory of victory' into doctrine and critical tasks occurs in small innovation group(s) created and managed by the product champion is strongly supported by historical evidence. To be accepted as a general explanation, a structural explanation would have to show that disruptive innovations have been caused by temporary structural changes in the organization that have made it more innovative. In particular, such an explanation would have to demonstrate that the product champion directly controlled a newly-created group in which the decision-making process was less formal, and that disruptive innovation resulted from this new structure.

An examination of this explanation using the case studies suggests that these changes did occur. Indeed, in all disruptive innovations examined, product champions created and managed small innovation groups, consistent with the predictions of disruptive theory. An emphasis on informality appears to have been a necessary prerequisite to disruptive change. The evidence to support this relationship is found by examining the circumstances under which each of the disruptive changes was initiated. For these reasons, the study concludes that small innovation groups have played a critical role in translating visions into tasks.

Disguising process

The prediction that disguising disruptive innovations as sustaining innovations eases disruptive organizational change was validated. Specifically, disguising held a dominant position in all the successful disruptive cases. Furthermore, disguising was not present in any great degree in the two disruptive cases that stagnated or failed.

The most interesting case also proved to be the one anomaly. In the surface land attack case, product champions achieved a successful disruptive innovation without having to disguise it as a sustaining innovation. Although this observation is at odds with the disguising model, it does not necessarily invalidate the proposition that dissimulation plays an important role in most cases. What the anomaly case warns of is the need to consider all conditions when predicting if a product champion will use disguising. It is worth noting that the Aegis and Tomahawk champions used disguising to develop these disruptive technologies, which eventually would be key components of the surface land attack disruptive innovation.

As discussed in the external cause section, disguising conditions are correlated with the level of interservice competition. High levels of interservice competition diminish the need to disguise. Can this inference be explained from a disguising perspective? The answer, based on case study evidence, is yes. As observed in the maneuver warfare and defensive sea control cases, the personal cost of championing a disruptive innovation can be very high. If a product champion fails to win the political battle to change the way the organization will fight, not only will the disruptive innovation fail, but those officers supporting the reform more than likely will have their careers ended as well. The most dramatic example of this in the study was Colonel Wyly, a strong maneuver warfare advocate under General Gray, who was forced to leave the Marine Corps just six months prior to his mandatory retirement at 30 years' service. To send a message to the rest of the Marine Corps, Wyly's superiors searched his desk for correspondence between Gray, the product champion of maneuver warfare, and Wyly. The meaning was clear: maverick maneuver warfare officers who risked engaging in a direct relationship with General Gray (outside the chain of command) would be dealt with harshly.

Thus, one purpose of disguising an innovation as sustaining is to protect the reform officers supporting the product champion. When interservice competition is high, however, reform officers do not have to be protected to the same degree. In the case of surface land attack warfare, for example, the product champion, Admiral Kelso, as well as the rest of the Navy, recognized that the Navy was in trouble. Consequently, the evidence shows an acting commanding officer of an air wing working in Kelso's small innovation group agreeing to have the surface community perform a mission previously done only by the air community. Arguably, under moderate interservice competition conditions such an agreement would have been career-ending.

In sum, the evidence strongly supports the proposition that, under certain conditions, product champions advance their disruptive innovations by disguising them as sustaining innovations. Clearly, however, when the external cause reaches a service-threatening threshold, disguising is not needed to effect the disruptive innovation.

Political process

As predicted, product champions in all the disruptive innovation cases ushered in a new vision of naval warfare by having a strategy for controlling the political process changing the organization. The evidence shows that each spent enormous effort building horizontal and vertical alliances and fought hard to ensure a senior officer who shared his new vision of war replaced him. This study concludes that generally they all were successful in building the support that would allow their innovations to be implemented. The evidence also strongly supports the political proposition that the product champion in the aborted case, Admiral Zumwalt, failed because he did not create a lasting power support base or a new junior officer path to flag rank.

Internal causes of *sustaining* innovation

As predicted, the sustaining innovation cases validate the small innovation group proposition. The evidence shows that product champions created and managed small groups for all the sustaining innovations.

The zealot product champion proposition seems to be case dependent and subject to other variables. In four of the cases, zealots furthered innovation. In the defensive advanced base case it was Major Pete Ellis, and in the continuous aim gunfire case it was Lieutenant William Sims. In the two sustaining technological innovation cases the zealots were Captain Locke for the Tomahawk innovation and Captain Meyer for the Aegis radar innovation. Conversely, zealot product champions did not promote sustaining innovation in three of the seven cases: MPF, MEU(SOC), and CVBG. The MPF and MEU(SOC) were direct results of civilian intervention (as directed by congressional mandate) and were, in turn, driven by the Commandant. The CVBG was a direct result of the CNO driving the innovation. Overall, zealot product champions may play a role in promoting sustaining innovation, but they certainly are not required.

That product champions promote and nurture vertical and horizontal support is validated. The study found this behavior in all the cases, most strongly when the product champions were officers in the broad middle ranks (0–3 to 0–6). Without advocacy and nurturing from Sims, Locke, and Meyer, for example, the continuous aim gunfire, Tomahawk, and Aegis innovations would have been stillborn.

An assessment of how senior naval leaders achieve disruptive innovation

This analysis of the generalized process of disruptive and sustaining innovations in the Navy and Marine Corps offers an answer to the first question of this study: how can senior naval leaders achieve a disruptive innovation when they are heavily engaged around the world and are managing sustaining innovations? The answer points to a direct dependency on the level on interservice competition.

Two important lessons emerge from this study. Generally, if interservice competition is moderate, product champions must use a disguising management approach to achieve a disruptive innovation. If they use a sustaining approach they are likely to fail. As demonstrated in the surface land attack case, however, product champions do not have to disguise the disruptive innovation as sustaining the old way of fighting when interservice competition remains high.

If a senior leader determines a new way of war is necessary to meet the challenges of a changing security environment, he first should form a small innovation group. All successful disruptive (and sustaining) innovations analyzed in this study, including the German Army development of mobile tank warfare, began with the product champion creating and monitoring small innovation groups. The next intellectual step is to create new tasks from the new vision of warfare. During this time, small innovation groups continue to play an important role while conducting experiments and simulations. If interservice competition is high, the next step is to institute a top-down directed change such as was observed in the surface land attack, maneuver warfare, and defensive sea control cases. If interservice competition is low or moderate, the critical task for the product champion is to disguise the new way of warfighting as a sustaining innovation.

Politically, the product champion should be promoting support alliances both vertically and horizontally to increase his power, the goal being to shift the distribution of power in favor of the product champion's new way of war. In doing so, the product champion should begin creating a new career path for junior officers committed to the new way of war. Because the innovation will take several years to develop, the product champion should shape the political battlefield so a senior officer with no allegiance to the old way of war replaces him.

What seems clear is that most of the innovations a senior naval leader will manage will be sustaining. Arguably, most senior naval leaders are in their current positions because they have demonstrated the requisite skills to manage sustaining innovations. It is apparent, however, that these skills are not suited for managing a disruptive innovation.

The challenge most senior product champions face in a moderate interservice rivalry environment is to manage a disruptive innovation as if it were a sustaining innovation. Admittedly, Admiral Kelso, Admiral Zumwalt, and General Gray all resorted to the same sustaining-management approach, but only Kelso succeeded, largely due to high interservice competition. Zumwalt and Gray fell short because they faced only moderate interservice competition; they would have had a greater chance of succeeding if they had used the disguising-managing approach of Admiral Moffett and General Russell.

Disruptive innovation theory versus Rosen's top-down theory and Davis's mid-level theory

The evidence presented in this study demonstrates the value of disruptive theory, which should be viewed as a reformulation of Rosen's top-down theory and Davis's mid-level theory. As discussed, senior naval leaders find themselves managing several different kinds of innovation simultaneously. In fact, this is the reason Rosen gives for breaking down the innovation phenomenon into technological and operational behavioral categories. He contends, 'that different kinds of innovation occur for different reasons in the same organization, and that different organizations will handle innovation very differently'.[5]

Davis – the politics of innovation

Whereas <u>Davis's mid-level theory predicts sustaining innovations</u> such as continuous aim gunfire, and <u>Rosen's top-down theory predicts disruptive innova-tions</u> such as amphibious warfare, <u>disruptive theory predicts both types of inno-vation</u>. Just as important, disruptive theory accounts for the impact the two types of innovation may have on each other. The distinction is not unimportant. Using disruptive theory, innovation advocates can understand a number of factors about disruptive innovation that are inexplicable under Rosen's or Davis's theory. For example, disruptive theory explains why civilian intervention using military mav-ericks does not lead to disruptive innovation and why it does lead to sustaining innovation, in contrast to the predictions of Rosen, which are limited to explain-ing new ways of fighting, and those of Davis, which are limited to explaining improving old ways of fighting.

In the same way, disruptive theory explains why product champions tend to be more successful in creating disruptive innovations when they manage the process differently from how they would if promoting sustaining innovations. Sustaining innovations often succeed even when those supporting them possess limited vision. The same cannot be claimed for disruptive innovations. Most disruptive innovations occur when existing components are combined in novel ways. Rarely does a single technological breakthrough create a new way of warfare. For the most part, all necessary components of a new way of fighting exist, but only the true visionary sees how they can be combined differently. That is why those opposing the change at first dismiss it (they see the components but fail to see the impact of the concept) and are then surprised when it changes everything they thought they knew.

This leads to the major point of this study. Disruptive theory explains why failure to achieve a disruptive innovation is the result of well-meaning product champions, often regarded as the most astute military leaders of their services, who improperly manage the disruptive innovation as if it were a sustaining inno-vation. For example, General Gray used civilian intervention (in the form of William Lind) to promote maneuver warfare, not understanding the central proposition of disruptive theory – that civilian intervention (either top-level or lower-level) is not the means to produce a disruptive innovation. He also did not protect and promote his young officers. This reasoning can be applied to Zumwalt's aborted innovation as well. Like Gray, Zumwalt failed to protect and promote his younger officers.

In short, disruptive theory embodies Rosen's top-down theory and Davis's mid-level theory. How a senior naval leader manages sustaining innovation is impor-tant because the majority of innovations are sustaining in nature. Occasionally, a vision of a new way of warfighting begins to emerge, and here the product champion must manage it differently. But how the product champion manages it depends on the external environment. By conceiving of innovation as being managed by two different processes, the study presents a more complete and accurate picture of how product champions have managed disruptive innovations successfully.

Prescribing a civilian strategy for promoting disruptive innovation

Chapter 2 concluded with a third question: what oversight role should civilians play in identifying the need for new military functions and capabilities and promoting disruptive innovations? One of the most important findings of this study is that civilian intervention did not provide the initiative for peacetime disruptive innovation. The evidence supports that if they use an indirect approach, civilians might be able to achieve greater success in spurring the military to examine new ways of war and perhaps even to adopt them. Indirect means civilians should disguise their intervention efforts. Instead of acting like a zealot, such as Bill Lind, they must be willing to couch their efforts so that the military product champions believe either that it is their own idea or vision they are embracing or that it is a sustaining innovation.

Second, civilians should consider ways to increase interservice competition to promote disruptive innovation. An important discovery in this study is that services tend to mask technology developments that have the potential to be disruptive. Both Aegis and Tomahawk developments are examples. The study found that when services are faced with high interservice competition, however, they tend to unmask these technologies with disruptive potential. Again, an increase in interservice competition should result in unmasking technology potential.

Third, civilians should choose senior military leaders carefully. Great effort by the Secretaries of the Navy and Defense went into selecting Admiral Zumwalt, the innovator. Indeed, Zumwalt had moved the Navy toward defensive sea control. If the civilian policymakers believed Zumwalt was going in the right direction, they should have ensured a surface warfare officer, such as Admiral Stan Turner, who was a strong Zumwalt supporter and could have been the intellectual architect of the disruptive innovation, relieved him. Instead, they chose an aviator, who began implementing a different warfighting vision.

Notes

1 Introduction

1 Rosen, *Winning the Next War*, 7.
2 Other examples of disruptive architectural innovation are carrier warfare and amphibious warfare.
3 According to Henderson and Clark, modest changes to components are acceptable for this definition. See Rebecca Henderson and Kim B. Clark, 'Architectural Innovation: The Reconfiguration of Existing Product Technologies and the Failure of Existing Firms', *Administrative Science Quarterly* 35 (March 1990), 10.
4 Christensen, *The Innovator's Dilemma*, xiv–xvi.
5 The author wishes to thank Bradd Hayes for his thoughtful comments in conceptualizing the three-point shot as a disruptive innovation.
6 The three-point shot was introduced in 1961 by the American Basketball League, which was founded by Abe Saperstein (who owned the Harlem Globetrotters) as a rival to the NBA. When the NBA started drafting black players in the early 1960s, the ABL talent base was drained and the league folded. In 1967, the ABA started up and it borrowed the three-point shot from the short-lived American Basketball League.
7 See Michael Murphy, 'The ABA Way: For Pure Entertainment, American Basketball Association Was a Slam Dunk', *Houston Chronicle* (4 February 1996), Sports 2 Section 17.
8 Deborah D. Avant, *Political Institutions and Military Change: Lessons From Peripheral Wars* (Ithaca, NY: Cornell University Press, 1994), 15.
9 Eliot Cohen, 'Defending America in the Twenty-first Century', *Foreign Affairs*, (November/December 2000), 41. Cohen states that 'a real transformation (military innovation), although entirely possible given the technology now available, will require either inspired leadership, or a stark threat from the outside'.
10 James Q. Wilson, *Bureaucracy* (New York: Basic Books, 1989), 221.
11 Clayton Christensen cites, as a classic case study of disruptive innovation, how steam-powered ships eventually replaced wind-powered ships in the transoceanic business. The first steamship operated on the Hudson River in 1819 and 'it under-performed transoceanic sailing ships on nearly every dimension of performance: It cost more per mile to operate; it was slower; and prone to frequent breakdowns'. Hence, it was not considered a serious threat to wind-powered transoceanic travel, even though it was well suited for inland waterways, where its performance was measured very differently. 'In rivers and lakes, the ability to move against the wind or in the absence of a wind was the attribute most highly valued by ship captains, and along that dimension, steam outperformed sail.' Sailing ship makers, however, continued to focus on different attributes (e.g., cost per mile) until finally steam ship performance surpassed them. The seeds of the demise of sailing ships were found in the sailing ship builders' reluctance to change strategy and build steam ships for the inland waterways. Why? The customers

of sailing ship manufacturers were transoceanic shippers and they did not demand the use of steam ships, until technological improvements made them economically competitive for transoceanic voyages. Here's the point. All makers of sailing ships ignored steam power and not one survived once steam-power ships dominated the high seas. See Christensen, *The Innovator's Dilemma*, 76.

12 See Andrew Krepinevich, *The Army and Vietnam* (Baltimore, MD: Johns Hopkins University Press, 1986), 34. Also see Rosen, *Winning the Next War*, 11.

13 As Andrew Krepinevich notes, 'How the Army responded to Kennedy's efforts to engineer a "revolution from above" in its approach to war says a lot concerning how strongly the Army Concept was embedded in the organizational psyche.' See Krepinevich, *The Army and Vietnam*, 27.

14 Richard Hundley, *Past Revolutions Future Transformations* (National Defense Research Institute, RAND, 1999), 28–9. Thomas Hone, Norman Friedman, and Mark Mandeles, *American and British Aircraft Development 1919–1941* (Annapolis, MD: United States Naval Institute Press, 1999), 92–4, 189. The British built carriers with armored flight decks so they could survive severe attacks from land-based air attack. Consequently, the hangars were relatively small because they were fully enclosed with hulls. The operational 'price' of armor was an aircraft capacity half that of Japanese and American carriers. Also, the British concept of operations did not include the 'deck park', the practice of stowing a major part of a carrier's complement of aircraft on the flight deck, and refueling and rearming there as well. The British stowed, refueled, and rearmed all their aircraft below, on the hanger deck. Thus, in 1939 a first-line British carrier carried only 24–30 aircraft, whereas American and Japanese carriers carried 80–100 aircraft because they designed their carriers to stow the majority of the planes on top of the flight deck. As it turned out, the key determinant of carrier warfare turned out to be the number of aircraft that could be launched for a strike and how quickly successive attacks could be mounted. Thus because the British could not generate 'pulses' of aircraft striking power comparable to those attainable from Japanese and American frontline carriers, the British never realized the full potential of carrier operations.

15 Christensen, *The Innovators Dilemma*, 98.

16 Ibid.

17 In contrast to the traditional doctrine-driven and technology-driven theorists who define military innovation differently, I suggest using Andy Ross' definition, which combines both doctrine and technology. Military innovation, he writes, 'includes not only the actual instruments or artifacts of warfare, but the means by which they are designed, developed, tested, produced, and supplied – as well as the organizational capabilities and processes by which hardware is absorbed and employed'. Andrew L. Ross, 'The Dynamics of Military Technology', in David Dewitt, David Haglund, and John Kirtland (eds), *Building a New Global Order: Emerging Trends in International Security* (Oxford, England: Oxford University Press, 1993), 111.

18 For a summary of the two perspective see Owen Cote, 'The Politics of Innovative Military Doctrine: The US Navy and Fleet Ballistic Missiles', Ph.D. diss., MIT, January 1996, 41.

19 Allison, Graham and Philip Zelikow, *Essence of Decision: Explaining the Cuban Missile Coisis* (2nd ed, New York: Longman Press, 1999), 143. Also see Emily Goldman, 'Institutional Learning under Uncertainty: Findings from the Experience of the US Military', unpublished manuscript, Department of Political Science, University of California, Davis, 1996. Goldman introduces a subdivision typology of the organizational approach, calling the approaches the 'institutionalist' and 'professionalist'.

20 Cote, 'The Politics of Innovative Military Doctrine', 6.

21 Williamson, Murray 'Innovation: Past and Future', *Joint Force Quarterly* (Summer 1994), 304.

22 Posen, *The Sources of Military Doctrine*, 212–13.

23 Rosen, 'New Ways of War', 39.
24 Ibid., 137.
25 Ibid., 135.
26 Rosen, *Winning the Next War*. Jeffrey Isaacson, Christopher Layne, and John Arquilla, *Predicting Military Innovation* (National Defense Research Institute, Santa Monica, CA: RAND, 1999), 18. See also K. M. Zisk, *Engaging the Enemy: Organizational Theory and Soviet Military Innovation, 1955–1991* (Ithaca, NY: Cornell University Press, 1993).
27 Rosen defines mainstream officers as those officers who support the values of the establishment to which they belong. See Rosen, 'New Ways of War', 136.
28 Ibid.
29 Ibid., 15.
30 Rosen, *Winning The Next War*, 14.
31 Ibid.
32 Rosen, 'New Ways of War', 139. Bradd Hayes and Douglas Smith (eds), *The Politics of Naval Innovation* (Strategic Research Department, Research Report 4–94, US Naval War College, 4–94), 97. Of note, Bradd Hayes has made the case for juxtaposing the Davis and Cote interservice competition model with Rosen's intraservice competition for explaining the external causal mechanism of innovation.
33 Vincent Davis, *The Politics of Innovation: Patterns in Navy Cases* (Monograph Series in World Affairs, Vol 4, No. 3, University of Denver, 1967), 21.
34 Cote, 'The Politics of Innovative Military Doctrine', 46.
35 Ibid., abstract.
36 Ibid., 77.
37 Ibid., 78.
38 Ibid., 28.
39 Ibid., 78.
40 Bradd Hayes argues that juxtaposing the intercompetition and intracompetition models into one framework explains more variation. See Hayes and Douglas (eds), *The Politics of Naval Innovation*, 97.
41 Isaacson, Layne, and Arquilla, *Predicting Military Innovation*, viii. Cote, 'The Politics of Innovative Military Doctrine', 47.
42 See Elizabeth Kier, *Imagining War: French and British Military Doctrine between the Wars* (Princeton, NJ: Princeton University Press, 1997). Also see Jeffrey Legro, *Cooperation under Fire: Anglo-German Restraint During World War II* (Ithaca, NY: Cornell University Press, 1995). Legro uses organizational culture to explain why belligerents in World War II cooperated even as they tried to destroy each other. For a critique of the use of organizational culture approaches in security studies, see M.C. Desch, 'Culture Clash: Assessing the Importance of Ideas in Security Studies', *International Security* (Summer 1998), 141–70.
43 See Douglas Porch, 'Military "Culture" and the Fall of France in 1940', *International Security* 24, 4 (Spring 2000), 148. Porch provides a critique of the viability of Kier's argument.
44 Kier, *Imagining War*, 144–5.
45 See Williamson Murray, *Experimentation in the Period Between the Two World Wars: Lessons for the Twenty-First Century* (Institute for Defense Analyses, November 2000). For a discussion of the gap between historians and political scientists in dealing with military innovation see Williamson Murray's Foreword in Gray *Strategy For Chaos*, ix–xii.
46 Williamson Murray, 'Armored Warfare', in Williamson Murray and Allan Millett (eds), *Military Innovation in the Interwar Period* (Cambridge, NY: Cambridge University Press, 1996), 18.
47 See Kier, *Imagining War* and Legro, *Cooperation Under Fire*. For a critique of the use of organizational culture approaches in security studies, see Desch, 'Culture Clash', 141–70.

48 A major stumbling block for innovation security scholars has been the ability to explain the impact of technological change upon naval warfighting. The problem, quite simply, is that a grand theory of innovation that explains technology advances and warfighting innovation does not exist. Stephen Rosen's 1991 book, *Winning The Next War*, and Williamson Murray and Allan Millett's 1996 book, *Military Innovation in the Interwar Period*, both conclude that a grand theory of innovation that explains technology advances and innovation does not exist. Without a causal mechanism to explain the link these two phenomena, the generally accepted proposition is that military organizations respond to technology advances by assimilating them into old doctrine. See Posen, *The Sources of Military Doctrine*, 55. As Posen notes this proposition is derived empirically by Bernard Brodie, 'Technological Change, Strategic Doctrine, and Political Outcomes', in Klaus Knorr (ed), *Historical Dimensions of National Security Problems* (Lawrence, KS: University of Kansas Press, 1976), 300; and Edward Katzenbach, 'The Horse Cavalry in the Twentieth Century: A Study in Policy Response', in Richard Head and Erwin Rokke (eds), *American Defense Policy* (Baltimore, MD: Johns Hopkins University, 1973). Relying on this approach, the best innovation theorists can do is to accept that new technologies are assimilated to an old doctrine, but hopefully 'somehow' stimulate a change to a new one. The proposition that a new technology will normally be assimilated to an old doctrine is posited by Posen, *The Sources of Military Doctrine*, 55.

49 Avant, *Political Institutions*, 14.

50 Rosen, 'New Ways of War', 141.

51 Ibid.

52 Commander Terry C. Pierce, USN, 'Teaching Elephants to Swim', *United States Naval Institute Proceedings* (May 1998), 26.

53 Rosen, 'New Ways of War', 141.

54 Ibid.

55 Ibid.

56 Ibid., 251.

57 Rosen, *Winning The Next War*, 52.

58 Ibid.

59 Ibid., 243.

60 Ibid., 221.

61 Ibid.

62 Rosen quotes from Raymond Isenson's, 'Project Hindsight: An Empirical Study of the Sources of Ideas Utilized in Operational Weapons Systems', in William Gruber and Donald Marquis (eds), *Factors in the Transfer of Technology* (Cambridge, MA: MIT University Press, 1969), 168.

63 Van Creveld, *Technology and War*, 220.

64 Jared Diamond, *Guns, Germs, and Steel* (New York: Norton Press, 1999), 242–4.

65 Rosen, *Winning The Next War*, 41; Diamond, *Guns, Germs, and Steel*, 242–4.

66 Rosen, *Winning The Next War*, 249–50.

67 Ibid., 224.

68 Of note, two studies have tested Davis inductive technology driven model. The first was by Ronald James Kurth, *The Politics of Technological Innovation in the United States Navy*, unpublished doctoral thesis, Harvard University, June 1970. Kurth tested of Davis hypothesis and found them to be valid. The second study was by the Strategic Research Department on the US Naval War College under the auspices of Captain Bradd Hayes, USN and Commander Douglas Smith, USN, who tested several of Davis' hypothesis in 1994 and found them also to be valid. See Hayes and Smith, *The Politics of Naval Innovation*. Note, for a general approach of building inductive theories from case study research see, Kathleen Eisenhardt, 'Building Theories from Case Study Research', *Academy of Management Review*, No. 4 (1989), 532–50.

69 Davis, *The Politics of Innovation*, 56.

70 Hayes and Smith, *The Politics of Naval Innovation*, 10.
71 Mathew Evangelista has proposed a five-step generalized stage model for tactical nuclear weapons innovation. Although Evangelista's generalized stages basically conform to Davis' model of innovation, investigating the sources of atomic weapons development and improvement is beyond the scope of this thesis. Rather, I am interested in the question that Davis studies: How does the Navy manage technology innovations that will support the use of atomic weapons or atomic power or other radical technical innovations? See Matthew Evangelista, *Innovation and the Arms Race: How the United States and the Soviet Union Develop New Military Technologies* (Ithaca, NY: Cornell University Press, 1988).
72 See Davis, *The Politics of Innovation*, 51–8, for a detailed discussion of his concluding hypotheses that I am discussing in this section.
73 See Avant, *Political Institutions*, 14–15. Other examples that use a similar definition of innovation are Posen's *The Sources of Military Doctrine*, Emily Goldman's organizational behavioral approach in *Mission Possible: Organization Learning in Peacetime* in Peter Trubowitz, Emily O. Goldman, and Edward Rhodes (eds), *The Politics of Strategic Adjustment: Ideas, Institutions and Interests* (New York: Columbia University Press, 1999), and two organizational culture studies by Elizabeth Kier, *Imagining War*, and Jeffery Legro's *Cooperation Under Fire*. Three excellent army studies include Andrew Krepinevich's *The Army and Vietnam*, David Johnson, *Fast Tanks and Heavy Bombers: Innovation in the US Army* (Ithaca, NY: Cornell University Press, 1998), and Kevin Sheehan's *Preparing for an Imaginary War? Examining Peacetime Functions and Changes of Army Doctrine* (unpublished Harvard PhD Thesis 1998).
74 Van Creveld, *Technology and War*.
75 Some advocates of this approach proclaim we are on the verge of a technology-driven Revolution in Military Affairs in which technology causes a paradigm shift in the nature and conduct of military operations. See Hundley, *Revolutions: Future Transformations*, xiii.
76 See Vice Admiral Harold G. Bowen, USN. retired, *Ships, Machinery and Mossbacks* (Princeton, NJ: Princeton University Press, 1954).
77 See Davis, *The Politics of Innovation*, 43.
78 Ibid., 46.
79 Kurth, *The Politics of Technological Innovation in the United States Navy*, 4.
80 Rosen, *Winning the Next War*, 7.
81 See Davis, *The Politics of Innovation: Patterns in Navy Cases*, 15.
82 Rosen, *Winning The Next War*, 8.
83 Karl Lautenschlager 'Technology and the Evolution of Naval Warfare', *International Security* (Fall 1983), 48.
84 Hayes and Smith, *The Politics of Naval Innovation*, 3.
85 Lieutenant Colonel Mark Chmar, Lieutenant Colonel Stephen Cullen, Lieutenant Colonel David Ralston, Lieutenant Colonel Michael Smith, *Peacetime Military Innovation: Getting Beyond Today* (National Security Program Occasional Paper, John F. Kennedy School of Government, Harvard University, 1999), 6.
86 Goldman, 'Mission Possible: Organizational Learning in Peacetime', 255–6.
87 Rosen, *Winning the Next War*, 5.
88 Ibid., 40.
89 I attribute the desegregating approach to Henry Chesbrough, Harvard Business Professor.
90 Figure 1 is based upon Rebecca Henderson and Kim Clark's architectural typology and was modified to reflect military innovations.
91 Henderson and Clark, 'Architectural Innovation', 11.
92 Christensen, *The Innovator's Dilemma*, 115.
93 Rosen, *Winning the Next War*, 5.
94 See Robert Alford, *The Craft of Inquiry: Theories, Methods, Evidence* (New York: Oxford Press, 1998), 107. Alford notes that dependent variables can be measured as either 'a continuous variable, a dichotomy, or a typology'.

2 Explaining disruptive innovations

1 Allison and Zelikow, *Essence of Decision: Explaining the Cuban Missile Crisis*, 9.
2 Rebecca Henderson, 'Managing Innovation in the Information Age', *Harvard Business Review* (January–February 1994), 100.
3 Continuous aim gunfire is a architectural innovation that Rosen does not consider a major innovation because it was sustaining in nature and not disruptive. Although it had a large impact on the Navy it did not create a new way to fight, but rather dramatically improved the existing way. I argue in this thesis that sustaining architectural innovations are important and should be considered, but I also argue that they are managed differently than Rosen's major military innovation, which I call disruptive architectural innovation.
4 Christensen, *The Innovator's Dilemma*, xii.
5 Henderson and Clark, 'Architectural Innovation'.
6 Ibid., 12.
7 Ibid., 11.
8 Posen, *The Sources of Military Doctrine*, 55.
9 Brodie, 'Technological Change, Strategic Doctrine, and Political Outcomes', 300; and Katzenbach, 'The Horse Cavalry in the Twentieth Century: A Study in Policy Response'.
10 This concept of expanding architectural subsystems to a higher level builds on the work of Clay Christensen who points out the usefulness of perceiving products as a *nested* set of components designed together into system architectures.
11 Henderson and Clark, 'Architectural Innovation', 14.
12 Williamson Murray, 'Innovation Past and Future', in Murray and Millett (eds), *Military Innovation in the Interwar Period*, 322–3.
13 In this section, I adapt conceptual and technical innovations originating in the business tradition to fit them into the political science tradition. For a meta-study of the impact of technological change on organizations see Henry W. Chesbrough, 'The Differing Impact of Technological Change upon Incumbent Firms: A Comparative Theory of Organizational Constraints and National Institutional Factors', Harvard Business School Working Paper, 98–110, April 1998.
14 Hundley, *Past Revolutions Future Transformations*, xiv. Generally speaking, scholars have not focused on radical innovations because the historical evidence shows that technology-driven transformations are usually brought about by combinations of technologies (architectural innovation), rather than individual technologies.
15 See Sheehan, 'Preparing For An Imaginary War? Examining Peacetime Functions and Changes of Army Doctrine'. Also see Wilson, *Bureaucracy*, 218–21.
16 Christensen, *The Innovator's Dilemma*.
17 Clay Christensen argues for this framework.
18 See Williamson Murray, 'Innovation: Past and Future', in Murray and Millet (eds), *Military Innovation in the Interwar Period*, 322.
19 See Murray, 'Armored Warfare', 14.
20 Ibid., 6.
21 Ibid.
22 Ibid.
23 Ibid., 43.
24 See John Ellis, *The Social History of the Machine Gun* (Baltimore, MD: Johns Hopkins University Press, 1996), 19.
25 Murray, 'Armored Warfare', 43.
26 Ibid., 40.
27 Barry Watts and Williamson Murray, 'Military Innovation in Peacetime', in Williamson Murray and Allan Millett (eds), *Military Innovation in the Interwar Period* (New York: Cambridge University Press, 1966), 372.
28 Murray, 'Armored Warfare', 72.

29 Rosen, *Winning the Next War*, 255–7; Cote, 'The Politics of Innovative Military Doctrine', 1–25.
30 Henderson and Clark, 'Architectural Innovation', 18.
31 Ibid.
32 Bradd Hayes, 'Transforming the Navy' (Report 00–3, Decision Strategies Department, Naval War College, 2000), 13.
33 In stating that different organizational structures lead to different types of innovation, I am not implying that the structure itself automatically leads to the innovation. Instead I am asserting that organizational arrangements can place significant obstacles in the way of innovators, and creating architectural innovation groups is a necessary but not sufficient variable for enabling the development of new architectural knowledge. Clearly, selecting the right people as member of these groups makes the greatest difference in the outcome. See Hayes, *The Politics of Naval Innovation*, 94.
34 Wilson, *Bureaucracy*, 232. Of note, Rosen in 'New Ways of War', 153, states that Admiral Moffett was 'initially very careful not take issue with any of the theories that discussed the use of carriers to support battleships'. Rosen, however, does not elaborate on the importance of dissimulation.
35 John Mosier, *The Myth of the Great War* (New York: Harper Collins Publishers, 2001), 3.
36 Ibid.
37 Murray, 'Armored Warfare', 21.
38 Ibid., 27.
39 Ibid.
40 J.P. Harris, *Men, Ideas and Tanks: British Military Thought and Armored Forces, 1903–1939* (Manchester, England: Manchester University Press, 1995), 249–51; James Corum, 'A Comprehensive Approach to Change: Reform in the German Army in the Interwar Period', in Harold Winton and David Mets (eds), *The Challenge of Change: Military Institutions and New Realities, 1918–1941* (Omaha, NE: University of Nebraska Press, 2000), 93–4.
41 Murray, 'Armored Warfare', 28.
42 Ibid., 29.
43 James S. Corum, *The Roots of the Blitzkrieg* (Lawrence, KS: Kansas University Press, 1992), 39.
44 Ibid.
45 Henderson and Clark, 'Architectural Innovation', 18.
46 Ibid., 17.
47 Murray, 'Armored Warfare', 26.
48 Ibid., 24.
49 Ibid., 41.
50 Ibid., 24.
51 Ibid., 33.
52 Ibid., 40.
53 Ibid.
54 Ibid., 24.
55 Ibid.
56 Henderson and Clark, 'Architectural Innovation', 13.
57 Rosen, *Winning the Next War*, 127. Also see Watts and Murray, 'Military Innovation in Peacetime', 381.
58 Murray, 'Armored Warfare', 55.
59 Williamson Murray, *Experimentation in the Period Between the Two World Wars: Lessons for the Twenty-First Century*, Institute for Defense Analyses (November 2000), 2.
60 Murray, 'Armored Warfare', 24.
61 Rebecca Henderson, 'Technological Change and Architectural Knowledge', in Michael D. Cohen and Lee S. Sproul (eds), *Organizational Learning* (Thousand Oaks, CA: Sage Publications, 1996), 369.
62 Henderson and Clark, 'Architectural Innovation', 17.

63 Corum, *The Roots of the Blitzkrieg*, 122–3.
64 Murray, 'Armored Warfare', 39.
65 Watts and Murray, 'Military Innovation in Peacetime', 43.
66 Henderson and Clark, 'Architectural Innovation', 17.
67 Williamson Murray, 'Armored Warfare', 40.
68 Ibid., 24.
69 Arguing against Posen's view of Hitler causing the Blitzkrieg are Williamson Murray who states, 'Hitler played little role in development of the panzer forces except to make a few favorable remarks to Guderian in 1934.' See Murray, 'Armored Warfare', 17. Echoing Murray's assertion is James S. Corum, who argues, 'The least acceptable interpretation of the postwar German doctrine of mobile warfare was proffered by Barry Posen in *The Sources of Military Doctrine*. Posen argued that the Reichswehr's adherence to a doctrine of maneuver is an example of organization theory, which states that military organizations like offensive doctrines and do not like to innovate. Posen missed the point. Postwar German military doctrine represents considerable innovation, and armies prefer the offense because it wins wars. Von Seeckt and the General Staff were extremely conservative in political matters, but the enormous postwar effort in establishing committees to critically examine army doctrine and organization refutes the image of the General Staff as military traditionalists. The Reichswehr adopted its doctrine of a quick war of maneuver leading to an early decision and rapid annihilation of the enemy force as simply the most sensible military doctrine for a future war.' See Corum, *The Roots of Blitzkrieg*, 66. The most useful synthesis of the entire question of German interwar armor warfare development and specifically the von Seeckt years is Robert M. Cinto's *The Path to Blitzkrieg: Doctrine and Training in the German Army 1920–1939* (New York: Lynne Rienner Publishers, 1999). Using largely unpublished sources Cinto concludes that 'the doctrine of the post-1935 army, emphasizing the close cooperation of mechanized, motorized, and air forces, did not required invention out of whole cloth after Hitler's declaration of rearmament. It was not an innovation of the Nazi period, as it was often characterized at the time. No one man – and certainly not Hitler – "invented" *Blitzkrieg*. Rather, the birth of this style of war was an evolutionary development, the result of 15 years of doctrinal experimentation that began during the Seeckt era...' See Cinto, *The Path to Blitzkrieg*, 244.
70 Williamson Murray, *The Change in the European Balance of Power, 1938–1939* (Princeton, NJ: Princeton University Press, 1984), 32.
71 Posen, *The Sources of Military Doctrine*, 184.
72 Ibid., 224–6.
73 Ibid., 191.
74 Ibid., 179.
75 Ibid., 225.
76 Ibid., 226.
77 Ibid.
78 See Johnson, *Fast Tanks and Heavy Bombers*, 9–11. Also see Hundley, *Past Revolutions Future Transformations*, 69.
79 Hundley, *Past Revolutions Future Transformations*, 69.
80 Johnson, *Fast Tanks and Heavy Bombers*, 9.
81 Williamson Murray and Allan R. Millett, *A War To Be Won: Fighting the Second World War* (Boston, MA: The Belknap Press of Harvard University Press, 2000), 23. Also see Murray, 'Armored Warfare', 38.
82 Citino, *The Path to Blitzkrieg*, 43.
83 Corum, *The Roots of the Blitzkrieg*, 37.
84 Ibid., 39.
85 Williamson Murray, 'Armored Warfare', 41.
86 Henderson and Clark, 'Architectural Innovation', 17.
87 Ibid., 28.

88 For example, Seeckt appointed Colonel Bruchmuller to the Artillery group; Major General von Below, who had commanded the German army at Caporetto, was placed on the Mountain Warfare Group; and Major General von Lettow-Vorbeck, who had fought brilliantly in German East Africa for four years, headed the group on colonial warfare.
89 Citino, *The Path to Blitzkrieg*, 43.
90 Ibid., 10–11.
91 Williamson Murray 'Armored Warfare', 40.
92 Ibid., 41.
93 Henderson and Clark, 'Architectural Innovation', 17. Also, a rich literature of the history of the Blitzkrieg describes its development and impact. Heinz Guderian, *Panzer Leader*, translated from the German by Constantine Fitzgibbon (New York: E.P. Dutton & Co., 1952), provides a subjective firsthand view; Corum, *The Roots of Blitzkrieg*.
94 Henderson and Clark, 'Architectural Innovation', 17.
95 Murray, *The Change in the European Balance of Power*, 36.
96 MacGregor Knox and Williamson Murray (eds), *The Dynamics of Military Revolution 1300–2050* (New York: Cambridge University Press, 2001), 158; Citino, *The Path to Blitzkrieg*, 228.
97 Knox and Murray (eds), *The Dynamics of Military Revolution 1300–2050*, 158; Citino, *The Path to Blitzkrieg*, 157–9.
98 Murray, *The Change in the European Balance of Power*, 36.
99 Murray and Millett, *A War To Be Won*, 23.
100 Ibid., 3.
101 Murray, 'Armored Warfare', 17.
102 Ibid.
103 Guderian, *Panzer Leader*, 62.
104 Ibid., 45–6.
105 Corum, *The Roots of Blitzkrieg*, 137.
106 Mathew Cooper, *The German Army, 1933–1945* (Chelsea, MI: Scarborough House Publishers, 1978), 38.
107 Murray, 'Armored Warfare', 44.
108 Guderian, *Panzer Leader*, 20.
109 Murray, *The Change in the European Balance of Power*, 35.
110 Guderian, *Panzer Leader*, 24.
111 Ibid.
112 Ibid., 25.
113 Ibid.
114 Ibid., 30.
115 Ibid.
116 Murray, *The Change of the European Balance of Power*, 35.
117 Corum, *The Roots of Blitzkrieg*, 140.
118 Murray, *The Change of the European Balance of Power*, 35.
119 Murray, 'Armored Warfare', 40.
120 Henderson and Clark, 'Architectural Innovation', 18.
121 Murray, 'Innovation: Past and Future', 316. For a more in-depth logistic discussion of the German *Blitzkrieg* see Martin van Creveld, *Supplying War, Logistics from Wallenstein to Patton* (New York: Cambridge University Press, 1977).
122 See Steve Rosen, 'New Ways of War: Understanding Military Innovation', *International Security*, 13, 1 (Summer 1988), 134–43 for a discussion of the differences between Barry Posen's intervention model and Rosen's intervention model.
123 Posen, *The Sources of Military Doctrine*, 225.
124 Ibid., 191.
125 Ibid., 206.

126 Ibid., 207.
127 Ibid., 207–9.
128 Corum, 'The Roots of Blitzkrieg', x.
129 Ibid., xvi.
130 See Murray, 'Armored Warfare', 106.
131 Timothy Lupfer in *Dynamics of Doctrine: The Changes in German Tactical Doctrine During the First World War* (Combat Studies Institute: US Army Command and General Staff College, July 1981), 41.
132 Ibid., 42.
133 Ibid.
134 Ibid., 57.
135 Williamson Murray, 'Armored Warfare', 17.
136 Williamson Murray, *The Change in the European Balance of Power*, 30–1.
137 Ibid., 30.
138 Ibid.
139 Ibid., 97–8.
140 Ibid., 84.
141 Ibid., 86.
142 Ibid., 85.
143 Ibid., 87.
144 Ibid., 91.

3 US Marine Corps innovation: the development of amphibious warfare

1 Stephen Rosen argues that officers within the Marine Corps developed new ideas about the way future wars would be fought and how they might be won not from a close study of the Japanese military, but rather in response to a changing security environment. See Rosen, *Winning the Next War*, 57.
2 Quoted in Jeter Isely and Philip Crowl, *The US Marines and Amphibious War: Its Theory and Its Practice in the Pacific* (Princeton, NJ: Princeton University Press, 1951), 6.
3 See Rosen, *Winning the Next War*, 159.
4 General Charles C. Krulak, USMC, 'Editorial', *Marine Corps Gazette* (July 1998), 19.
5 Isely and Crowl, *The US Marines and Amphibious War*, 4.
6 See Lieutenant Colonel Jon T. Hoffman, USMCR, Marine Corps Command and Staff College Brief, 28 August 2000, held by Marine Corps Historical Center, Navy Yard, Washington DC.
7 As an illustration, some 400 Marines from the First Defense Battalion in December 1941 finally surrendered Wake Island after 15 days of facing insurmountable odds.
8 See David J. Ulbrich, 'Clarifying the Origins and Strategic Mission of the US Marine Corps Defense Battalion, 1898–1941', *War & Society* 17, 2 (October 1999), 81–109. Also see Charles D. Melson, *Condition Red: Marine Defense Battalions in World War II* (Washington DC: Washington Marine Corps Historical Center, 1966). Both Ulbrich and Melson make compelling arguments that the advanced based defense theory dominated Marine Corps thinking right up until 1940.
9 Marine Corps Board convened by Commandant of the Marine Corps, General Lemuel C. Shepherd, Jr, *The Evolution of Modern Amphibious Warfare* (Quantico, VA: Unpublished Paper Breckinridge Library Quantico: 4 April 1959).
10 Ibid., 1.
11 Ulbrich, 'Clarifying the Origins', 84.
12 Rosen, *Winning the Next War*, 64.
13 Dirk Ballendorf and Merrill Bartlett, *Pete Ellis: An Amphibious Warfare Prophet 1880–1923* (Annapolis, MD: United States Naval Institute Press, 1997), 155.

14 Lieutenant Colonel Kenneth Clifford, USMCR, *Progress and Purpose: A Developmental History of the United States Marine Corps: 1900–1970* (Washington DC: Marine Corps Historical Center Publication).

15 Ibid., 3.

16 Prior to the Spanish-American War, the United States did not have a central advisory authority for determining naval policy. Afterwards, with the increased responsibilities of defending newly-acquired territories, Navy Secretary John D. Long established the General Board with no executive functions but merely with an advisory capacity. Soon, however, the General Board became in reality the spokesman for the Secretary of Navy as it became a consistent source of long-term planning for mobilization and construction. See Clifford, *Progress and Purpose*, 6.

17 Ibid., 8.

18 Victor Krulak, *First To Fight, An Inside View of the US Marine Corps* (Annapolis, MD: United States Naval Institute Press, 1999), 73.

19 Reports submitted were 'General Principles Governing the Selection and Establishment of Advanced Bases, The Composition of an Advanced Base Outfit', and 'Additional Notes on Field Work Construction for Advanced Bases'. See Clifford, *Progress and Purpose*, 14.

20 Brigadier General Edwin Howard Simmons, *The United States Marines: A History* (Annapolis, MD: United States Naval Institute Press, 1992), 111.

21 Major J.H. Russell, 'The Preparation of War Plans for the Establishment and Defense of a Naval Advance Base' (Lecture given at the US Naval War College, 1910, Naval War College Archives, Newport, RI).

22 Clifford, *Progress and Purpose*, 17.

23 Ibid.

24 David Moy, *War Machines: Transforming Technologies in the US Military, 1920–1940* (College Station, TX: Texas A&M University Press, 2001), 23.

25 Krulak, *First To Fight*, 73.

26 Ibid.

27 Allan R. Millett, *Semper Fi: The History of the United States Marine Corps* (New York: The Free Press, 1991), 277.

28 Of significance, as product champion of amphibious warfare, Commandant of the Marine Corps, General Russell would be responsible for creating the *Tentative Landing Doctrine* which served as the foundation for all amphibious assaults during WW II. See Major John H. Russell, 'A Plea For a Mission and Doctrine', *Marine Corps Gazette* (June 1916), 13.

29 Ballendorf and Bartlett, *Pete Ellis*, 109.

30 Ibid.

31 Ulbrich, *Clarifying the Origins*, 86.

32 See John Lejeune, 'The United States Marine Corps', *United States Naval Institute Proceedings* (October 1955), 863–6. Ben Fuller, 'The Mission of the Marine Corps', *Marine Corps Gazette* (November 1930), 33.

33 Earl Ellis, *Advanced Base Operations in Micronesia*, 1921 (Quantico, VA: Marine Corps Research Center) Breckinridge Library Historical Amphibious File, File 165, Breckinridge Library, Quantico, VA.

34 Holland Smith and Percy Finch, *Coral and Brass* (New York: Scribner's, 1949), 57.

35 Ibid., 59.

36 Rosen, *New Ways of War*, 155.

37 Clifford, *Progress and Purpose*, 92.

38 Ibid., 29.

39 Allan R. Millett, *In Many a Strife: General C. Thomas and the US Marine Corps: 1917–1956* (Annapolis, MD: United States Naval Institute Press, 1993), 75.

40 See Rosen, *New Ways of War*, 164.

41 First Lieutenant Anthony A. Frances, *History of the Marine Corps Schools* (Quantico, VA: Unpublished Paper, Archives, Marine Corps Research Center, 1945), 24.

42 Merrill Bartlett, *Lejeune: A Marines Life 1867–1942* (Annapolis, MD: Naval Institute Press, 1991), 52–3. Also see, Colonel James W. Hammond, Jr, USMC, 'Lejeune of the Naval Service', *United States Naval Institute Proceedings* (November 1981), 33.

43 Frances, *The History of the Marine Corps Schools*, 27.

44 Ibid., 28.

45 Ibid., 34.

46 Ibid., 36.

47 Ibid., 40.

48 Ulbrich, 'Clarifying the Origins', 81–109.

49 Millett, *Semper Fi*, 342.

50 Memo From Commandant of the Marine Corps, Russell, dated 18 August 1933 (Historical Amphibious File, Box 12 A, Field Officers Schools, Marine Corps 1932–1933, Breckinridge Research Facility, Quantico, VA).

51 Smith and Finch, *Coral and Brass*, 60.

52 Ballendorf and Bartlett, *Pete Ellis*, 111.

53 Ibid.

54 Victor Krulak, *First To Fight*, 18.

55 Ibid., 80.

56 Ibid.

57 Lieutenant Colonel Kenneth Clifford, *Progress and Purposes: A Developmental History of the US Marine Corps: 1900–1970* (Washington, DC: History and Museums, 1973), 43.

58 Ibid., 44.

59 Lieutenant General Holland Smith, 'The Development of Amphibious Tactics in the US Navy', *Marine Corps Gazette* (August 1946), 44.

60 Smith and Finch, *Coral and Brass*, 60.

61 Millett, *In Many a Strife*, 110.

62 Major J.H. Russell, 'The Preparation of War Plans for the Establishment and Defense of a Naval Advance Base', Lecture given at the US Naval War College, 1910, Naval War College Archives, Newport, R.I. Also see Krulak, *First To Fight*, 75.

63 Colonel Robert Heinl, 'The US Marine Corps: Author of Modern Amphibious Warfare', in Lieutenant Colonel Merrill L. Bartlett, USMC (ed), *Assault From the Sea* (Annapolis, Maryland, 1983), 188.

64 Rosen, *Winning the Next War*, 64.

65 Clifford, *Progress and Purposes*, 46.

66 The result was FTP 167 and FM 31–5.

67 Krulak, *First To Fight*, 75.

68 Ibid., 73.

69 Ibid., 76.

70 Ibid., 74.

71 Millett, *Semper Fidelis*, 303. Millet notes that the Marines' combat skill and the great German losses had been exaggerated impressions by the Association Press writers. Millett writes that 'Belleau Wood thus was a dearly bought but stunningly successful public relations coup for the Corps'.

72 Millett, *In Many a Strife*, 100.

73 Memo From Commanding General Quantico Schools to Commandant of the Marine Corps (Quantico, VA: Marine Corps Research Center, Amphibious Archive Files, 1936).

74 Memo from Major General John Russell to the Chief of Naval Operations, Subject: *Expeditionary Force*, 17 August 1933 (Quantico, VA: Marine Corps Research Center, Amphibious Archive Files, 1933).

75 The staff numbered about 35 officers. Krulak, *First To Fight*, 80.
76 Ibid., 81.
77 Memorandum from Colonel E.B. Miller to General Russell, Commandant of the Marine Corps, 12 May 1934 (Washington, DC: National Archives 1520-30-120).
78 Ibid.
79 Millett, *Semper Fi*, 335.
80 Ibid.
81 Ibid., 336.
82 Dr Donald Bittner, *Major General John Russell, United States Marine Corps: The Statesman Commandant* (Quantico, VA: unpublished paper, 12 February 1989).
83 Ibid., 19.
84 Moy, *War Machines*, 65. As Moy notes, the list of requirements presented a daunting challenge in naval engineering and architecture. The most difficult task by far was to design a hull that would exhibit the desired beaching characteristics in surf. Most Pacific atolls had shoreline that continuously pounded by heavy surf that made landing and retracting extremely difficult. For example, rudders and propellers were driven deep into the sand and the silt clogged water-cooled engines. The state of naval architecture after World War I had not advanced to the point where engineers could design and model shallow draft craft for laboratory testing. Consequently, the only reliable method of determining a new boat's beaching characteristics was to build and take it to a beach and test it.
85 Krulak, *First To Fight*, 90.
86 Millett, *Semper Fi*, 337.
87 Krulak, *First to Fight*, 91.
88 Ibid., 90.
89 Clifford, *Progress and Purpose*, 48.
90 The first amphibious ships designed from the keel up were delivered in 1942. See Millett, *Semper Fi*, 83.
91 Krulak, *First To Fight*, 91.
92 Millett, 'Assault From the Sea', 82.
93 This is one of the key teaching points of Marine Corps Historian Jon Hoffman in his lecture to Marine Corps Command and College on the 'Development of Amphibious Doctrine'.
94 Vice Admiral Daniel Barbey, *MacArthur's Amphibious Navy: Seventh Amphibious Force Operations 1943–1945* (Annapolis, MD: United States Naval Institute Press, 1969), 43.
95 See Commander Terry Pierce, USN, 'Operational Maneuver From the Sea', *United States Naval Institute Proceedings* (August 1994), 34. Also see Captain Richard S. Moore, USMC, 'Blitzkrieg From the Sea', *Naval War College Review* (November–December 1983), 42–3.
96 Barbey, *MacArthur's Amphibious Navy*, x. Of particular note, on 15 September 1950, American troops led by General MacArthur surprised the North Koreans by conducting a major amphibious landing at Inchon. Immediately afterwards General MacArthur sent a message to Admiral Barbey, MacArthur's amphibious admiral during World War II, and stated, 'The landing was made in the best Barbey tradition.' The Inchon landing during the Korean War was conducted in accordance with the doctrine Barbey and MacArthur had used during World War II.
97 Smith, 'The Development of Amphibious Tactics in the US Navy', 47.
98 Ibid., 33.
99 Ibid., 38.
100 Rosen, *Winning the Next War*, 76.
101 Wilson, *Bureaucracy*, 220–1.
102 Ballendorf and Bartlett, *Pete Ellis*, 67.

4 Post-World War II Marine Corps disruptive innovations: (I) helicopter warfare

1 See Rosen, 'New Ways of War', 151.
2 Lieutenant Colonel Eugene W. Rawlins, USMC, *Marines and Helicopters 1946–1962* (Washington DC: History and Museums Division, 1976), v.
3 Clifford, *Progress and Purpose*, 71.
4 As cited in Joseph H. Alexander and Merrill L. Bartlett, *Sea Soldiers in the Cold War: Amphibious Warfare 1945–1991* (Annapolis, MD: United States Naval Institute Press, 1995), 11.
5 Ibid., 1–2.
6 Robert Heinl, Jr Col. USMC, Retired, *Victory at High Tide: The Inchon-Seoul Campaign* (New York: The Nautical & Aviation Publishing Company of America, 1979), 7.
7 Lieutenant General Roy S. Geiger ltr to CMC, dated 21 August 1946 (Serial 0265–46, Box 11, Accession No. 14051, Record Group 127, WNRC, Suitland, Md.). Also see Clifford, *Progress and Purpose*, 71.
8 Several years later Vandegrift provided this reaction to the Geiger's report: 'I refused to share the atomic hysteria familiar to some ranking officers. The atomic bomb was not yet adapted for tactical employment, nor would this happen soon…I did feel obliged to study the problem in all its complexity. For if we believed the basic mission of the Marine Corps remained unchanged in the atomic age, we knew the conditions surrounding the mission would change and change radically.' Robert Asprey, *Once a Marine: The Memoirs of General A.A. Vandergrift as Told to Robert Asprey* (New York: Robert & Company, 1964), 319.
9 Ibid., 67.
10 Clifford, *Progress and Purpose*, 72. 'At the time, the seaplane in the immediate future was the Martin "Mars" with an empty weight of 75,000 pounds and a cargo and fuel load of 63,000 pounds. It was to have a troop-carrying capacity of 133 equipped men in seats. Howard Hughes, millionaire aircraft designer, was building a prototype eight-motored transport seaplane, which was designed to carry a 44-ton tank as part of its payload of 120,000 pounds.' The architectural innovation board, however, did not seem very optimistic about the outlook for either the Hughes or Martin seaplane being delivered in adequate numbers with the next half decade.
11 Robert D. Heinl, *Soldiers of the Sea: The United States Marine Corps, 1775–1962* (Annapolis, MD: United States Naval Institute Press, 1962), 513.
12 The first commanding officer of HMX-1 was Colonel Edward Dyer. James Ginther, *Keith Barr McCutcheon: Integrating Aviation Into the United States Marine Corps, 1937–1971* (unpublished disseration, Texas Tech University, 1999), 73.
13 The *Amphibious Operations – Employment of Helicopters (Tentative)* was also called PHIB 31. It was named PHIB 31 because it was the thirty-first in a series of manuals the school's architectural group developed to teach amphibious war concepts in the post-war era. James Ginther, *McCutcheon*, 74.
14 Ibid., 59.
15 See Clifford, *Progress and Purpose*, 75. In the exercise, HMX-1 was simulated landing one regimental combat team. During the actual landing, a total of 66 men and a considerable amount of equipment were transported to the beach by helicopter. For the entire operation, a total of 29 hours were flown and a total of 103 carrier landings and take-offs were made.
16 See Clifford, *Progress and Purpose*, 78.
17 Keith B. McCutcheon, 'Employment of Helicopters in the Marine Corps', draft copy in File 7, Box 7, McCutcheon Papers. Ginther, *McCutcheon*, 86.
18 The unit consisted of four helicopters employed as part of a Marine Corps Observation Squadron (VMO) 6 whose mission consisted of tactical air reconnaissance, artillery spotting, and other flight operations within the capabilities of assigned aircraft in support of ground units.

19 During Operation Switch in November 1951, McCutcheon's squadron was tasked to replace a front-line battalion with a reserve battalion. All movement of troops in and out of the combat zone was by helicopter. The squadron's 12 helicopters flew 262 flights, transported 950 combat equipped men and successfully completed the switch in 10 hours. James Ginther, *McCutcheon*, 97.

20 See Clifford, *Progress and Purpose*, 83. Nearly 10,000 Marines were evacuated by helicopter during the Korean War.

21 Lynn Montross, *Cavalry of the Sky: The Story of US Marine Combat Helicopters* (New York: Harper & Brothers Press, 1954), 178. Ginther, *McCutcheon*, 100.

22 Ginther, *McCutcheon*, 53.

23 Ibid., 139.

24 See Clifford, *Progress and Purpose*, 86.

25 'Report of the Fleet Marine Force Organization and Composition Board', 7 January 1957, Archives Branch, MCRC. Hereafter referred to as the Hogaboom Board Report, Section II, 13. Ginther, *McCutcheon*, 142.

26 Ginther, *McCutcheon*, 136.

27 As an illustration, Major General Vernon Megee, a senior fixed winged aviator wrote, '...as the "devil's advocate" regarding the helicopter...There were occasions when I felt that the helicopter enthusiasts had not properly evaluated the inherent vulnerability of the their magic aircraft to ground fire and air attack....Our operations in Korea...unchallenged by enemy air, and due to the covered approaches our mountainous front line positions were immune to effective enemy ground fire.' James Ginther, *McCutcheon*, 124.

28 Ibid., 134–6. As a result only 30 of 180 medium lift helicopters were built before this program was cancelled. The follow-on programs were the CH-46 and CH-53.

29 Ibid., 136–8.

30 Ibid., 137.

31 Millet, *Semper Fi*, 456. Ginther, *McCutcheon*, 124. The field artillery officers had every right to be paranoid of the close air support mission eliminating the need for field artillery. This is exactly what happened to the artillery arm once the Cobra helicopter was employed.

32 Millet, *Semper Fi*, 456. Victor Krulak, a proponent of airmobility, used the sustaining tactic quite effectively.

33 Although the Marine Corps had based development of its air arm on the close air support mission, a consistent and effective method of employing aircraft in this capacity had yet to be developed. The major problems were in the realm of communications and coordination between air and ground units. Ginther, *McCutcheon*, 40.

34 Ibid., 77.

35 Keith Barr McCutcheon to Clayton Jerome, 15 January 1952, and Keith Barr McCutcheon to Jack Beighle, 16 January 1952, both in Folder 5, Box 6, McCutcheon Papers. Ginther, *McCutcheon*, 101.

36 Ibid., 103.

37 As Ginther notes, 'As in the case of Marine fixed-wing aviation in the past, McCutcheon had to temper his faith in the combat capabilities offered by his helicopters against the ignorance of ground commanders of their capabilities and the politics of the service to accomplish his mission.' Ibid., 102.

38 Ibid., 40.

39 Ibid., 234.

5 Post-World War II Marine Corps disruptive innovations: (II) MAGTF warfare – combined arms operations

1 For the most in-depth discussion of MAGTF development see Ginther, *McCutcheon*.

2 Ibid., 60.

3 Ibid., 16.
4 During operations in the Solomon and Marshall Islands new methods were perfected which enabled Marine aviation to function as part of a close-knit team and provide exceptional support for Army and Marine units on Peleliu, in the Philippines, and on Okinawa. See memo from E.H. Simmons, Director of Marine Corps History and Museums to Deputy Chief of Staff for Aviation, 31 July 1979 (MAGTF File, Headquarters Marine Corps, History Division, Washington Navy Yard, Washington, DC).
5 See memo from E.H. Simmons, Director of Marine Corps History and Museums to Deputy Chief of Staff for Aviation, 31 July 1979 (MAGTF File, Headquarters Marine Corps, History Division, Washington Navy Yard, Washington, DC).
6 MCDP 1, *Warfighting* (Quantico, VA; USMC, 1997), 94.
7 Ibid., 95.
8 See Clifford, *Progress and Purpose*, 109.
9 Ibid.
10 See Clifford, *Progress and Purpose*, 110.
11 Ginther, *McCutcheon*, 7.
12 Ibid., 6.
13 Ibid., 8.
14 McCutcheon, 'Employment of Helicopters in the Marine Corps'.
15 Ginther, *McCutcheon*, 121.
16 Ibid., 125.
17 Ibid., 7.
18 Ibid., 126.
19 Ibid., 8.
20 Ibid., 153.
21 Ibid., 154.
22 Ibid.,155.
23 Ibid., 8.
24 Keith Barr McCutcheon to Frank Lamson, 31 January 1952, Folder 6, Box 5, McCutcheon Papers (Quantico, VA: Research Center Archives).
25 Ginther, *McCutcheon*, 8.
26 Ibid., 7.
27 Keith Barr McCutcheon to Frank Lamson, 31 January 1952, Folder 6, Box 5, McCutcheon Papers (Quantico, VA: Research Center Archives).

6 US Marine Corps inchoate disruptive innovation: maneuver warfare

1 The Marine Corps defines maneuver warfare as 'a warfighting philosophy that seeks to shatter the enemy's cohesion through a variety of rapid, focused, and unexpected actions which create a turbulent and rapidly deteriorating situation with which the enemy cannot cope'. See Marine Corps Doctrine Publication 1, *Warfighting*, 1997, 73.
2 Rosen, *Winning the Next War*, 96.
3 William Lind, *What Great Victory? What Revolution?* (Unpublished Article written in 1991). The following note appears on the title page; '(Note: The following article was accepted for publication by the editorial board of the *Marine Corps Gazette*; subsequently, its publication was blocked at a higher level. The author has given permission for unlimited reproduction and distribution.)'
4 Ibid., 2.
5 Statement by Lieutenant General Bernard E. Trainor, USMC (ret), 16 June 2003. Author holds original statement.
6 Ibid.
7 Ibid.
8 Ibid.
9 Ibid.

10 Major Michael N. Peznola, USMC, *A Matter of Trust? Maneuver Warfare in the Marine Corps: A 10 Year Assessment* (Unpublished Masters Thesis, Marine Corps University, 1999), 2.

11 Bill Lind is the President of the Military Reform Institute. Largely civilian intellectuals, not military officers with the exception of retired Air Force Colonel John Boyd, drive the Military Reform Movement. Boyd is noted for his OODA theory.

12 William S. Lind, *Maneuver Warfare Handbook* (Boulder, CO: Westview Press, 1985), 1.

13 Major Richard Hooker, Jr (ed), *Maneuver Warfare: An Anthology* (Navato, CA: Presidio Press, 1993), xiii.

14 David Evans, 'Marines have the last word on one who did it his way', *Chicago Tribune* (12 April 1991), 27.

15 John Scharfen's interview with Major General Alfred Gray, Commanding General 2nd Marine Division, 'Tactics and Theory of Maneuver Warfare', *Amphibious Warfare Review*, July 1984.

16 Ibid., 11.

17 Letter from Bill Lind to Terry Pierce, dated 25 May 1999.

18 Statement by Lieutenant General Bernard E. Trainor, USMC (ret), 16 June 2003.

19 John Fialka, 'A Very Old General May Hit the Beach With Marines', *The Wall Street Journal* (9 January 1991), 1.

20 The OODA Loop or Boyd Cycle is named after John Boyd who pioneered the concept in his lecture, 'The Patterns of Conflict'. Boyd identified a four-step mental process: observation, orientation, decision, and action. He theorized that each party to a conflict first observes the situation. On the basis of the observation, he orients; that is, he makes an estimate of the situation. On the basis of orientation, he makes a decision. Finally, he implements the decision – he acts. Because the action has created a new situation, the process begins anew. Boyd argued that the party who consistently completes the cycle faster gains an advantage that increases with each cycle. His enemy's reactions become increasingly slower by comparison and therefore less effective until, finally, he is overcome by events. 'A Discourse on Winning and Losing: The Patterns of Conflict', unpublished lecture notes and diagrams, August 1987. Also see MCDP 1, *Warfighting*, 1997, 102. Also see Fialka, 'A Very Old General May Hit the Beach With Marines', 1.

21 The Marine Corps officially accepted Boyd's OODA (Observe-Orient-Decide-Act) loop model as the theory of maneuver warfare and adopted it as the basis for their capstone operational philosophy in FMF 1 *Warfighting*, 1989. For an outstanding summary of the Boyd Theory, see Major Robert Polk, USA, *A Critique of the Boyd Theory – Is It Relevant to the Army?* (Monograph, School of Advanced Military Studies, United States Army Command and General Staff College, Fort Leavenworth, Kansas, 1999).

22 Gary Hart and William S. Lind, *America Can Win* (New York: Alder and Alder Press, 1986), 5.

23 Interview with Colonel G.I. Wilson, 5 February 2000.

24 Captain Kevin Clover, 'Maneuver Warfare: Where Are We Now?', *Marine Corps Gazette* (February 1988), 54.

25 Gray's tasking of the group was to determine how to teach maneuver warfare to the 2nd Marine Division, to include developing a series of large-scale exercises to the test that new-found knowledge. These exercises were called the Fort Pickett maneuver warfare exercises. Interview with Wilson.

26 Interview with G.I. Wilson.

27 Scharfen, 'Tactics and Theory of Maneuver Warfare', 14.

28 Ibid., 13.

29 Lind, *Maneuver Warfare Handbook*, 1.

30 Ibid., 2. Interview with Lieutenant Colonel Scott Moore, 9 March 2001, who stated that he had signed a contract with Bill Lind to be co-author of the book, but when

Lind's boss, Senator Gary Hart decided to run for president, Scott removed his name as co-author so it would not appear that an active duty military officer was supporting a candidate for political office.

31 Although Gray assigned Schmitt, who was a member of the Warfighting Center's Doctrine Branch, to draft the Maneuver Warfare volume, other key members of the group that Schmitt consulted included Brigadier General Paul K. Van Riper, Colonel Mike Wyly, Mr. William S. Lind, Lieutenant Colonels Raymond Cole and Gary I. Wilson, Brigadier General (Select) Anthony Zinni, and Brigadier General James M. Myatt. See Colonel John E. Greenwood, USMC (retired), 'FMFM 1: The Line of Departive', *United States Naval Institute Proceedings* (May 1990), 155.

32 The four major chapters include: the nature of war, the theory of war, preparing for war, and the conduct of war. The nature of war includes: friction, uncertainty, fluidity, disorder, and complexity. The theory of war includes: spectrum of conflict, levels of war, initiative and response, styles of warfare, centers of gravity, and critical vulnerabilities. Preparing for war includes: doctrine, training, professional education, and equipping. The conduct of war includes: maneuver warfare, orientating of the enemy, philosophy of command, shaping the action, decision-making, mission tactics, commander's intent, main effort, surfaces and gaps, and combined arms.

33 Marine Corp Doctrine Publication 1, *Warfighting*, 1997, Preface.

34 From General Gray to the President of FY01 USMC Lieutenant Colonel Selection Board, dated 7 October 1999.

35 Captain Tim Jackson, *Warfighting Skills Program Abstract* (Washington D.C.: Marine Corps Institute, 28 September 1989).

36 Ibid. Of particular importance is that Jackson invited Colonel Boyd to give a series of lectures to the 'Young Turks' in the Washington D.C. area. Boyd accepted and gave the lectures to the officers of Marine Barracks. See Letter from Captain Tim Jackson to Colonel Boyd, dated 21 April 1988, Letter found in Box 21 of the Boyd file, held at Archives in Quantico, VA.

37 Interview with Colonel G.I. Wilson, 5 March 2001.

38 The purpose of the Command's War Fighting Lab was outlined in the Marine Corps Home page at http://www.mcwl.org/mcwl-new/

39 Bill Lind Memorandum to Terry Pierce, dated 27 September 1999.

40 For a detailed discussion of the Scharnhorst changes see Charles Edward White, *The Enlightened Soldier* (New York: Praeger 1989).

41 Gary Hart and William S. Lind, Detailed Outline of *Military Reform*, 14 December 1984, 9 (William Lind's personal files).

42 Bill Lind Memorandum to Terry Pierce, dated 27 September 1999. Original memo held by author.

43 Ibid.

44 Ibid.

45 Bill Lind Memorandum to Terry Pierce, dated 27 September 1999. A Lind anecdote illustrates this point. He writes, 'When I was working for [Senator] Gary Hart, sometime in the late 1979s as I recall, an Air Force four-star with some intellectual ambitions asked to meet with Hart to discuss military reform (we weren't yet calling it that yet, of course). Hart invited me, and we met for lunch in the Senate dining room. Before Hart joined us, the general, whose name I forget, said to me, "You know, we all write think pieces. The problem is getting anyone to read them. You first get a Senator to give a speech on the Floor saying that we should cut the Air Force budget by $10 billion, then you put out your think piece. And then, everybody reads it." I replied, "You are exactly right." We may not have gotten the mule to go where we wanted it to, but we did learn how to get its attention.'

46 Letter from Bill Lind to Terry Pierce, dated 25 May 1999.

47 Statement by Lieutenant General Bernard E. Trainor, USMC (ret), 16 June 2003.

48 Ibid.

49 Memorandum from Bill Lind to Senator Gary Hart entitled, *Report on Trip to Marine Corps Base 29 Palms*, 23–27 October 1978.
50 Ibid., 8.
51 William S. Lind to Senator Gary Hart, *Report on Trip to Marine Corps Base 29 Palms*, 30 May–1 June 1979.
52 William S. Lind to Senator Gary Hart, *Report on Trip to Marine Corps Base 29 Palms*, 30 June–2 July 1981.
53 Gary Hart and William S. Lind, Detailed Outline of *Military Reform* (14 December 1984), 9.
54 Rosen, *Winning the Next War*, 11. Rosen writes that the 'classical example of such a reaction can be found in the failure of the US Army to develop army-wide capabilities for counterinsurgency even after being personally ordered to do so by the president'.
55 Undersecretary of Navy, Seth Cropsey, memo to Secretary of Navy, dated 1 April 1985.
56 Ibid.
57 William S. Lind and Jeffrey Record, 'The Marines' Brass is Winning the Battle But Losing the Corps, *The Washington Post* (28 July 1985), B1.
58 Ibid.
59 Ibid.
60 Ibid.
61 Ibid.
62 Interview with Bill Lind on 17 December 1999 and 8 March 2001.
63 Statement by Lieutenant General Bernard E. Trainor, USMC (ret), 16 June 2003.
64 In his guidance, General Al Gray stated, 'My intent in PME [professional military education] is to teach military judgement rather than knowledge. Judgement that will lead to a maneuver warfare style of war.' See Letter from Commandant of the Marine Corps to the Commanding General, Marine Corps Combat Development Command. Subject: Training and Education dated 1 July 1989 (Quantico VA: Research Center).
65 Peznola, *A Matter of Trust?*, 64.
66 Ibid., 69.
67 Ibid., 70.
68 Ibid., 72.
69 Scharfen, 'Tactics and Theory of Maneuver Warfare', 10.
70 Ibid.
71 Ibid., 30.
72 Ibid., 30.
73 Statement by Lieutenant General Bernard E. Trainor, USMC (ret), 16 June 2003.
74 William S. Lind, 'Defining Maneuver Warfare for the Marine Corps', *Marine Corps Gazette* (March 1980), 55.
75 In Lieutenant General Trainor's approach he identifies six factors of modern war for the high intensity modern battlefield to include: Intelligence, Electronics, Maneuver, Combined Arms, Flexible Logistics and C3. Scharfen, 'Tactics and Theory of Maneuver Warfare', 10.
76 Ibid., 12.
77 See memo from Major M.E. Williams, Marine Corps Liaison Officer, Infantry School, Fort Benning, Georgia, to Commanding General, MCDEC, Quantico, VA, copy to Commanding General, 2nd Marine Division, Subject Maneuver Concept Paper, 16 March 1983, Found in General Gray Box 39 II, Archives (Quantico, VA: Marine Corps Research Center).
78 Scharfen, 'Tactics and Theory of Maneuver Warfare', 11.
79 Ibid.
80 Interview with Colonel G.I. Wilson, 4 March 2000.
81 Interview with Colonel G.I. Wilson, 5 March 2001.

82 Statement by Lieutenant General Bernard E. Trainor, USMC (ret), 16 June 2003.

83 Interview with Colonel G.I. Wilson, 4 March 2000.

84 A memo from Franklin Charles Spinney to William F. Buckley, Jr 28 April 1991 Boyd Box 21, Archives (Quantico, VA.: Marine Corps Research Center).

85 Evans, 'Marines have the last word on one who did it his way', 27.

86 Ibid.

87 Ibid.

88 Ibid.

89 Peter Cary, 'The fight to change how America fights', *US News & World Report*, (6 May 1991), 31.

90 Tim Jackson recently defeated a passover on the strength of Gray endorsements, but this occurred ten years after Gray was Commandant and the career damage had already been done. Author holds original Gray letters to promotion board.

91 Interview with Lieutenant General Paul van Ripper, 18 February 2001.

92 Interview with Colonel T.X. Hammes, USMC, G-5, I MEF, 26 February 2001.

93 William S. Lind, 'Missing the Boat: A Response to Generals Knutson, Hailston, and Bedard', *The Marine Corps Gazette* (October 2000), 34.

94 Lind, *Maneuver Warfare Handbook*, 1.

95 Michael Gordon and General Bernard Trainor, *The Generals' War* (New York: Little Brown, 1995), 164.

96 Lind, *What Great Victory? What Revolution?*

97 For a complete explanation of the revolutionary development of tactics in the last years of World War I see Tim Lupfer's *Dynamics of Doctrine* and Bruce Gudmunsson, *Stormtrooper Tactics* (New York: Praeger Publishers, 1989).

98 Statement by Lieutenant Colonel Paul (Lester) Kuckuk, USMC, 16 June 2003. Author holds original statement.

99 Statement by Lieutenant General Bernard E. Trainor, USMC (ret), 16 June 2003. Author holds original statement.

100 Statement by Lieutenant General Wallace (Chip) Gregson, USMC, 12 June 2003. Author holds original statement.

101 See Gordon and Trainor, *The Generals' War*, 163–4.

102 Interview with Major Chris Yunker, USMC (ret), 19 February 2001.

103 See Wilson, *Bureaucracy*, 231.

104 Interview with Colonel Jim Lasswell, USMC (ret), 20 February 2001.

7 US Marine Corps sustaining innovations and summary of disruptive Marine Corps cases

1 Prior to 1988, Marine Expeditionary Units (MEUs) were called Marine Amphibious Units (MAUs). General Gray changed the name from MAUs to MEUs in 1988 in order to reflect more accurately Marine Corps' missions and capabilities. The point Gray was trying to make was that MAGTFs were not limited to amphibious operations alone. Rather, they are capable of projecting sustained, combined-arms combat power ashore in order to conduct a wide range of missions. To reduce confusion, I use the term MEU to refer to MAUs that existed prior to 1988.

2 Marine Corps Doctrine MCDP *Expeditionary Operations* (Quantico,VA: Marine Corps Doctrine Division), 75.

3 Stephen Rosen, 'New Ways of War', 151.

4 Memorandum from General Gray, Commanding General FMF, Atlantic, to Commandant of the Marine Corps, 26 March 1986, *Report of Examination of Marine Corps Special Operations Enhancements*.

5 A centralized effort by the Executive Branch of the US Government to develop the procedures and methods to combat terrorism resulted in NSDD-138, Presidential Guidance and Policy on Combating Terrorism, and NSDD-221, Presidential

Guidance and Policy of US Capabilities for Conducting Special Operations. Following the President's order, the Secretary of Defense directed the Military Departments to organize, train, and equip 'Special Operations Forces (SOF) capable of conducting the full range of special operations on a worldwide basis...not later than the end of Fiscal Year 1990'. It further stated that, 'Each Service will assign SOF and related activities sufficient resource allocation priority...'.

6 Jon Hoffman, assistant director of Marine Corps historical center, Hoffman's undated SOC operations Binder, *Historical Review of MAU/MEU(SOC)*.

7 Jon Hoffman, *Historical Review of MAU/MEU(SOC)* (lecture notes) (Washington, DC: Marine Corps Historical Center).

8 Millett, *Semper Fi*, 630.

9 General P.X. Kelly, Commandant of the Marine Corps, for the Joint Chiefs of Staff, 22 July 1985.

10 General C.E. Mundy, Jr, Commandant of the Marine Corps, Memorandum for the Commanding General, Marine Corps Combat Development Command, Subject *Special Operations Capabilities in Fleet Marine Forces*, 3 November 1992 (Washington, DC: Marine Corps Historical Center).

11 Ibid.

12 Ibid.

13 Rosen, 'Winning the Next War', 21.

14 Allan Millett, *Semper Fi*, 617.

15 See declassified Secret Document from Commandant of the Marine Corps with Subject Title: Compendium of Major Decisions in the Evolution of the Maritime Prepositioning Ships (MPS) Program, dated 23 December 1983 (Washington, DC: Marine Corps Historical Center).

16 Ibid.

17 See Commandant Marine Corps White Letter 2-81, 'Amphibious Operations and Maritime/Near Term Prepositioning Ships'. Located in Slide and Reports, December 1983, Maritime Prepositioning Force Study Background Material, Breckinridge Archive Files, (Quantico, VA: Marine Corps Research File).

18 See declassified Secret Document from Commandant of the Marine Corps with Subject Title: *Compendium of Major Decisions in the Evolution of the Maritime Prepositioning Ships (MPS) Program*, dated 23 December 1983.

19 Memorandum from the Commandant of the Marine Corps to Commanding General, Marine Corps Development and Education Command, 21 January 1986, Subject 'Maritime Prepositioning Force (MPF), Operations Study'. From Studies and Reports, Box 1985–86, Quantico (Quantico, VA: Research Center).

20 See Jeffrey Record, 'The Military Reform Caucus', *Washington Quarterly* 6 (Spring 1983)', 125–9; James W. Reed, 'Congress and the Politics of Defense Reform', in Asa Clark *et al.* (eds), *The Defense Reform Debate* (Baltimore, MD: Johns Hopkins University Press, 1984), 230–49. These articles contend that the Military Reform Movement lost momentum when it began to address several different issues simultaneously rather than focusing on the Army's air-land battle doctrine and the Marine Corps maneuver warfare doctrine.

21 Millet, *Semper Fi*, 335.

8 US Navy sustaining innovation: continuous aim gunfire

1 The background of this case is taken from a series of three lectures delivered by Elting E. Morison at the California Institute of Technology in 1950. It was printed as Chapter 2 of Morison's *Men, Machines and Modern Times* (Boston, MA: MIT Poers, 1966), 17–44.

2 In 1899, gun captains aboard five US Navy warships took a hulk under fire for five minutes per ship at a range of only 1,600 yards. After 25 minutes of fire, two hits had been scored. Morison, *Men, Machines and Modern Times*, 22.

3 Elting Morison, *Admiral Sims and the Modern American Navy* (Boston, MA: Houghton Mifflin, 1942), 86.

4 Michael Tushman and Charles O'Reilly, *Winning Through Innovation* (Boston, MA.: Harvard Business School, 1997), 4.

5 Morison, *Admiral Sims and the Modern American Navy*, 83–90.

6 Ibid.,107.

7 Ibid.,120.

8 See Kurth, *The Politics of Technological Innovation in the United States Navy*, 13.

9 See Rosen, *Winning the Next War*, 10.

10 Richard Neustadt, *Presidential Power* (New York: The Free Press, 1991), 18–24. Also see Rosen, *Winning the Next War*, 10.

11 Neustadt, *Presidential Power*, 24.

12 Rosen illustrates this point by using President Lyndon Johnson's order to innovate in Vietnam, but because he did not know exactly what type of disruptive innovation he wanted he could not give unambiguous orders. See Rosen, *Winning the Next War*, 10–11.

9 US Navy disruptive innovation: carrier warfare

1 Rosen, 'New Ways of War', 151.

2 Ibid.

3 The battle line was a tactical concept for fleet actions. The battle line served two purposes: (1) it allowed admirals to command several ships, and (2) it allowed the admiral to concentrate the fleet's firepower. The battleship with the big gun was the decisive weapon. Building a better battleship meant bigger guns and thicker ship's armor. At the Battle of Jutland in 1916, 250 warships of the British and German fleets clashed in a line-of-battle encounter. Both sides claimed victory with several ships sunk on both sides. In US Navy preparations for the next war, the capital ship remained the battleship, as the admirals planned on refighting the Battle of Jutland somewhere in the Pacific against Japanese battleships.

4 In 1921, Navy and Army aircraft conducted joint bombing tests and sank the anchored *Ostfriesland*. Based on this event, Mitchell told a Select House Committee that 'it is a very serious question whether air power is auxiliary to the Army and the Navy, or whether armies and navies are not actually auxiliary to air power'. He asserted, 'The Navy would and could not develop aviation at sea.' Mitchell's criticism fired a fierce debate over the Navy's efforts in aviation. In the long run, he spurred the Navy to conduct tactical experiments with the aircraft carrier *Langley* well before it might have on its own. Mitchell as quoted in Andrew Marshall, *Historical Innovation: Carrier Aviation Case Study*, memorandum from the Office of Net Assessment, Secretary of Defense (27 June 1994), 9. Hone, Friedman, and Mandeles generally agree with this assessment, but they would tie in the importance of the simulation games at the Naval War College and importance of Reeves ability to get more striking power (through more aircraft) and the impact of this on changing the carrier aviation rules for future games. Thomas C. Hone, Norman Friedman, and Mark D. Mandeles, *American and British Aircraft Development: 1914–1944* (Annapolis, MD: United States Naval Institute Press, 1999), 42.

5 Hone, Friedman, and Mandeles, *American and British Aircraft Development*, 165.

6 Andrew Marshall, *Historical Innovation: Carrier Aviation Case Study*, Memorandum from the Office of Net Assessment, Secretary of Defense, 27 June 1994, 1.

7 Tremendous credit must go to Mahan and Luce who instilled a spirit of intellectual curiosity as well as a willingness to experiment at the Naval War College with the impact of new technologies on naval warfare. Besides the advent of the airplane, the Navy was also championing new technologies such as steam turbines, long-range guns, fire control calculators, submarines, radio communications, and the shift to oil-fired engines.

8 Marshall, *Historical Innovation*, 7.

9 Ibid., 7.

10 Hone, Friedman, and Mandeles, *American and British Aircraft Development*, 33–4; Barry Watts and Williamson Murray, 'Military Innovation in Peacetime', in Williamson Murray and Millett (eds), *Military Innovation in the Interwar Period* (New York: Cambridge Press, 1996), 392.

11 Peter P. Perla, *The Art of Wargaming: A Guide for Professionals and Hobbyists* (Annapolis, MD: United States Naval Institute Press, 1990), 71.

12 Watts and Murray, 'Military Innovation in Peacetime', 393.

13 Marshall, *Historical Innovation*, 5.

14 Hone, Friedman, and Mandeles, *American and British Aircraft Development*, 39

15 For a detailed discussion of the carrier warfare innovation case, see Rosen, 'New Ways of War', 151–8. For a detailed study of Admiral Moffett's career, see William Trimble, *Admiral William A. Moffett: Architect of Naval Aviation* (Washington, DC: Smithsonian Institution Press, 1994).

16 The evidence is unclear what factors were the most important in Moffett being appointed to head naval aviation. Steve Rosen has cited evidence supporting it was a senior naval leadership decision (see Rosen, *Winning the Next War*, 77). Clark G. Reynolds, on the hand, suggests Moffett energized political connections to secure the job (See Clark G. Reynolds, 'William A. Moffett: Steward of the Air Revolution', James C. Bradford (ed), *Admirals of the New Steel Navy: Makers of the American Naval Tradition, 1880–1930* (Annapolis, MD: United States Naval Institute Press, 1990), 378.

17 See Watts and Murray, 'Military Innovation in Peacetime,' in Murray and Millett, *Military Innovation in the Interwar Period*, 394.

18 Trimble, *Moffett: Architect of Naval Aviation*, 65.

19 House Committee on Naval Affairs, *Title of Hearings*, 67th Cong., 1st sess., 1921, 88–90.

20 Trimble, *Moffett: Architect of Naval Aviation*, 78.

21 Ibid., 80.

22 Marshall, *Historical Innovation*, 8.

23 Ibid., 25.

24 Hone, Friedman, and Mandeles, *American and British Aircraft Development*, 184.

25 Ibid., 39.

26 Marshall, *Historical Innovation*, 10.

27 Hone, Friedman, and Mandeles, *American and British Aircraft Development*, 41.

28 Hone, Friedman, and Mandeles, *American and British Aircraft Development*, 42; Marshall, *Historical Innovation*, 12.

29 Hone, Friedman, and Mandeles, *American and British Aircraft Development*, 42; Marshall, *Historical Innovation*, 42.

30 See Marshall, *Historical Innovation*, 10.

31 It would be headed by Rear Admiral Montgomery Taylor and was called the Taylor board. See Trimble, *Moffett: Architect of Naval Aviation*, 203.

32 Ibid.

33 Marshall, *Historical Innovation*, 34.

34 As quoted in Trimble, *Moffett: Architect of Naval Aviation*, 203.

35 Hearings before the General Board of the Navy, 1927 (micro. Roll 7), 98, 108, 131. Also see Trimble, *Moffett: Architect of Naval Aviation*, 206.

36 *Saratoga* and *Lexington* each carried about 70 aircraft. *Ranger*, at just 13,800 tons, could carry about 72 aircraft because her hangar was significantly larger. The problem with *Ranger* was that because of her small size, she lacked the protection, speed, range, and seakeeping qualities of larger ships. To his credit, Moffett recognized this shortcoming and increased the tonnage of the next carrier to about 20,000 tons. The 20,000-ton carriers with higher speeds, more aircraft, and enhanced protection turned

out to be about the right size for World War II engagements. See Trimble, *Moffett: Architect of Naval Aviation*, 15.

37 Ibid.
38 Report to Board to Consider and Recommend upon Present Aeronautic Policy, 11 May 1927 (3rd endorsement, 11 November 1927), G.B. 449 (Seriel No. 1353), box 191, General Board, General Records of the Department of the Navy, Record Group 80, National Archives. See Trimble, *Moffett: Architect of Naval Aviation*, 204.
39 Trimble, *Moffett: Architect of Naval Aviation*, 204.
40 Marshall, *Historical Innovation*, 17.
41 Ibid., 18.
42 House Committee on Naval Affairs, *Hearings on Sundry Legislation Affecting the Naval Establishment*, 1927–1928, 70th Cong., 1st sess., 1928, 837, 842, 845. See Trimble, *Moffett: Architect of Naval Aviation*, 208.
43 J.J. Clark and Clark G. Reynolds, *Carrier Admiral* (New York: McKay Press, 1967), 44. Also see Rosen, 'New Ways of War', 156.
44 Eugene E. Wilson, *Gift of Foresight: The Reminiscences of Commander Eugene E.Wilson*, A Naval History Project, The Oral History Research Officer (Columbia University, 1962), typescript, 315.
45 Rosen, 'New Ways of War', 152.
46 Ibid., 153.
47 Trimble, *Moffett: Architect of Naval Aviation*, 81.
48 Ibid., 163.
49 Hone, Friedman, and Mandeles, *American and British Aircraft Development*, 40.
50 Ibid., 47.
51 Thomas Wildenberg, *All The Factors of Victory: Admiral Joseph Mason Reeves and the Origins of Carrier Airpower* (Dulles, VA: Brassey's, Inc., 2003), 116.
52 Ibid., 153–4.
53 Reeve's chief of staff Captain E.E. Wilson in his memoirs writes, 'The Navy had created the first American strategic air force.' Hone, Friedman, and Mandeles write, 'In brief, in the climax to Problem IX, *Saratoga*, under the command of Reeves as commander of the "enemy" force, left the main force of battleships and, accompanied by one light cruiser, made a high-speed run from the west in order to launch a 70-plane strike against the locks of the Panama Canal from the range of 140 miles. On her approach, *Saratoga* used her own aircraft to scout, and her light attack planes dove on the canal's locks from an altitude of 10,000 feet. They caught "defending" forces by surprise, thus demonstrating – supposedly – the potential of carrier aviation.' (Hone, Friedman, and Mandeles, *American and British Aircraft Development*, 48–9).
54 Reynolds, 'Moffett: Steward of the Air Revolution', 379.
55 Reynolds, 'Moffett: Steward of the Air Revolution', 379; and Clark G. Reynolds, *Admiral John H. Towers: The Struggle for Naval Air Supremacy* (Annapolis, MD: United States Naval Institute Press, 1991), 199.
56 Reynolds, 'Moffett: Steward of the Air Revolution', 379–80.
57 Hone, Friedman, and Mandeles, *American and British Aircraft Development*, 39.
58 As quoted in Rosen, 'General Board, Serial Number 1554, Moffett to the President of the General Board, Design of Future Aircraft Carriers', 12 November 1931, Naval War College Library, Microfilm Collection. See Rosen, 'New Ways of War', 153 Footnote 47.
59 Trimble, *Moffett: Architect of Naval Aviation*, 198.
60 See Watts and Murray, 'Military Innovation in Peacetime', 155 Footnote 53. Also See Hone, Friedman, and Mandeles, *American and British Aircraft Carrier Development*, 109.
61 As quoted in Rosen, 'New Ways of War', 155, n. 53, General Board, File 421, Serial Number 1298, 'Policy on Naval Aviation Personnel', 26 September 1925, National Archives.
62 See Rosen, 'New Ways of War', 156. Also see Edward Arpee, *From Frigates to Flat-Tops: The Story of the Life and Achievements of Rear Admiral William Adger*

Moffett, U.S.N., 'the Father of Naval Aviation', October 31, 1869–April 4, 1933 (Lake Forest, IL: 1953), 114–15, 118–19.

63 Wildenberg, All the Factors of Victory, 152–3.

10 Disruptive innovation: Japanese carrier warfare

1 Mark R. Peattie, *Sunburst: The Rise of Japanese Naval Air Power, 1909–1941* (Annapolis, MD: United States Naval Institute Press, 2001), 152 [fn 63] and 169.

2 David C. Evans and Mark R. Peattie, *Kaigun: Strategy, Tactics, and Technology in the Imperial Japanese Navy 1887–1941* (Annapolis, MD: United States Naval Institute Press, 1997), 282.

3 Peattie, *Sunburst*, 1.

4 Ibid., 4.

5 Mark Peattie writes, 'Viewing the air operations at Tsingtao as a whole, Charles Burdick, the acknowledge authority on the siege, has asserted that 'the sophistication of Japanese aircraft employment – i.e., coordination with land forces, bombing equipment, and general mobility – was well ahead of any other country.' ' See Peattie, *Sunburst*, 7–9.

6 Ibid., 9.

7 Evans and Peattie, *Kaigun*, 168.

8 Evans and Peattie, *Kaigun*, 11. Peattie, *Sunburst*, 16.

9 Peattie, *Sunburst*, 10.

10 Evans and Peattie, *Kaigun*, 301. Peattie, *Sunburst*, 19–20.

11 Rosen, *Winning The Next War*, 96–7.

12 Mark Mandeles advocates a departmental organizational theory to promote innovation. See Thomas C. Hone and Mark D. Mandeles, 'Interwar Innovation in Three Navies: US Navy, Royal Navy, Imperial Japanese Navy', *Naval War College Review* (Spring, 1987).

13 Ibid., 69 [fn 26].

14 See Peattie, *Sunburst*, 26. While interservice rivalry spurred naval technological development, there was a down side. As Mark Peattie notes, 'While the Naval Aviation Department was the source of a good deal of innovative thinking about naval air doctrine and technology, effective aircraft design and production was continually hampered by one of the inherent and critical flaws in the prewar Japanese government: the debilitating and corrosive rivalry between the armed services. The ongoing failure of the army and navy to cooperate, as well as the inadequate organizational integration of the navy itself, eventually led to serious aircraft production problems that could have been avoided had there been more opportunities for inter- and intraservice discussion of critical technological issues.' Peattie, *Sunburst*, 28–9.

15 For a detailed discussion see Peattie, *Sunburst*, 21–51.

16 Peattie, *Sunburst*, 28.

17 Ibid., 89.

18 Ibid., 91.

19 Ibid., 58.

20 Ibid., 53.

21 Ibid., 58.

22 Rosen, 68.

23 See Gordon W. Prange, *At Dawn We Slept: The Untold Story of Pearl Harbor* (New York: Penguin Books, 1981), 39.

24 By 1941, Japan was poised for further expansionist adventures into Southeast Asia – Malaya, the Philippines, and the Netherlands East Indies. The Japanese convinced themselves that necessity and self-protection demanded they take over the vast resources of these promised lands to break through real or imagined encirclement and beat off the challenge of any or a combination of their international rivals – the United States, Great Britain, and Soviet Russia. See Prange, *At Dawn We Slept*, 4.

25 Evans and Peattie, *Kaigun*, 201.
26 Prange, *At Dawn We Slept*, 12.
27 Evans and Peattie, *Kaigun*, 475.
28 Ibid.
29 Peattie, *Sunburst*, 74.
30 Shigeru Fukudome, 'Hawaii Operation', in Paul Stiwell (ed), *Air Raid: Pearl Harbor!* (Annapolis, MD: United States Naval Institute Press, 1981), 62.
31 Ibid., 59, and Evans and Peattie, *Kaigun*, 473. Note: The battleship umpires of the 1927 Naval Staff College tabletop wargame ruled that the planes caused minimal damage during the Pearl Harbor attack and they ruled that Lieutenant Commander Kaku Tomeo's plan had been rash launching the air strike.
32 Mark R. Peattie, *Sunburst*, 72.
33 Ibid., 72.
34 Ibid., 74.
35 Ibid., 75.
36 Evans and Peattie, *Kaigun*, 282.
37 Prange, *At Dawn We Slept*, 24.
38 Peattie, *Sunburst*, 147.
39 Evans and Peattie, *Kaigun*, 347.
40 Peattie, *Sunburst*, 148–9.
41 Ibid., 151.
42 In December 1940, Admiral Fukudome was Yamamoto's Chief of Staff Combined Fleet. He was the only person acquainted with detailed plans of the operation from the moment of conception (Fukudome, 'Hawaii Operation', 61).
43 See Fukudome, 'Hawaii Operation', 61. Yamamoto outlined his bold concept to Rear Admiral Onishi Takijiro, then Chief of Staff of the Eleventh Air Fleet, and instructed him to study its practicability.
44 Masatake Okumiya and Jiro Horikoshi with Martin Caidin, *Zero: The Story of Japan's Air War in the Pacific* (Ibooks – Simon and Schuster, 1956), 22. Prange, *At Dawn We Slept*, 23–24. In 1934 Genda reported as an instructor to the Yokosuka Air Corps. There he expanded his ideas on the use of fighters and carriers in combat – theories that were to become known as Gendaism.
45 Prange, *At Dawn We Slept*, 26.
46 Ibid., 98.
47 Ibid., 99–100. Prange discusses that Yamamoto entrusted Kuroshima (strange guy) and Wantanbe with the detailed work and with solving the torpedo challenge.
48 Ibid., 101.
49 Peattie, *Sunburst*, 151.
50 Ibid., 152, 161.
51 Steve Rosen, 'Theories of Victory: Understanding Military Innovation', *International Security* 13 (1988), 136.
52 Peattie, *Sunburst*, 83.
53 Ibid., Peattie notes that occasionally, Yamamoto did try to convince select members of the Gun Club's upper echelon who admired him of the folly of building several new super-battleships. But when he represented the Japanese Navy during Naval Conferences among the big three Navies, Yamamoto followed instructions and made the case for Big Gun orthodoxy.
54 For a similar discussion of US carrier development, see Williamson Murray, *Experimentation in the Period Between the Two World Wars: Lessons for the Twenty-First Century*, Institute for Defense Analyses, IDA Document D-2502, 5.
55 For a discussion of the pitfalls of event-based experiments in US carrier development, see Williamson Murray, *Experimentation in the Period Between the Two World Wars*, 28.
56 Edwin P. Hoyt, *Yamamoto: The Man Who Planned The Attack On Pearl Harbor* (Guilford, CT: The Lyons Press, 1990), 2.

57 H. Agawa, *The Reluctant Admiral: Yamamoto and the Imperial Navy* (New York. Kodansha International, 1979), 97.

58 Hoyt, *Yamamoto*, 90–1, 101.

59 Agawa, *The Reluctant Admiral*, 160, 163.

60 Ibid., 163.

61 Awaga, 165.

62 Hoyt, *Yamamoto The Reluctant Admiral,* 103.

63 Evans and Peattie, *Kaigun*, 262.

64 Mark R. Peattie writes, 'Ozawa saw in the potential of the navy's scattered carriers a revolutionary means to achieve formidable offensive air power. Led by his subordinate, Lieutenant Commander Fuchida Mitsuo, the air group commander of *Akagi*, who continually preached to him the importance of carrier concentration and urged that the First and Second carrier divisions should train and operate together. Thus, Ozawa himself came to see the need not only to reorganize the carrier divisions but to harness all the navy's air power under a single command.' See Peattie, *Sunburst*, 151.

65 Ibid.

66 Hoyt, *Yamamoto*, 12.

67 Ibid.

68 Peattie, *Sunburst*, 151.

69 Ibid., 150.

70 Ibid., 151.

71 Prange, *At Dawn We Slept*, 102.

72 Peattie, *Sunburst*, 152.

73 Prange, *At Dawn We Slept*, 102.

74 Peattie, *Sunburst*, 167, 195.

75 Prange, *At Dawn We Slept*, 14.

76 See Peattie, *Sunburst*, 152. Prange ask the question, 'Did Yamamoto organize the First Air Fleet for the express purpose of attacking the US Pacific at Pearl Harbor? No, Yamamoto did not know in December 1940, when he approved the air fleet concept (that gave rise to the First Air Fleet) that there would be a need for sure to attack. Yes, however, because if events forced him to lead the Combined Fleet into battle with Kimmel's ships, Yamamoto determined that he must strike that first blow, and he could not do so without the *nucleus of a carrier tasks force in being*. See Prange, *At Dawn We Slept*, 106.

77 Peattie, *Sunburst*, 152.

78 Evans and Peattie, *Kaigun*, 352.

79 Ibid.

80 Evans and Peattie, *Kaigun*, 483–5. Peattie, *Sunburst*, 167, 195.

81 Peattie, *Sunburst*, 206.

82 Hoyt, *Yamamoto*, 109.

83 Ibid., 51.

84 Agawa, *The Reluctant Admiral*, 76.

85 Arthur J. Marder, *Old Friends, New Enemies: The Royal Navy and the Imperial Japanese Navy: Strategic Illusions 1936–1941* (New York: Oxford Press 1981), 305.

86 Ibid.

87 Prange, *At Dawn We Slept*, 158.

88 Hoyt, *Yamamoto*, 13.

89 Evans and Peattie, *Kaigun*, 475.

11 US Navy disruptive innovation: CWC – naval combined arms warfare

1 Frederick H. Hartmann, *Naval Renaissance: The US Navy in the 1980s* (Annapolis, MD: United States Naval Institute Press, 1990), 23.

2 Naval Doctrine Publication 6, *Naval Command and Control*, Navy and Marine Corps Joint Publication (19 May 1995), 6. Command and control can be compared to the functioning of the central nervous system. Sensory nerves detect what is happening, inside and outside the body, and their input is processed by the brain. The brain (the commander) interprets the sensations, and sends the appropriate instructions to the muscles via the motor nerves.

3 Naval Doctrine Publication 6, *Naval Command and Control*, 6.

4 Captain P.J. Doerr, USN, 'CWC Revisited', *United States Naval Institute Proceedings* (April 1986), 39.

5 Hartmann, *Naval Renaissance*, 28.

6 Vice Admiral James H. Doyle Jr, USN (Ret), Oral History, by Naval Historical Center 1997, 55, 56.

7 Dr Schneiderman worked in Doyle's tactical development and experiment section (TAC D&E). Other member's of Schneiderman's small innovation group came from TAC D&E. The CWC tactical memo was COMTHIRDFLT TACMEMO 510-1-76, Composite Warfare Commander. Written comments to author from Captain Stuart Landersman, USN (Retired), 28 March 2001.

8 Phone interview with Captain Stuart Landersman, USN (Retired), 22 March 2001.

9 The hunter-killer group's mission was to locate Soviet diesel-electric submarines and hold them until they were forced to surface from lack of battery power or oxygen. The primary ship of the group was a World War II aircraft carrier equipped with S-2 aircraft and ASW helicopters. In a typical practice scenario, an aircraft some distance from the carrier and destroyers would get contact by either visual, electronic intercept, or magnetic anomaly. Manual plots maintained on the carrier and the destroyers would provide an area of probability of the submarine. The rear admiral commanding all the ships would direct the screen commander to detach two destroyers from the screen as a surface attack unit, which would speed off to the area of probability and attempt to gain sonar contact.

10 Landersman came on the CWC scene in February 1977, after the concept had been promulgated and tried at sea in a few exercises.

11 Captain Landersman notes, 'Like a pouting kid I started to pick up my toys to go home. Strutzer [a commander from the aviation staff] stopped me, went back to his seniors, and I was permitted to participate on my own terms.' He continues, 'Often Stutzer and I would reflect on how the development of CWC was determined with my threat to "pick up my toys and go home", because it determined that the subordinate warfare coordinators (later changed to commanders) would be separate subordinate commands.' Phone interview with Captain Landersman, 26 March 2001.

12 The school's first name was Office of Executive Director for Pacific Fleet Tactical Training, later Tactical Training Group Pacific.

13 By this time the CWC concept had moved up the tactical development process to become Pacific Fleet doctrine and was promulgated under the technical name CINC-PACFLT TACNOTE 510-1-78.

14 Landersman believed in Clausewitz's concept of economy of force, which he interpreted to mean, 'Don't waste any asset in the battle group.' With that in mind, Landersman taught his students that in coordinated battle group ASW every resource of the battle group (that the ASW commander could control or influence in any way) had a role in ASW. This was a paradigm shift in naval ASW.

15 Phone interview with Captain Landersman, 25 March 2001.

16 As Landersman predicted, the submarine community used its new role of direct support to the battle group to receive congressional funding for the new *Los Angeles* class submarine.

17 Phone interview with Captain Landersman, 26 March 2001. The amphibious surface warfare community also objected to CWC. Leading the amphibious opposition was Rear Admiral Dave Ramsey, who was not convinced that the aviation community

under CWC would give amphibious commanders the resources needed to protect the landing force. Based on World War II battles in the Pacific, Ramsey believed the carriers would hoard air resources to protect the carrier. Landersman, however, did not have much to fear from the amphibious community because within the pecking order of surface warfare, amphibious forces fell behind the cruiser-destroyer club, of which Landersman was a part.

18 Phone interview with Captain Landersman, 26 March 2001.
19 Ibid.

12 US Navy sustaining innovation: carrier battle group concept

1 The Commander-in-Chief, Atlantic Fleet, Admiral Paul David Miller, experimented with putting Marines on carriers in the 1990s, but the concept was never institutionalized. See Peter Swartz, *Origins and Development of the US Navy Carrier Battle Group Concept*, unpublished paper held by author, 31 August 1998.
2 The specialized attack carriers were CVAs (conventional propulsion) and CVANs (nuclear propulsion).
3 The ASW carriers were CVSs.
4 Hartmann, *Naval Renaissance*, 21.
5 Ibid., 20.
6 Admiral James L. Holloway III, 'The Aircraft Carrier: An Overview', *Wings of Gold*, special report (Summer 1987), 5–7. See Hartmann, *Naval Renaissance*, 26.
7 See Swartz, *Origins and Development of the US Navy Carrier Battle Group Concept.*
8 Two or more CVBGs operating together made up what Holloway called a carrier battle force (CVBF).
9 See Swartz, *Origins and Development of the US Navy Carrier Battle Group Concept.*

13 US Navy disruptive innovation – aborted: Project 60 – defensive sea control warfare

1 I am contrasting Zumwalt's defensive sea control innovation with the Hayward offensive sea control innovation. In the former, sea control is considered a higher priority than power projection ashore. In contrast, offensive sea control considered the two roles as equal.
2 Amazingly, Zumwalt after being CNO for only 72 days delivered his new warfighting innovation to Laird, and this included hand circulating the Project 60 study to all senior officers in OPNAV and the fleet for comments and obtaining the Secretary of the Navy's approval. See Jeffrey Sands, *On His Watch: Admiral Zumwalt's Efforts to Institutionalize Strategic Change*, Center for Naval Analysis, CRM 93–22, July 1993, 19.
3 Sands, *On His Watch*, 1 Also see Admiral Elmo Zumwalt Jr, *On Watch* (Arlington, VA: Elmo Zumwalt and Associates, 1976), Appendix C, 519–29. Here Zumwalt list unclassified summaries of the growth in Soviet naval capabilities.
4 The Project 60 briefing was classified Secret but has been declassified. See CNO's Project 60. Presentation to SECDEF, 9 September 1970, 4.
5 Ibid., 4.
6 Ibid., 27.
7 Zumwalt writes, 'For the last quarter-century or more there have been three powerful "unions", as we call them, in the Navy – the aviators, the submariners, and the surface sailors – and their rivalry has played a large part in the way the Navy has been directed. (The submariners have not had a CNO in recent times, but they have had the aforesaid Admiral Rickover, who most of the time is more than a match for most CNOs)...Whichever union such a commander comes from, it is hard for him not to favor fellow members, the men he has worked with most closely, when he constructs a staff or passes out choice assignments.' Zumwalt, *On Watch*, 63–4.

8 Ibid., 35.
9 Sands, *On His Watch*, 23.
10 Ibid.
11 Zumwalt, *On Watch*, 66.
12 Sands, *On His Watch*, 24.
13 Project 60 emphasized six basic precepts necessary to meet the specific threats the Soviet Navy posed: reprioritize naval missions; retire forces early to fund modernization; modernize following the high-low mix; pursue new initiatives in research and development; reduce support costs; and pursue people programs. See CNO's Project 60 Presentation to SECDEF. Also see, Zumwalt, *On Watch*, 67–84 and Sands, *On His Watch*, 25.
14 I do not discuss the personnel initiatives, which arguably demanded a great deal of time from Zumwalt.
15 Assured Second Nuclear Strike was the Navy's top nuclear strategic mission. However, I am referring to non-nuclear naval missions. See CNO's Project 60 Presentation to SECDEF, 3.
16 E.T. Wooldridge, *Into the Jet Age* (Annapolis, MD: United States Naval Institute Press, 1995), 221.
17 George W. Baer, *One Hundred Years of Sea Power: The US Navy, 1890–1990* (Stanford, CA: Stanford University Press, 1994), 390. Zumwalt writes that Moorer's emphasis on 'high end' campaigns degraded the two 'low end' campaigns of coastal control and river patrol. See Elmo Zumwalt, Jr and Elmo Zumwalt III, with John Pekkanen, *My Father, My Son* (New York: Macmillan, 1986), 44. Also see Baer, *One Hundred Years of Sea Power*, 390.
18 Zumwalt and Zumwalt, *My Father, My Son*, 44.
19 Baer, *One Hundred Years of Sea Power*, 393.
20 Ibid., 393.
21 Zumwalt asserted that for a NATO war in the mid-1970s, JCS plans called for moving seven million tons of military dry cargo and five million tons of military POL in the first six months. Of this total only 6 per cent could be moved by air. Of note, Zumwalt states these figures are consistent with the Navy's experience in Southeast Asia, where 96 per cent was moved in ships. CNO's Project 60 Presentation to SECDEF, 6.
22 Ibid., 28.
23 Ibid., 6.
24 Ibid., 8.
25 Ibid., 9.
26 Such a strategy builds his work as a systems analyst in 1966, when he authored a report entitled 'Major Fleet Escort Study' – establishing what type of replacement destroyer would be needed for sea control missions. See Norman Friedman, 'Elmo Russell Zumwalt Jr', Robert Love, Jr (ed), *The Chiefs of Naval Operation* (Annapolis, MD: United States Naval Institute Press, 1980), 366–7.
27 Zumwalt, *On Watch*, 76.
28 Ibid., 76–7.
29 Zumwalt became familiar with the surface missiles when he conducted a study of surface-to-surface missiles which led directly to the development of the Harpoon anti-ship missile. Friedman, 'Elmo Russell Zumwalt Jr', 367.
30 Zumwalt, *On Watch*, 72.
31 Zumwalt first articulated this high-low concept as a captain with his article, 'A Course for Destroyers', US Naval Institute Press *Proceedings* (November 1962), 27–39. His central theme is 'destroyer trends of the present era tend to confirm that it is no longer feasible to construct in adequate numbers a true general purpose [war]ship' (35). See also the discussion in Zumwalt's book, *On Watch*, 71–7.
32 Zumwalt, *On Watch*, 72.
33 CNO's Project 60 Presentation to SECDEF, 11.

34 Besides building and operating a dual-mission carrier for both power projection and sea control, Zumwalt also planned on placing existing helicopters on destroyers for the purpose of antisubmarine warfare. This program eventually evolved into the highly-successful LAMPS helicopter program. Zumwalt also proposed deploying one hydrofoil gunboat (PHG) to the Mediterranean to test its suitability in tracking Soviet missile ships, which were trailing US carriers and other major combatants. Another Project 60 concept was the employment of nuclear attack submarines as surface task group escorts with the submarine under the control of the battle group commander, supported by an embarked submarine officer. Finally, Zumwalt proposed beefing-up surface ship construction. To enhance the surface ship capability for the sea control mission, in the face of the Soviet anti-ship missile, Zumwalt proposed building surface ships that were air-capable. The goal of building these 'Surface Effect Ships' was to develop ships up to 10,000 tons and greater, with sustained speed ranges of 80 to 150 knots, by the 1980s. See Sands, *On His Watch*, 109.

35 Ibid., 25.

36 Ibid., 20.

37 Ibid., 36.

38 Zumwalt recounts in his book that he lobbied 485 senators and congressmen for the 1973 Trident vote. See Zumwalt, *On Watch*, 160–3.

39 Sands, *On His Watch*, 46.

40 A fourth non-warfighting baron was created for logistics support. See Ibid., 47.

41 Zumwalt, *On Watch*, 44–5.

42 Ibid., 45.

43 Sands, *On His Watch*, 69.

44 Stansfield Turner, *Secrecy and Democracy* (New York: Houghton Mifflin, 1985), 11. Also see John Lehman's discussion of 'Anti-Zumwalt Bloodshed Following Zumwalt's retirement', in John F. Lehman Jr, *Command of the Seas* (New York: Charles Scribner's Sons, 1998). Also see Sand's discussion of anti-Zumwalt purges, *On His Watch*, 70.

45 Zumwalt, *On Watch*, 53.

46 Sands, *On His Watch*, 76.

47 Ibid., 70.

48 Captain W. Lewis Glenn Jr, USN, *Reminiscences by Staff Officers of Admiral Elmo Zumwalt, Jr, US Navy*, vol. 1, US Naval Institute, Oral History, 1989.

49 Zumwalt, 'A Course for Destroyers', 27–39.

50 Sands, *On His Watch*, 20.

51 Ibid., 19. Sands write in footnote 35, 'David Rosenberg compares the effort by Admiral Zumwalt with that of Admirals Ernest J. King in 1942 and Forrest Sherman in 1949. Both entered the office wishing to change the course of the Navy, and both faced challenges at least and probably greater than that faced by Admiral Zumwalt. Yet neither laid out and publicized what he attempted to do to the extent that Admiral Zumwalt did.' Also see David Rosenberg, *Project 60: Twelve Years Later*, Enclosure (2) 'History of the Project 60 Effort.' (Chief of Naval Operations Special Project, 23 July 1982).

52 Sands, *On His Watch*, 23.

53 Admiral Worth Bagley, *Reminiscences by Staff Officers of Admiral Elmo Zumwalt, Jr, US Navy*, 315.

14 US Navy disruptive innovation: maritime action groups – surface land attack warfare

1 Generally speaking there are two types of Tomahawk missiles – nuclear and conventional. The focus of this case study is the tactical Tomahawk, which is a cruise missile that carries a 1,000 pound conventional warhead for a distance of up to 1,400 miles. It can be fired from almost any type of ship including submarines. Its speed is

subsonic, but because it is small and flies close to the ground, radar detection is difficult. Because of its subsonic speed, defenders can shoot it down, but unlike manned aircraft, it leaves no pilot or crewman to become a prisoner of war. See Frank Uhlig Jr, *How Navies Fight: The US Navy and Its Allies* (Annapolis, MD: US Naval Institute Press, 1994), 390. Also see letter from Rear Admiral Locke to Captain Bradd Hayes dated 28 August 2001.

2 Bradd Hayes paraphrases the Maritime Action Group argument by noting 'how six destroyers and cruisers could fire enough missiles to destroy a division of 750 enemy tanks and still have enough left to ward off an enemy aircraft or missile response'. See Hayes and Smith, *The Politics of Naval Innovation* 90, fn 7, who cites Thomas E. Ricks, 'How Wars Are Fought Will Change Radically, Pentagon Planner Says', *Wall Street Journal* (15 July 1994), 1.

3 See Hayes and Smith, *The Politics of Naval Innovation*, 86.

4 Simply asserting that TLAM and Aegis ships can outperform strike aircraft would be misleading since from a cost-benefit perspective each TLAM missile cost $1,000,000 while an equivalent 500-pound bomb from a strike aircraft cost $500. Also, the pilot of the strike aircraft can choose not to drop the bomb and return to the carrier and to use it at another time.

5 See Hayes, *The Politics of Naval Innovation*, 88.

6 Edward Smith notes that...*From the Sea* provided a new strategic concept, not a force plan, a new maritime strategy, or a new naval doctrine. 'It [... *From the Sea*] defined the post-Cold War need for a flexible littoral strategy, outlined the types of capabilities naval forces would require to implement it in a joint context, and called for development of appropriate naval strategies and tactics.' See Edward Smith, 'From the Sea: The Process of Defining a New Role for Naval Forces in the post-Cold War World', in Peter Trubowitz, Emily O. Goldman, and Edward Rhodes (eds), *Politics of Strategic Adjustment: Ideas, Institutions, and Interests* (New York: Columbia Press, 1999), 267. Colonel Gary Anderson, *Beyond Mahan: A Proposal for a US Naval Strategy in the Twenty-First Century* (Newport Paper #5, Naval War College, 1993), vii.

7 Robert J. Art and Stephen Ockenden, 'The Domestic Politics of Cruise Missile Development, 1970–1980', in Richard K. Betts (ed), *Cruise Missiles: Technology, Strategy, and Politics*, (Washington, DC: Brookings Institution, 1981), 360.

8 See Hayes, *The Politics of Naval Innovation*, 88.

9 Admiral William A. Owens, USN, *High Seas* (Annapolis, MD: United States Naval Institute Press, 1995), 115.

10 See Smith, 'From the Sea', 267.

11 Anderson, *Beyond Mahan,* vii.

12 See Smith, 'From the Sea', 272.

13 Gregory Engel, 'Cruise Missiles and the Tomahawk', in Bradd Hayes and Douglas Smith, *The Politics of Naval Innovation* (Newport, RI: Naval War College Press, 1994).

14 These air-to-surface missiles included the AS-4 as well as the more capable AS-6 and AS-9 missiles. Also see Thomas Hone, Douglas Smith, Roger Smith, and Roger Easton, 'Aegis-Evolutionary or Revolutionary Technology?', in Hayes and Smith (eds), *The Politics of Naval Innovation*, 45.

15 Ibid., 46.

16 Ibid., 47.

17 Ibid., 50.

18 In 1974, the Naval Ship Systems Command merged with Naval Ordnance to become the Naval Sea Systems Command (NAVSEA). The Navy promoted Captain Meyer to rear admiral and he managed both Aegis and NAVSEA's Surface Combat Systems Division. These new responsibilities were important for Meyer because he not only managed Aegis, but also the design of a destroyer-size Aegis ship, and the other Aegis cruiser. Hone, Smith, Smith, and Easton, 'Aegis', 52.

19 Ibid., 52.
20 Ibid., 60.
21 Steve Meyer, as quoted in Hone, Smith, Smith, and Easton, 'Aegis', 52.
22 Hone, Smith, Smith, and Easton, 'Aegis', 71.
23 Engel, 'Cruise Missiles and the Tomahawk', 18.
24 Zumwalt's efforts eventually lead to the development of the Harpoon ship-to-ship cruise missile that had a general range of 60 or so nautical miles. See Engel, 'Cruise Missiles and the Tomahawk', 17.
25 The Harpoon has an advertised range of 60 or so miles, while Tomahawk has an advertised range four times. See Engel, 'Cruise Missiles and the Tomahawk', 18. Interview with Admiral Zumwalt, former Chief of Naval Operations, Washington DC, 28 May 1993.
26 Source is Rear Admiral Locke letter and comments to Captain Bradd Hayes dated 28 August 1994. This letter/comments clarified an interview Locke had with Commander Engel on May 5, 1993, cited in Hayes, *The Politics of Naval Innovation*.
27 An ICBM is a ballistic missile. They are ballistic from the way they fly. The flight of a ballistic missile has two parts. During the first part, the missile's rocket engine blasts it onto its planned course and gives it the desired speed. After a short time, the engine shuts off. The missile then coasts through the second part of the flight until it drops down on its target. A ballistic missile is guided only during the first part of flight. The Tomahawk cruise missile, however, flies the entire course under power from its engine and under the control of its guidance system. It does not fly in a ballistic path, but usually at a low fixed altitude.
28 The product champions of the Sea Launched Cruise Missile argued that it did not have the basing vulnerability of the ICBM and that it had an advantage over the mobile land-based ICMB then being considered because of its unique flight profile and smaller launch signature. See Engel, 'Cruise Missiles and the Tomahawk', 19.
29 Ibid., 21.
30 Source is Rear Admiral Locke letter and comments to Captain Bradd Hayes dated 28 August 1994.
31 Locke comments to Hayes, 28 August 1994, 16. Sea-launched cruise missiles with nuclear warheads were not new and would only be a small part of the US nuclear arsenal. As Locke notes, 'The threat of low flying cruise missiles would make the Soviet air defense system obsolete but would only marginally improve our strategic deterrence. An anti-ship version would not be new either. Its principal advantage over the Navy's existing anti-ship weapon, the Harpoon, was greater range', 25.
32 Locke comments to Hayes, 28 August 1994, 17.
33 Ibid.
34 Gregory Engel paraphrasing Rear Admiral Locke's testimony to Congress, Engel, 'Cruise Missiles and the Tomahawks', 28.
35 Ibid., 28.
36 Ibid., 29.
37 Ibid., 29.
38 Ibid., 30.
39 Ibid., 31.
40 Locke comments to Hayes, 28 August 1994, 42.
41 Ibid., 43.
42 The consequence of Watkins missing an earlier opportunity to develop Tomahawk with GPS resulted in no Tomahawks with GPS available for Desert Storm, In comparison, the Air Force used conventional air-launched cruise missiles with GPS in that campaign. Locke comments to Hayes, 28 August 1994, 44.
43 Engel, 'Cruise Missiles and the Tomahawks', 42.
44 Other warfighting concepts contained in...*From the Sea* that the Naval Services still are working on institutionalizing are Operational Maneuver From the Sea and Ship to Objective Maneuver.

45 Smith, 'From the Sea', 269.
46 Hayes, *The Politics of Naval Innovation*, 90.
47 Smith, 'From the Sea', 271.
48 As quoted in Smith, 'From the Sea', 271.
49 Author interviews with Bradd Hayes, 28 March 2001.
50 Hayes, *The Politics of Naval Innovation*, 91.
51 Art and Ockenden, 'The Domestic Politics of Cruise Missile Development', 360, 388–9, 392.

15 US Navy disruptive innovation: tactical collaborative network and a summary of disruptive Navy cases

1 Naval Doctrine Publication (NDP 1), *Naval Warfighting*, is a naval version of the Marine Corps maneuver warfare doctrine manual, *Warfighting*.
2 Marine Corps Doctrine MCDP, *Warfighting* (Quantico, VA, 1997), is the Marine Corps version of maneuver warfare.
3 Ronald O'Rourke coined 'speed of command' in his CRS Report for Congress, 'Navy Network-Centric Warfare Concept: Key Programs and Issues for Congress', 6 June 2001.
4 Admiral Dennis C. Blair, 'We Can Fix Acquisition', *United States Naval Institute Proceedings* (May 2002).
5 Rear Admiral Phil Balisle, 'CEC Provides Theater Air Dominance', *Naval Institute Proceedings* (May 2002), 4 (http://www.usni.org/Proceedings/Articles02/PRObalisle05.htm).
6 Cost figure is a Pentagon estimate. See Sandra Erwin, 'Stakes Are High in Competition for Naval Air-Defense Program', *Defense News* (21 June 2001), 2.
7 Edward J. Walsh, 'Industry and the Navy Push Technology for Future Air Defense', *Military Information Technology Online Archives*, 3 (www.mit-kmi.com/Archives/4_3_MIT/4_3_Art8.cfm).
8 Ronald O'Rourke, 'Navy Network-Centric Warfare Concept: Key Programs and Issues for Congress', CRS Report for Congress, 6 June 2001, 4.
9 See Erwin, 'Stakes Are High in Competition for Naval Air-Defense Air-Defense Program'.
10 See Walsh, 'Industry and the Navy Push Technology for Future Air Defense'.
11 Erwin, 'Stakes Are High in Competition for Naval Air-Defense Program', 3.
12 Ibid.
13 Walsh, 'Industry and the Navy Push Technology for Future Air Defense', 5.
14 Warren Citrin, *A Brief History of the Tactical Component Network* (15 May 2003, original held by author).
15 Ibid.
16 Ibid.
17 But technical requirements were not enough. A network, by its nature, is a community endeavor and must be structured in a way that encourages participation and interoperability. The DOD's inattention to this aspect is famously responsible for the nearly total lack of interoperability within and among the military services. Additionally, a network foundation, which TCN® was intended to be, must support third-party development of compatible applications. So a set of business and acquisition principles had to be followed to complete the network vision. These were as follows:

   • The network foundation applications should be provided to the DOD as complete products. That is, software should not be acquired in the same way as an aircraft carrier, with big proposal efforts that reward nonsense such as Capability Maturity Model (CMM) levels and other big-industry buy-ins that have nothing to do with product quality and are designed to eliminate small business form competing for defense contracts. Rather, the product should be funded by the vendor and presented as a complete item to the customer for test and evaluation against documented requirements.

- The foundation software products should be sold on a licensed basis, as is done in the commercial world. This would encourage small innovative businesses to become involved in the process because of the build once, sell many times license approach. The large defense contractors, with their purchased CMM levels and underbid pricing, have a perfectly abysmal record in providing software.
- Design the software products around an independent, component-based model that does not encroach on the schedule and programmatic interests of the individual combat system engineering organizations. Do not require each network client to adhere to the operational requirements of the network, as in CEC. Rather, make the network flexible enough to adhere to the operational requirements of the clients.

18 Citrin, 'A Brief History of the Tactical Component Network'.
19 See Greg Schneider, 'Scuttled by the Process', *The Washington Post* (29 August 2001), 3.
20 Ibid., E01.
21 Ibid., E02.
22 Interview with Mun Fenton, Office of Naval Research, Missile Defense Program Director for Future Naval Capabilities, 14 October 2001.
23 Senator Inouye from Hawaii as well as Maryland Congressman Hoyer have been strong proponents of TCN®. Interview with Mun Fenton, 21 October 2001.
24 Mark Trainer, *Notes on TCN® Development* (25 May 2003. Original held by author).
25 Ibid.
26 Ibid.
27 Ibid.
28 Schneider, 'Scuttled by the Process', E01.
29 Milestone 3 is the coveted authorization from the DOD for full rate production.
30 Secretary of Defense Donald Rumsfeld, 'DOD Acquisition and Logistics Excellence Week Kickoff – Bureaucracy to Battlefield', Remarks as Delivered at the Pentagon, 10 September, 2001 (www.defenselink.mil/cgi-bin/dlprint.cgi).

16 Conclusion

1 Edward Marolda and Robert Schneller Jr, *Shield and Sword: The United States Navy and the Persian Gulf War* (Annapolis, MD: United States Naval Institute Press, 2001), 258–359.
2 Ibid., 359.
3 See Table 3 for a summary of all the variable results.
4 Causality for manipulating interservice competition to cause doctrinal innovation was demonstrated by Owen Cote. See Cote, 'The Politics of Innovative Doctrine: The US Navy and Fleet Ballistic Missiles', 4.
5 Rosen, *Winning the Next War*, 5.

Bibliography

Document collections

Archives of the United States Marine Corps Combat Development Command, Marine Corps Research Center, Quantico, VA.

Archives of the United States Marine Corps, Marine Corps Historical Center, Navy Yard, Washington, DC.

Archives of the United States Navy, Naval Historical Center, Navy Yard, Washington, DC.

Modern Military Branch, National Archives, Washington, DC.

Books

Agawa, H., *The Reluctant Admiral: Yamamoto and the Imperial Navy* (New York: Kodansha International, 1979).

Alexander, Joseph H. and Merrill L. Bartlett, *Sea Soldiers in the Cold War: Amphibious Warfare 1945–1991* (Annapolis, MD: United States Naval Institute Press, 1995).

Alford, Robert, *The Craft of Inquiry: Theories, Methods, Evidence* (New York: Oxford Press, 1998).

Allison, Graham and Philip Zelikow, *Essence of Decision: Explaining the Cuban Missile Crisis* (2nd edn, New York: Longman Press, 1999).

Arpee, Edward, *From Frigates to Flat-Tops: The Story of the Life and Achievements of Rear Admiral William Adger Moffett, USN* (Lake Forest, IL: 1953).

Art, Robert J. and Stephen Ockenden, 'The Domestic Politics of Cruise Missile Development, 1970–1980', in Richard K. Betts (ed), *Cruise Missiles: Technology, Strategy, and Politics* (Washington, DC: Brookings Institution, 1981).

Asprey, Robert, *Once a Marine: The Memoirs of General A.A. Vandergrift as Told to Robert Asprey* (New York: Robert & Company, 1964).

Avant, Deborah D, *Political Institutions and Military Change: Lessons From Peripheral Wars* (Ithaca, NY: Cornell University Press, 1994).

Baer, George W, *One Hundred Years of Sea Power: The US Navy, 1890–1990* (Stanford, CA: Stanford University Press, 1994).

Ballendorf, Dirk and Merrill Bartlett, *Pete Ellis: An Amphibious Warfare Prophet 1880–1923* (Annapolis, MD: United States Naval Institute Press, 1997).

Barbey, Vice Admiral Daniel, *MacArthur's Amphibious Navy: Seventh Amphibious Force Operations 1943–1945* (Annapolis, MD: United States Naval Institute Press, 1969).

Bartlett, Merrill, *Lejeune: A Marine's Life 1867–1942* (Annapolis, MD: United States Naval Institute Press, 1991).

Bennett, Andrew, *Condemned to Repetition: The Rise, Fall, and Reprise of Soviet-Russian Military Interventionism, 1973–1996* (Cambridge, MA: MIT Press, 1999).

Betts, Richard (ed), *Cruise Missiles: Technology, Strategy, Politics* (Washington, DC: The Brookings Institute Press, 1981).

Bowen, Harold G, *Ships, Machinery and Mossbacks* (Princeton, NJ: Princeton University Press, 1954).

Brodie, Bernard, 'Technological Change, Strategic Doctrine, and Political Outcomes', in Klaus Knorr (ed), *Historical Dimensions of National Security Problems* (Lawrence, KS: University of Kansas Press, 1976).

Christensen, Clay, *The Innovator's Dilemma: When New Technologies Cause Great Firms to Fail* (Cambridge, MA: Harvard Business School, 1997).

Cinto, Robert M, *The Path to Blitzkrieg: Doctrine and Training in the German Army 1920–1939* (New York: Lynne Rienner Publishers, 1999).

Clark, J.J. and Clark G. Reynolds. *Carrier Admiral* (New York: McKay Press, 1967).

Clifford, Lieutenant Colonel Kenneth, *Progress and Purposes: A Developmental History of the US Marine Corps* (Washington, DC: History and Museums, 1973).

Cooper, Mathew, *The German Army, 1933–1945* (Chelsea, MI: Scarborough House Publishers, 1978).

Corum, James S., *The Roots of Blitzkrieg* (Lawrence, KS: University of Kansas Press, 1992).

Corum, James S., 'A Comprehensive Approach to Change: Reform in the German Army in the Interwar Period', in Harold Winton and David Mets (eds), *The Challenge of Change: Military Institutions and New Realities, 1918–1941* (Omaha, NE: University of Nebraska Press, 2000).

Creveld, Martin van, *Supplying War, Logistics from Wallenstein to Patton* (New York: Cambridge University, 1977).

Creveld, Martin van, *Technology and War* (New York: The Free Press, 1989).

Diamond, Jared, *Guns, Germs, and Steel* (New York: Norton Press, 1999).

Ellis, John, *The Social History of the Machine Gun* (Baltimore, MD: Johns Hopkins University Press, 1996).

Engel, Gregory, 'Cruise Missiles and the Tomahawk', in Bradd Hayes and Douglas South (eds), *The Politics of Naval Innovation* (Newport, RI: Naval War College Press, 1994).

Evangelista, Matthew, *Innovation and the Arms Race: How the United States and the Soviet Union Develop New Military Technologies* (Ithaca, NY: Cornell University Press, 1988).

Evans, David C. and Mark R. Peattie, *Kaigun: Strategy, Tactics, and Technology in the Imperial Japanese Navy 1887–1941* (Annapolis, MD: United States Naval Institute Press, 1997).

Friedman, Norman, *Carrier Air Power* (New York: The Routledge Press, 1981).

Friedman, Norman, 'Elmo Russell Zumwalt Jr', in Robert Love Jr (ed), *The Chiefs of Naval Operation* (Annapolis, MD: United States Naval Institute Press, 1980).

Fuchida, Mitsuo, 'I Led the Air Attack on Pearl Harbor', in Paul Stillwell (ed), *Air Raid: Pearl Harbor!* (Annapolis, MD: United States Naval Institute Press, 1981).

Fukudome, Admiral Shigeru, 'Hawaii Operation', in Paul Stillwell (ed), *Air Raid: Pearl Harbor!* (Annapolis, MD: United States Naval Institute Press, 1981).

Genda, Minoru, 'Evolution of Aircraft Carrier Tactics of the Imperial Japanese Navy', in Paul Stillwell (ed), *Air Raid: Pearl Harbor!* (Annapolis, MD: United States Naval Institute Press, 1981).

Gibson, James William, *The Perfect War: Technowar in Vietnam* (New York: The Atlantic Monthly Press, 1986).

Goldman, Emily, 'Mission Possible: Organizational Learning in Peacetime', in Peter Trubowitz, Emily O. Goldman, and Edward Rhodes (eds), *The Politics of Strategic Adjustment: Ideas, Institutions, and Interests* (New York: Columbia University Press, 1999).

Gordon, Michael and General Bernard Trainor, *The Generals' War* (New York: Little Brown, 1995).

Gray, Colin S., *Strategy For Chaos: Revolutions in Military Affairs and the Evidence of History* (London: Frank Cass Publishers, 2002).

Gudmundsson, Bruce I., *Stormtroop Tactics: Innovation In The German Army, 1914–1918* (New York: Praeger Publishers, 1989).

Guderian, Heinz, *Panzer Leader*, trans. Constantine Fitzgibbon (New York: First Da Capo, 1996).

Harris, J.P., *Men, Ideas and Tanks: British Military Thought and Armoured Forces, 1903–1939* (Manchester, England: Manchester University Press, 1995).

Hart, Gary and William S. Lind, *America Can Win* (New York: Alder and Alder Press, 1986).

Hartmann, Frederick H., *Naval Renaissance: The US Navy in the 1980s* (Annapolis, MD: United States Naval Institute Press, 1990).

Hayes, Bradd, 'Conclusions', in Bradd Hayes and Douglas Smith (eds), *The Politics of Naval Innovation* (Strategic Research Department, Research Report 4–94, US Naval War College, 4–94, 1994).

Heinl, Robert Jr, *Soldiers of the Sea: The United States Marine Corps, 1775–1962* (Annapolis, MD: United States Naval Institute Press, 1962).

Heinl, Robert Jr, *Victory at High Tide: The Inchon-Seoul Campaign* (New York: The Nautical & Aviation Publishing Company of America, 1979).

Heinl, Robert Jr, 'The US Marine Corps: Author of Modern Amphibious Warfare', in Merrill L. Bartlett (ed), *Assault From the Sea* (Annapolis, MD: United States Naval Institute Press, 1983).

Hooker, Richard Jr, (ed), *Maneuver Warfare: An Anthology* (Novato, CA: Presidio Press, 1993).

Hone, Thomas C., Norman Friedman, and Mark D. Mandeles. *American and British Aircraft Development: 1919–1944* (Annapolis, MD: United States Naval Institute Press, 1999).

Hone, Thomas, Douglas Smith, Roger Smith, and Roger Easton, 'Aegis – Evolutionary or Revolutionary Technology?', in Bradd Hayes and Douglas Smith (eds), *The Politics of Naval Innovation* (Newport, RI: Naval War College Press, 1994).

Hoyt, Edwin P., *Yamamoto: The Man Who Planned The Attack On Pearl Harbor* (Guilford, CT: The Lyons Press, 1990).

Hundley, Richard, *Past Revolutions Future Transformations* (National Defense Research Institute: RAND, 1999).

Ienaga, Saburo, *The Pacific War: 1931–1945: A Critical Perspective On Japanese Role In World II By a Leading Japanese Scholar* (New York: Pantheon Books, 1968).

Isely, Jeter and Philip Crowl, *The US Marines and Amphibious War: Its Theory and Its Practice in the Pacific* (Princeton, NJ: Princeton University Press, 1951).

Isenson, Raymond, 'Project Hindsight: An Empirical Study of the Sources of Ideas Utilized in Operational Weapons Systems', in William Gruber and Donald Marquis (eds), *Factors in the Transfer of Technology* (Cambridge, MA: MIT University Press, 1969).

Johnson, David E, *Fast Tanks and Heavy Bombers: Innovation in the US Army, 1917–1945* (Ithaca, NY: Cornell University Press, 1998).

Katzenbach, Edward, 'The Horse Cavalry in the Twentieth Century: A Study in Policy Response', in Richard Head and Erwin Rokke (eds), *American Defense Policy* (Baltimore, MD: Johns Hopkins University Press, 1973).

Knox, MacGregor and Murray, Williamson (eds), *The Dynamics of Military Revolution 1300–2050* (New York: Cambridge University Press, 2001).

Kier, Elizabeth, *Imagining War: French and British Military Doctrine Between the Wars*, (Princeton, NJ: Princeton University Press, 1997).

Krepinevich, Andrew, *The Army and Vietnam* (Baltimore, MD: Johns Hopkins University Press, 1986).

Krulak, Victor, *First To Fight, An Inside View of the US Marine Corps* (Annapolis, MD: United States Naval Institute Press, 1999).

Layman, R.D., *Before the Aircraft Carrier: The Development of Aviation Vessels: 1849–1922* (Annapolis, MD: United States Naval Institute Press, 1989).

Layman, R.D., *Naval Aviation in the First World War: Its Impact and Influence* (Annapolis, MD: United States Naval Institute Press, 1996).

Legro, Jeffrey, *Cooperation Under Fire: Anglo-German Restraint During World War II* (Ithaca, NY: Cornell University Press, 1995).

Lehman, John F. Jr, *Command of the Seas* (New York: Charles Scribner's Sons, 1988).

Levine, Alan J., *The Pacific War: Japan versus the Allies* (New York: Praeger Publishers, 1995).

Lind, William S, *Maneuver Warfare Handbook* (Boulder, CO: Westview Press, 1985).

Marder, Arthur J., *Old Friends, New Enemies: The Royal Navy and the Imperial Japanese Navy: Strategic Illusions 1936–1941* (New York: Oxford Press 1981).

Marolda, Edward and Robert Schneller Jr, *Shield and Sword: The United States Navy and the Persian Gulf War* (Annapolis, MD: United States Naval Institute Press, 2001).

McBride, William M, *Technological Change and the United States Navy, 1865–1945* (Baltimore, MD: Johns Hopkins University Press, 2000).

Miller, David S., *War Plan Orange: The US Strategy to Defeat Japan: 1897–1945* (Annapolis, MD: United States Naval Institute Press, 1991).

MCDP 1 (originally FMF 1), *Warfighting* (Quantico, VA: USMC, 1989).

MCDP 1, *Warfighting* (Quantico, VA: USMC, 1997).

Melson, Charles D., *Condition Red: Marine Defense Battalions in World War II* (Washington, DC: Washington Marine Corps Historical Center, 1996).

Millett, Allan R., *Semper Fi: The History of the United States Marine Corps* (New York: The Free Press, 1991).

Millett, Allan R., *In Many a Strife: General C. Thomas and the US Marine Corps: 1917–1956* (Annapolis, MD: United States Naval Institute Press, 1993).

Millett, Allan, 'Assault From The Sea', in Williamson Murray and Allan Millett (eds), *Military Innovation in the Interwar Period* (New York: Cambridge Press, 1996).

Montross, Lynn, *Cavalry of the Sky: The Story of US Marine Combat Helicopters* (New York: Harper & Brothers Press, 1954).

Morison, Elting, *Admiral Sims and the Modern American Navy* (Boston, MA: Houghton Mifflin, 1942).

Morison, Elting, *Men, Machines and Modern Times* (Boston, MA: MIT Press, 1966).

Mosier, John, *The Myth of the Great War* (New York: Harper Collins Publishers, 2001).

Moy, David, *War Machines: Transforming Technologies in the US Military, 1920–1940* (College Station, TX: Texas A&M University Press, 2001).

Murray, Williamson, *The Change in the European Balance of Power, 1938–1939: The Path to Ruin* (Princeton, NJ: Princeton University Press, 1984).

Murray, Williamson, 'Innovation: Past and Future', in Williamson Murray and Allan Millett (eds), *Military Innovation in the Interwar Period* (New York: Cambridge University Press, 1996).

Murray, Williamson and Allan Millet, *Military Innovation in the Interwar Period* (New York: Cambridge University Press, 1996).

Murry, Williamson, 'Armored Warfare', in Williamson Murray and Allan Millett (eds), *Military Innovation in the Interwar Period* (Cambridge, NY: Cambridge University Press, 1996).

Murray, Williamson and Allan R. Millett, *A War To Be Won: Fighting the Second World War* (Boston, MA: Belknap Press of Harvard University Press, 2000).

Neustadt, Richard, *Presidential Power* (New York: The Free Press, 1991).

Nomura, Kichisaburo, 'Stepping-Stones to War', in Paul Stillwell (ed), *Air Raid: Pearl Harbor!* (Annapolis, MD: United States Naval Institute Press, 1981).

O'Hanlon, Michael, *Technological Change and the Future of Warfare* (Washington, DC: Brookings Institution, 2000).

Okumiya, Masatake and Jiro Horikoshi with Martin Caidin, *Zero: The Story of Japan's Air War in the Pacific* (New York: ibooks – Simon & Schuster, 1956).

Owens, William A., *High Seas* (Annapolis, MD: United States Naval Institute Press, 1995).

Peattie, Mark R., *Sunburst: The Rise of Japanese Naval Air Power, 1909–1941* (Annapolis, MD: United States Naval Institute Press, 2001).

Perla, Peter P., *The Art of Wargaming: A Guide for Professionals and Hobbyists* (Annapolis, MD: United States Naval Institute Press, 1990).

Posen, Barry, *The Sources of Military Doctrine: France, Britain, and Germany between the World Wars* (Ithaca, NY: Cornell University Press, 1984).

Prange, Gordon W., *At Dawn We Slept: The Untold Story of Pearl Harbor* (New York: Penguin Books, 1981).

Quinn, James Brian, *Strategies For Change: Logical Incrementalism* (New York: Richard Irwin Press, 1980).

Reynolds, Clark G., 'William A. Moffett: Steward of the Air Revolution', in James C. Bradford (ed), *Admirals of the New Steel Navy: Makers of the American Naval Tradition, 1880–1930* (Annapolis, MD: United States Naval Institute Press, 1990).

Reynolds, Clark G., *Admiral John H. Towers: The Struggle for Naval Air Supremacy* (Annapolis, MD: United States Naval Institute Press, 1991).

Rosen, Stephen P., *Winning the Next War: Innovation and the Modern Military* (Ithaca, NY: Cornell University Press, 1991).

Ross, Andrew L., 'The Dynamics of Military Technology', in David Dewitt, David Haglund, and John Kirtland (eds), *Building a New Global Order: Emerging Trends in International Security* (Oxford, England: Oxford University Press, 1993).

Sakai, Saburo, with Martin Caidin and Fred Saito, *Samurai!* (New York: ibooks – Simon & Schuster, 1957).

Sakaida, Henry, *Imperial Japanese Navy Aces 1937–45* (Oxford, GB: Military Book Club/Osprey Aerospace, 1998).

Smith, Edward, 'From the Sea: The Process of Defining a New Role for Naval Forces in the post-Cold War World', in Peter Trubowitz, Emily O. Goldman, and Edward Rhodes (eds), *Politics of Strategic Adjustment: Ideas, Institutions, and Interests* (New York: Columbia Press, 1999).

Smith, Holland and Percy Finch, *Coral and Brass* (New York: Chaoles Scribner's, Sons, 1949).

Simmons, Edwin Howard, *The United States Marines: A History* (Annapolis, MD: United States Naval Institute Press, 1992).

Spector, Ronald, *Eagle Against the Sun: The American War With Japan* (New York: First Vintage Books Edition, November 1985).

Stanley II, Colonel Roy, USAF, *Prelude to Pearl Harbor: War in China, 1937–41 Japan's Rehearsal for World War II* (New York: Charles Scribner's Sons, 1982).

Till, Geoffrey, 'Adopting The Aircraft Carrier: The British, American, and Japanese Case Studies', in Williamson Murray and Allan Millett (eds), *Military Innovation in the Interwar Period* (New York: Cambridge University Press, 1996).

Trimble, William, *Admiral William A. Moffett: Architect of Naval Aviation,*(Washington, DC: Smithsonian Institution Press, 1994).

Turner, Stansfield, *Secrecy and Democracy* (New York: Houghton Mifflin, 1985).

Tushman, Michael and Charles O'Reilly, *Winning Through Innovation* (Boston, MA: Harvard Business School, 1997).

Uhlig, Frank Jr, *How Navies Fight: The U.S. Navy and Its Allies* (Annapolis, MD: United States Naval Institute Press, 1994).

Wagner, Ray, *Prelude to Pearl Harbor: The Air War in China: 1937–1941* (San Diego, CA: San Diego Aero-Space Museum, 1991).

Walt, Stephen, *The Origins of Alliances* (Ithaca, NY: Cornell University Press, 1987).

Watts, Barry and Williamson Murray, 'Military Innovation in Peacetime', in Williamson Murray and Allan Millet (eds), *Military Innovation in the Interwar Period* (New York: Cambridge University Press, 1996).

White, Charles Edward, *The Enlightened Soldier* (New York: Praeger, 1989).

Wildenberg, Thomas, *Destined For Glory: Dive Bombing, Midway, and the Evolution of Carrier Airpower* (Annapolis, MD: United States Naval Institute Press, 1998).

Wildenberg, Thomas, *All The Factors of Victory: Admiral Joseph Mason Reeves and the Origins of Carrier Airpower* (Dulles, VA: Brassey's, 2003).

Wilson, James Q., *Bureaucracy* (New York: Basic Books, 1989).

Winton, Harold and David Mets (eds), *The Challenge of Change: Military Institutions and New Realities, 1918–1941* (Omaha, NE: University of Nebraska Press, 2000).

Wooldridge, E.T., *Into the Jet Age* (Annapolis, MD: United States Naval Institute Press, 1995).

Zisk, K.M., *Engaging the Enemy: Organizational Theory and Soviet Military Innovation, 1955–1991* (Ithaca, NY: Cornell University Press, 1993).

Zumwalt Elmo, Jr, *On Watch* (Arlington, VA: Zumwalt & Associates, 1976).

Zumwalt Elmo Jr, and Elmo Zumwalt III, with John Pekkanen, *My Father, My Son* (New York: Macmillan, 1986).

Articles and reports

Anderson, Gary, 'Beyond Mahan: A Proposal for a US Naval Strategy in the Twenty-First Century', *Newport Paper Number 5* (1993).

Balisle, Rear Admiral Phil, 'CEC Provides Theater Air Dominance', *United States Naval Institute Proceedings* (May 2002).

Boyd, Carl, 'Japanese Military Effectiveness: The Interwar Period', in Allan Millett and Williamson Murray (eds), *Military Effectiveness*, vol. 2, *The Interwar Period* (Winchester, MA: Unwin Hyman, Inc., 1988).

Cary, Peter,'The fight to change how America fights', *US News & World Report* (6 May 1991).

Clover, Kevin, 'Maneuver Warfare: Where Are We Now?', *Marine Corps Gazette* (February 1988).

Cohen, Eliot 'Defending America in the Twenty-first Century', *Foreign Affairs* (November/December 2000).

Davis, Vincent, *The Politics of Innovation: Patterns in Navy Cases* (Monograph Series in World Affairs, Vol 4, No. 3, University of Denver, 1967).

Desch, M.C. 'Culture Clash: Assessing the Importance of Ideas in Security Studies', *International Security* (Summer 1998).

Doerr, Captain P. J., 'CWC Revisited', *United States Naval Institute Proceedings* (April 1986).

Eisenhardt, Kathleen, 'Building Theories from Case Study Research', *Academy of Management Review No. 4* (1989).

Erwin, Sandra, 'Stakes Are High in Competition for Naval Air-Defense Program', *Defense News* (21 June 2001).

Evans, David, 'Marines have the last word on one who did it his way', *Chicago Tribune* (12 April 1991).

Evans, David and Mark Peattie, 'Ill Winds Blow', *United States Naval Institute Proceedings* (October 1997).

Fialka, John, 'A Very Old General May Hit the Beach With Marines', *The Wall Street Journal* (9 January, 1991).

Ferris, John, 'A British Unofficial Aviation Mission and Japanese Naval Developments 1919–1929', *Journal of Strategic Studies* (September 1982), 416–39.

Fioravanzo, Vice Admiral Giuseppe, 'The Japanese Military Mission to Italy in 1941', *United States Naval Institute Proceedings* (January 1956), 24–31.

Fuller, Ben, 'The Mission of the Marine Corps', *Marine Corps Gazette* (November 1930).

Genda, Minoru, 'Tactical Planning in the Imperial Japanese Navy', *Naval War College Review* (October 1969), 45–50.

Greenwood, Colonel John E., USMC (ret), 'FMFM 1: The Line of Departure', *United States Naval Institute Proceedings* (May 1990).

Hammond, Jr, James W. 'Lejeune of the Naval Service', *United States Naval Institute Proceedings* (November 1981).

Henderson, Rebecca, 'Managing Innovation in the Information Age', *Harvard Business Review* (January–February 1994).

Henderson, Rebecca and Kim B. Clark, 'Architectural Innovation: The Reconfiguration of Existing Product Technologies and the Failure of Existing Firms', *Administrative Science Quarterly*, 35 (March 1990).

Hirama, Rear Admiral Yoichi, 'Japanese Naval Preparations for World War II', *Naval War College Review* (Spring 1991).

Holloway, James L. III, 'The Aircraft Carrier: An Overview', *Wings of Gold* (special report, summer 1987).

Hone, Thomas C. and Mark D. Mandeles, 'Interwar Innovation in Three Navies: U.S. Navy, Royal Navy, Imperial Japanese Navy', *Naval War College Review* (Spring 1987), 63–83.

Hughes, Captain Wayne P. USN (ret), 'Naval Tactics And Their Influence on Strategy', *Naval War College Review* 39, (January–February 1986).

Isaacson, Jeffrey, Christopher Layne, and John Arguilla, *Predicting Military Innovation* (National Defense Research Institute, Santa Monica, CA: RAND 1999).

Isom, Dallas Woodbury, 'The Battle of Midway', *Naval War College Review* (Summer 2000).

Jackson, Tim, *Warfighting Skills Program Abstract* (Washington, DC: Marine Corps Institute, 28 September 1989).

Koda, Yoji, 'A Commander's Dilemma: Admiral Yamamoto and the "Gradual Attrition" Strategy', *Naval War College Review* (Autumn 1993).

Krulak, Charles C., 'Editorial', *Marine Corps Gazette* (July 1998).

Lautenschlager, Karl, 'Technology and the Evolution of Naval Warfare', *International Security* (Fall 1983).

Lejeune, John, 'The United States Marine Corps', *United States Naval Institute Proceedings* (October 1955).

Lind, William S., 'Defining Maneuver Warfare for the Marine Corps', *Marine Corps Gazette* (March 1980).

Lind, William S., 'Missing the Boat: A Response to Generals Knutson, Hailston, and Bedard', *The Marine Corps Gazette* (October 2000).

Lind, William S. and Jeffrey Record, 'The Marines' Brass is Winning the Battle But Losing the Corps', *The Washington Post* (28 July 1985).

Lupfer, Timothy, in *Dynamics of Doctrine: The Changes in German Tactical Doctrine During the First World War* (Combat Studies Institute: US Army Command and General Staff College, July 1981), 41.

Moore, Lynne Lucius, '*Shinano*: The Jinx Carrier', *United States Naval Institute Proceedings* (February 1953), 142–9.

Moore, Richard S., USMC, 'Blitzkrieg From the Sea', *Naval War College Review* (November–December 1983).

Murphy, Michael, 'The ABA Way: For Pure Entertainment, American Basketball Association Was a Slam Dunk', *Houston Chronicle* (4 February 1996).

Murray, Williamson, 'Innovation: Past and Future', *Joint Force Quarterly* (Summer 1994).

Murray, Williamson, *Experiment in the Period Between the Two World Wars: Lessons for the Twenty-First Century*, Institute for Defense Analyses (November 2000).

Ohmae, Toshikazu and Roger Pineau, 'Japanese Naval Aviation', *United States Naval Institute Proceedings* (December 1972), 70–7.

O'Rourke, Ronald, 'Navy Network-Centric Warfare Concept: Key Programs and Issues for Congress', Washington, DC: United States Congress, CRS Report for Congress, 6 June 2001.

Pierce, Terry C., 'Operational Maneuver From the Sea', *United States Naval Institute Proceedings* (August 1994).

Pierce, Terry C., 'Teaching Elephants to Swim', *United States Naval Institute Proceedings* (May 1998).

Pierce, Terry C., 'Sunk Cost Sink Innovation', *United States Naval Institute Proceedings* (May 2002).

Polk, Major Robert, USA, *Critique of the Boyd Theory – Is It Relevant to the Army?*, Monograph, School of Advanced Military Studies (United States Army Command and General Staff College, Fort Leavenworth, KS, 1999).

Porch, Douglas, 'Military "Culture" and the Fall of France in 1940,' *International Security* 24, 4 (Spring 2000).

Ricks, Thomas E., 'How Wars Are Fought Will Change Radically, Pentagon Planner Says', *Wall Street Journal* (15 July 1994).

Romanelli, Elaine and Michael Tushman, 'Organizational Transformation as Punctuated Equilibrium: An Empirical Test', *Academy of Management Journal*, 37 (1994).

Rosen, Stephen P., 'New Ways of War: Understanding Military Innovation', *International Security*, 13, 1 (Summer 1988).

Rosenberg, David, *Project 60: Twelve Years Later*, Enclosure (2) 'History of the Project 60 Effort'. (Chief of Naval Operations Special Project) (23 July 1982).

Russell, Major John H., 'The Preparation of War Plans for the Establishment and Defense of a Naval Advance Base', Lecture given at the US Naval War College, 1910, (Naval War College Archives, Newport, RI).

Russell, Major John H., 'A Plea For a Mission and Doctrine', *Marine Corps Gazette* (June 1916).

Sands, Jeffrey. *On His Watch: Admiral Zumwalt's Efforts to Institutionalize Strategic Change*, Center for Naval Analysis, CRM 93–22 (July 1993).

Scharfen, John, interview with Major General Alfred Gray, Commanding General 2nd Marine Division, 'Tactics and Theory of Maneuver Warfare', *Amphibious Warfare Review* (July 1984).

Schmitt, John, 'A Critique of the Hunter Warrior Operational Concept', *Marine Corps Gazette* (June 1998).

Seno, Sadao, 'A Chess Game with No Checkmate: Admiral Inoue and the Pacific War', *Naval War College Review* (January–February 1974).

Smith, Holland, USMC, 'The Development of Amphibious Tactics in the US Navy', *Marine Corps Gazette* (November 1946).

Till, Geoffrey, 'Adopting the Aircraft Carrier: The British, American, and Japanese case Studies', in Allan Millett and Williamson Murray (eds), *Military Innovation in the Interwar Period* (New York: Cambridge University Press, 1996).

Tushman, Michael, and Elaine Romanelli, 'Organizational Evolution: A Metaphorphosis Model of Convergence and Reorientation', *Research in Organizational Behavior*, 7 (1985).

Ulbrich, David J., 'Clarifying the Origins and Strategic Mission of the US Marine Corps Defense Battalion, 1898–1941', *War & Society*, 17, 2 (October 1999).

Walsh, Edward J., 'Industry and the Navy Push Technology for Future Air Defense', *Military Information Technology Online Archives*, [www.mit-kmi.com/Archives/4_3_MIT/4_3_Art8.cfm].

Zumwalt, Elmo, Jr, 'A Course for Destroyers', *U.S. Naval Institute Press Proceedings*, (November 1962), 27–39.

Documents

Citrin, Warren, *A Brief History of the Tactical Component Network*, 15 May 2003, original held by author.

Chmar, Lieutenant Colonel Mark, Lieutenant Colonel Stephen Cullen, Lieutenant Colonel David Ralston, and Lieutenant Colonel Michael Smith, *Peacetime Military Innovation: Getting Beyond Today*, (National Security Program Occasional Paper), John F. Kennedy School of Government, Harvard University, 1999.

Clifford, Kenneth, *Progress and Purpose: A Developmental History of the United States Marine Corps: 1900–1970*, Washington, DC, Marine Corps Historical Center Publication, 1973.

Commandant Marine Corps, declassified Secret Document from Commandant of the Marine Corps with Subject Title: *Compendium of Major Decisions in the Evolution of the Maritime Prepositioning Ships (MPS) Program*, dated 23 December 1983. (Washington, DC: Marine Corps Historical Center).

Commandant Marine Corps White Letter 2–81, 'Amphibious Operations and Maritime/Near Term Prepositioning Ships'. Located in Slide and Reports, Dec 1983, Maritime Prepositioning Force Study Background Material, Breckinridge Archive Files, (Quantico, VA: Marine Corps Research File).

Cropsey, Seth, Undersecretary of Navy, memo to Secretary of Navy, dated 1 April 1985.

Ellis, Earl, Advanced Base Operations in Micronesia 1921 (Marine Corps Research Center, Breckenridge Library Historical Amphibious File, File 165, Breckenridge Library, Quantico, VA).

Geiger, Lieutenant General Roy S., letter to CMC, dated 21 August 1946 (Serial 0265–46, Box 11, Accession No. 14051, Record Group 127, WNRC, Suitland, MD).

Hart, Gary and William S. Lind. Detailed Outline of *Military Reform*, 14 December 1984.

Hayes, Bradd, *Transforming the Navy, Report 00–3, Decision Strategies Department*, (Naval War College, 2000).

Hearings before the General Board of the Navy, 1927, (micro. Roll 7), Record Group 80, National Archives.

Hoffman, Jon T., Marine Corps Command and Staff College Brief, held by Marine Corps Historical Center, Navy Yard, Washington, DC, 28 August 2000.

Hoffman, Jon T., assistant director of Marine Corps historical center, Hoffman's undated SOC operations Binder, *Historical Review of MAU/MEU(SOC)*.

Hoffman, Jon T., *Historical Review of MAU/MEU(SOC)* (lecture notes) (Washington, DC: Marine Corps Historical Center).

Kelly, General P.X., Commandant of the Marine Corps, Memorandum for the Joint Chiefs of Staff, 22 July 1985.

Letter from Commandant of the Marine Corps to the Commanding General, Marine Corps Combat Development Command. Subject: Training and Education dated 1 July 1989, (Quantico, VA: Research Center).

Marine Corps Doctrine MCDP, *Warfighting* (Quantico, VA: 1997).

Marine Corps Doctrine MCDP, *Expeditionary Operations* (Quantico, VA: 1998).

Marshall, Andrew, *Historical Innovation: Carrier Aviation Case Study*, memorandum from the Office of Net Assessment, Secretary of Defense, 27 June 1994.

McCutcheon, Keith B., 'Employment of Helicopters in the Marine Corps', draft copy in File 7, Box 7, McCutcheon Papers. (Quantico, VA: Research Center Archives).

McCutcheon, Keith Barr to Frank Lamson, 31 January 1952, Folder 6, Box 5, McCutcheon Papers, (Quantico, VA: Research Center).

Memorandum from Bill Lind to Senator Gary Hart entitled, *Report on Trip to Marine Corps Base 29 Palms*, 23–27 October 1978.

Memorandum from Bill Lind to Senator Gary Hart entitled, *Report on Trip to Marine Corps Base 29 Palms*, 23–27 October 1978.

Memorandum from Bill Lind, William S. to Senator Gary Hart, *Report on Trip to Marine Corps Base 29 Palms*, 30 May–1 June 1979.

Memorandum from Bill Lind to William S. to Senator Gary Hart, *Report on Trip to Marine Corps Base 29 Palms*, 30 June–2 July 1981.

Memorandum from Bill Lind to Terry Pierce, dated 25 May 1999.

Memorandum from Bill Lind to Terry Pierce, dated 27 September 1999.

Memorandum from the Commandant of the Marine Corps to Commanding General, Marine Corps Development and Education Command, 21 January 1986, Subject 'Maritime Prepositioning Force (MPF) Operations Study'. From Studies and Reports, Box 1985–86, Quantico. (Quantico, VA: Research Center).

Memorandum From E.H. Simmons, Director of Marine Corps History and Museums to Deputy Chief of Staff for Aviation, 31 July 1979, (MAGTF File, Headquarters Marine Corps, History Division, Washington Navy Yard, Washington, DC).

Memorandum From Commanding General Quantico Schools to Commandant of the Marine Corps, Quantico, VA. Marine Corps Research Center, Amphibious Archive Files, 1936.

Melson, Charles D., *Condition Red: Marine Defense Battalions in Word War II*, (Washington DC: Washington Marine Corps Historical Center, 1966).

Mundy, Jr, General C.E., Commandant of the Marine Corps, Memorandum for the Commanding General, Marine Corps Combat Development Command, Subject *Special Operations Capabilities in Fleet Marine Forces*, 3 November 1992 (Washington, DC: Marine Corps Historical Center).

Naval Doctrine Publication 6, *Naval Command and Control* (Navy and Marine Corps Joint Publication, 19 May 1995).

Polk, Robert, *A Critique of the Boyd Theory – Is It Relevant to the Army?* Monograph, School of Advanced Military Studies, United States Army Command and General Staff College, Fort Leavenworth, Kansas, 1999.

Rawlins, Eugene W. *Marines and Helicopters 1946–1962* (Washington, DC: History and Museums Division, 1976).

Rumsfeld, Secretary of Defense Donald, 'DOD Acquisition and Logistics Excellence Week Kickoff – Bureaucracy to Battlefield'. Remarks as Delivered at the Pentagon, 10 September, 2001 [www.defenselink.mil/cgi-bin/dlprint.cgi].

Trainer, Mark, *Notes on TCN Development*, 25 May 2003. Original held by author.

Unpublished material

Bittner, Donald. *General John Russell, United States Marine Corps: The Statesman Commandant*, Quantico, Virginia, unpublished paper (12 February 1989).

Chesbrough, Henry W. 'The Differing Impact of Technological Change upon Incumbent Firms: A Comparative Theory of Organizational Constraints and National Institutional Factors, Harvard Business School Working Paper (April 1998), 98–110.

Chesbrough, Henry *Assembling the Elephant: A Review of Empirical Studies of the Impact of Technical Change upon Incumbent Firms*, Harvard Working Paper 99–104, (Harvard Business School, 1999).

Cote, Owen, 'The Politics of Innovative Military Doctrine: The US Navy and Fleet Ballistic Missiles', PhD Thesis, MIT (January 1996).

Frances, Anthony, *The History of the Marine Corps Schools* (Quantico, Virginia, Marine Corps Research Center, unpublished Paper, 1945).

Goldman, Emil, 'Institutional Learning under Uncertainty: Findings from the Experience of the US Military', unpublished manuscript, Department of Political Science, University of California, Davis, 1996.

Ginther, James, *Keith Barr McCutcheon: Integrating Aviation Into the United States Marine Corps, 1937–1971*, unpublished PhD Thesis, Texas Tech University (1999).

Hoffman, Jon, *Historical Review of MAU/MEU(SOC)* (lecture notes), Washington, DC, Marine Corps Historical Center (undated).

Kurth, Ronald James, *The Politics of Technological Innovation in the United States Navy*, unpublished doctoral PhD thesis, Harvard University (June 1970).

Lind, William, *What Great Victory? What Revolution?*, unpublished article written in 1991.

Marshall, Andrew, *Historical Innovation: Carrier Aviation Case Study*, memorandum from the Office of Net Assessment, Secretary of Defense, (27 June 1994).

Peznola, Michael N., *A Matter of Trust? Maneuver Warfare in the Marine Corps: A 10 Year Assessment*, unpublished Masters Thesis, Marine Corps University (1999).

Ralston, David and Michael Smith, *Peacetime Military Innovation: Getting Beyond Today*, (National Security Program Occasional Paper), John F. Kennedy School of Government, Harvard University (1996).

Russell, Major J.H., 'The Preparation of War Plans for the Establishment and Defense of a Naval Advance Base'. Lecture given at the US Naval War College, Naval War College Archives, Newport, RI (1910).

Sheehan, Kevin, *Preparing for an Imaginary War? Examining Peacetime Functions and Changes of Army Doctrine*, unpublished Harvard PhD Thesis, Harvard, (1988).

Shepherd, General Lemuel C. Jr, Marine Corps Board convened by Commandant of the Marine Corps, *The Evolution of Modern Amphibious Warfare*, unpublished paper: Breckinridge Library, Quantico (4 April 1959).

Swartz, Peter, *Origins and Development of the US Navy Carrier Battle Group Concept*, (31 August 1998).

Yunker, Chris, Young Turk maneuver warfare conference entitled 'Maneuver Warfare at 10', held at Quantico 5 December 1998.

Oral histories

Doyle, Vice Admiral James H. Jr, USN (Ret.), Oral History, by Naval Historical Center 1997.

Glenn Jr, Captain W. Lewis, USN, *Reminiscences by Staff Officers of Admiral Elmo Zumwalt, Jr*, US Navy, vol. 1, US Naval Institute, Oral History, 1989.

Reminiscences by Staff Officers of Admiral Elmo Zumwalt, Jr, US Navy, Vol. 1, Oral History, US Naval Institute, 1989, Admiral Worth Bagley.

Wilson, Eugene E. *Gift of Foresight: The Reminiscences of Commander Eugene E.Wilson*, A Naval History Project, The Oral History Research Officer, Columbia University, 1962.

Interviews

Gray, Alfred. Interview with Terry Pierce, 4 March 2001.

Gray, Alfred. Interview with John Scharfen, 'Tactics and Theory of Maneuver Warfare', *Amphibious Warfare Review* (July 1984).

Gudmundsson, Brian. Interview with Terry Pierce, 2 March 2001.

Hammes, T.X., Interview by Terry Pierce, 26 February 2001.

Landersman, Stuart. Interview with Terry Pierce, 22 March 2001

Landersman, Stuart. Interview with Terry Pierce, 26 March 2001.

Landersman, Stuart. Interview with Terry Pierce, 28 March 2001.

Lasswell, Jim. Interview by Terry Pierce, 20 February 2001.

Lind, William. Interview by Terry Pierce, 17 December 1999 and March 8, 2001.

Murray, Williamson. Interview by Terry Pierce, 17 January 2001

Ripper, Paul Van. Interview by Terry Pierce, 18 February 2001.

Schrieber, Nick. Interview by Terry Pierce, 3 March 2001.

Trainor, Lieutenant General Bernard E., USMC (retired). Interview by Terry Pierce, 16 June 2003.

Wilson, G.I. Interview by Terry Pierce, 5 February 2000.

Yunker, Chris. Interview by Terry Pierce, 19 February 2001.

Index

AAW (Anti-air Warfare) 145, 146, 147,
149, 167; intermediate range 168; new
generation 166
accuracy 117, 169; unheard-of 170
ACM (advanced cruise missile) 168;
strategic 169; tactical 169
acquisition system 184–5
advance base operations 51–8, 60, 64,
109; defensive 59, 112, 113; offensive
112, 113
advantage 10; decisive 2, 36, 37;
significant 26
Aegis Project 153, 164, 165, 166–8,
173, 177, 179, 183, 197, 198; CEC and
178, 180
Afghanistan 101, 102; Soviet invasion of
108
after-exercise reports 33
Agile Sword (exercise) 109
air base shortages 157
Air Club (Japan) 139, 140
aircraft 16, 29, 114; aircraft assault
support 80; ASW 153, 159; attack
134, 135, 137, 140, 148, 152, 166;
carrier-based, nuclear weapon delivery
11; communications relay 177; fighter
6, 75, 127, 135, 148, 153, 158; fitted
with floats 129; high-speed 81;
land-based, surviving severe attacks
from 130; launching and landing 125;
maritime patrol 147; military use of
132; naval 135, 164; number in the air
125; operating specifications of 123;
reconnaissance 134; role in naval power
projection 133; scouting 127, 128, 134;
sea-based 122; self-sufficiency in
design and manufacture 134; some of
the finest in the world 134; specialized
134, 152; spotter 122, 127, 128, 129;
strike 128, 153, 159, 164; see also
bombers; helicopters
aircraft carriers 8, 67, 80, 147, 156; battle
group concept 152–4; big-deck 158, 159;
development 133–5; escort 74; invention
of 3; modern, prototype of 133; need to
continue to build 126; nuclear-powered
153; nuclear weapon delivery 11; only
naval aviators be given command of 128;
small 94; unfinished ships converted into
122; see also carrier warfare
air defense 6
airfields (naval) 128
Air-Land Battle doctrine 24, 88
air power 122, 142
Air Power and Maneuver Warfare (van
Creveld) 110
airships 128
air-to-air combat/missions 76, 80
Akagi (Japanese battle cruiser) 134,
135, 143
Akita, Cmdr. Sasaki 138
Aleutians 59
all arms concept 33
Allies 23, 27, 40, 48, 67; stunning victory
(1918) 36
alligator concept 66
amphibious assault 152; air 73, 74, 75;
vertical and surface 70
amphibious operations 71, 109; air support
an integral part of 81; most successful
95; viability in atomic warfare 80
*Amphibious Operations – Employment of
Helicopters (Tentative)* 73
amphibious warfare 24, 71, 106, 111, 113,
182–3; atomic age 72; development of
51–69; helicopters 73; new way of
conducting 112

Amphibious Warfare School 89, 95
Anderson, Col. Gary 91, 99, 113
annihilation 48
Ansel, Lt. Walter C. 60
anti-guerrilla operations 74
antisubmarine patrol 133; *see also* ASW
API (applications program interface)
    179–80
APL (Applied Physics Laboratory) *see*
    Johns Hopkins University
Arabs 87
architectural innovation 16, 17, 20, 21–3,
    28, 52, 61, 65, 66, 109; all military
    doctrinal changes are 24; disruptive 26,
    71, 160; importance of 15;
    understanding the impact of different
    types of 24–5
architectural knowledge 22, 37, 57;
    applying 30, 41; building 30, 36, 40,
    46; embedded 34, 40; requires explicit
    management by senior military leaders
    30; uprooted 23
Ardennes 46
ARG (Amphibious Ready Group) 183
armored vehicles 41, 42, 44
armored warfare 19–20, 23, 24; close air
    support of 44; development 27, 32–50;
    disguising 43–6; exploitation potential
    on operational level of war 27; mobile
    50; promoted 46
Art, Robert J. 164
artillery 63, 75, 80, 93, 114; friendly fire
    36; prolonged preliminary
    bombardment 32; supporting 48
Asia: Japan's desire to expand empire
    throughout 135–6; Southeast 50, 136;
    Southwest 109
assassination 139, 140
ASUW (antisurface warfare) 146–7, 149
ASW (antisubmarine warfare) 145,
    146–7, 149, 152, 153, 159, 170
Atlantic 149, 155, 158
atolls 66, 67
atomic bombs 71, 72, 73, 112;
    introduction into US naval aviation
    strike force 24
ATP-1 (navy tactics manual) 149
attack-in-depth 48
attack squadrons 75
attrition warfare 87, 88, 89, 93, 96, 97,
    98, 136; key players in challenging
    111
Australia 67
Avant, Deborah 11

aviation 8, 126, 156; advent of 132;
    Army 128; civil 133–4; control and
    coordination of 134; established as a
    permanent combat arm 80; helicopter
    71; Japanese Army progress in 133;
    Marine 79, 80, 82; political techniques
    to successfully introduce into Navy 12;
    sea-based 122; split among services 7;
    tactical 146; *see also* naval aviation

Baer, George 157
Baghdad 101
Bagley, Rear-Adml. Worth 153, 157
Bahamas 106
Baker, Rear-Adml. Ted 172, 173
balance of power theory 4, 19, 130
Balisle, Rear-Adml. Phil 177, 182
ballistic missiles 169; fleet 11;
    submarine-launched 6; targeting silos
    170; *see also* ICBMs
Banana Wars (1903–33) 86, 106
bandwidth 178
Bangladesh 106
Barbey, Vice-Adml. Daniel 67
Barrow, Gen. Robert H. 96
baseball 2, 116
basketball 2
battalions 101; defense 52, 53, 58;
    motorized 44
battleships 8, 28–9, 123, 124, 127, 132,
    133; air superiority over 129;
    alternative to carrier aviation that could
    be carried on board 128; carrier
    aviation disguised as sustaining 139;
    combat among 122; defending 127;
    dreadnought 21, 121; made subordinate
    to carrier task forces 142; modern, cult
    of 13; obsolescence of 125, 129, 136;
    scouting for 135; spotting 122, 129;
    support for 128
beaches 57, 66; air superiority over
    landing 80; Army not interested in
    seizing 56; defended 53, 67, 68;
    offensive operations 58
beaching-type ships 67
Beck, Gen. Ludwig 32, 34–5, 39, 41,
    42, 86
Beirut 86
Belleau Wood, battle of (1918) 51, 57, 61
Berkeley, Brig.-Gen. Randolph C. 60
'Big Gun Club' (Japan) 133, 134, 136,
    139–43
Bigley, Vice-Adml. Tom 149
Bikini 71, 112

Blair, Adml. Dennis 176
Blatt, Capt. Wallace 74
Blitzkrieg 16, 17, 25, 28, 32, 42, 49, 79,
    86, 176; Allied response to 23;
    development of 5, 39, 40, 43; entire
    theoretical groundwork for 41;
    incorrect primary credit for innovation
    38; intellectual bases for 36; logistics
    failure 46–7; mobile-war doctrines
    gradually transformed into 48;
    outperformed 1, 26; primary creator of
    39; supply lines inadequate for
    supporting 47
Blomberg, Gen. Field Marshal Werner
    von 45
'body count' mentality 87
bombardment 32, 68
bombers 39, 42, 124, 127, 169; 'Backfire'
    166; firepower to destroy 135; medium
    134–5; torpedo 134
bombings 105
Boorda, Adml. Mike 164, 173
boot camp 96
Bowen, Rear-Adml. Harold 11–12
box-formation concept 138
Boyd, Lt.-Col. John 88, 90, 91, 98
Bradley, Gen. Omar 71
breakbulk cargo ships 108
Breckinridge, Brig.-Gen. James 58, 62
Bright Star 85 (exercise) 109
Britain *see* British Army; RAF; Royal
    Navy
British Army 57; armored warfare/tanks
    19, 20, 24, 27, 32–8, 41, 44, 46;
    culture 50
Brodie, Bernard 23
Brooke, Field Marshal Alan 34
Brown, Harold 108
brown-water riverine force 157
'Bubbas' 97, 98
bureaucracies 2–3, 14; civilian leaders
    and 111; self-initiated change 1
Burke, Adml. Arleigh 160
Burnett-Stuart, Gen. Sir John 33
Busch, Capt. Dan 184
Bush, George Sr 164

California 128
Cambodia 106
Cambrai, Battle of (1917) 32
Camp Lejeune 87, 89
capital 22
Caribbean 79, 106; American interests
    in 51

Carolines 59
carrier warfare 29, 121–31; developing 3;
    Japanese 3, 17, 18, 132–44; specialized
    task forces 152; *see also* CVBGs
casualties 67, 98
Caulfield, Maj.-Gen. Matt 172, 173
cavalry 33, 41, 45; coupling tanks with
    43; reconnaissance units 44; tanks
    replacing 50
CEC (Cooperative Engagement
    Capability) 177–85
Central America 56, 106
centralized control/planning 30, 100
Chambers, Capt. Washington Irving 12
change *see* engines of change; social
    change; technological change; throttles
    of change
China 56, 60, 66, 74, 79, 118; Boxer
    Rebellion (1900) 106; State
    Department's Open Door Policy in 121
China War (1937) 134, 137, 140
Chosin operations (1951) 74
Christensen, Clayton 17, 25, 26
CincPac (Commander-in-Chief,
    Pacific) 83
Citino, Robert 38
Citrin, Warren 179–81, 184
civilian intervention 3, 4, 19, 21, 27–8,
    38–9, 47–9, 91–4, 97, 111–12, 186,
    190, 193; can be effective in promoting
    innovation 6; causing sustaining
    innovation 107, 110; military
    organizations innovate only in response
    to 9; opportunity to assess the value of
    116; perceived 98; produces military
    innovation 5; retarded maneuver
    warfare development 85; shifts in 87
civil-military operations 106
Clark, Adml. J. J. 127
Clark, Kim 15–16, 22, 34
Clark, Adml. Vern 173
Clements, William 170
close air support 44, 80, 81
CNA (Center for Naval Analyses) 108,
    156
coaling stations 53
Cobra gunship 76, 114
cohesion 48
Cold War 99, 164, 194
collaborative tracking 179
Collier, Capt. Eugene F. C. 63
Colombo 132
colonial territories 136
combat arms 14; primary 1

combined-arms warfare 33, 38, 40, 41, 42, 46, 79–84, 112, 176; classic dilemma 158; mobile 48; naval 145–51; new doctrine that emphasized 36; penetration relying on surprise 48; poses the enemy with no-win situation 80; tank a key component of 50; true operating force 70–1
command and control 91, 148; adequate 147; effective 145; limited means and facilities 146; shared 176
Commander's Intent 101
Command and Staff College 95
Commodore-64-era technology 179
communications 180; disruption of 48; effective, lack of 36; two-way 160
competition: civil-military 4–5, 193; head-to-head 26; internecine 8; interservice 3, 6–7, 92, 110, 193; intraservice 5, 19–20, 110, 193
component-linkage approach 15, 20, 23, 24
component technology 17
compromise 129, 135
conflict: ancient way of looking at 97; civil-military 4; conventional 87; head-to-head 28; interservice 4, 7; low-intensity 106
Connelly, Adml. Richard L. 68
continuous aim gunfire 1, 13, 16, 23, 24, 63, 116–20, 159–60
convoys 158
Conway, Gen. James 101
Coolidge, Calvin 128
Cooper, Matthew 43
core components 22, 23, 24
Correspondence School 57
Corum, James 38, 47–8
Cote, Owen 3, 6, 7
cotton gin invention 10
counterinsurgency capability 3
counterterrorism 105
coup d'état (Haiti 1992) 106
Craven, Capt. Thomas T. 124
Creveld, Martin van 9–10, 11, 110
crossing-the-beach operations 74, 112
cruise missiles 166; advanced 168, 169; anti-ship 169–70; deployment of 158; expanded capability 169; launched from SSNs 170; long-range 158; strategic 168, 169, 170; submarine-launched 169, 170, 173; tactical 169; *see also* Tomahawk

cruisers 122, 127, 147; Aegis 173, 179; converting to carriers 134; heavy, construction limits on 134; light 153; nuclear-powered 167
Cuba 51, 52, 62
Culebra 53, 65
culture 3, 19, 49–50; decentralized 86; institutional 85–6; *see also* organizational culture
Cunningham, Lt. Alfred 80
Curtiss, Glenn 12
Custer, Gen. George Armstrong 27
CVBGs (carrier battle groups) 145, 152–4, 173, 167, 191, 198; command centers for operations 167
CV/CVN carriers 153, 159
CWC (Composite Warfare Commander) concept 145–51

dangerous waters 158
Davis, Vincent 6, 10–11, 13, 14, 15, 19, 20, 24, 26, 199–200
Davison, Lt.-Cmdr. Ralph 127
DDS (data distribution system) 177, 180
death threats 140
decentralization 30, 41, 88
decision-making 91, 95; authority delegated 101; tactical 92
defensive operations 7, 52, 55, 57, 58; sea control warfare 155–63
demolition operations 104
dependent variables 1–2, 11, 116
deployment 10, 79, 87, 108; cruise missile 158; forward 104, 105, 106; rapid 109, 110
DePoy, Phil 156
destroyers 127, 147, 153; Aegis 167; new, high-pressure boilers and high-temperature steam systems in 12; sinking of 168; World War I, APDs converted from 67
Dewey, Adml. George 12, 53
Diamond, Jared 10
Diego Garcia 108
Dill, Sir John 35
disarmament 40; rush to achieve 59
disguising process 113–14, 118–19, 148–9, 159–60, 167, 169–71, 188–9, 196–7
disguising propositions 37–8, 127–30, 183–4; armored warfare 43–6; maneuver warfare 40, 95–7
dismissal 140
dispersion 73, 74, 81

disruptive innovation 1, 11, 14, 18, 104, 107, 116, 118, 192, 200; aborted 155–63; amphibious warfare 51–69; assessment of how senior naval leaders achieve 198–9; carrier warfare 121–44; championed 4; CWC 145–51; described 2; disguised as sustaining innovation 64; explaining 19–50; external causes 8, 193–5; helicopter warfare 70–8; incentives to achieve 5; inchoate 29, 35, 85–103, 110, 111, 176; internal causes 8, 196–8; MAGTF warfare 79–84; maneuver warfare 85–103; maritime action groups/surface land attack warfare 164–75; prescribing civilian strategy for promoting 201; resisted 3; Rosen's top-down theory and Davis's mid-level theory versus 199–200; tactical 129, 176–91

dissent 30–1

dissimulation 40, 43, 45

Doctrine Division 100

Dominican Republic 106

DOT (Division of Operations and Training) 55

Dowding, Air Marshal Hugh 6

Doyle, Vice-Adml. James Jr 146, 150

dreadnoughts 21, 121, 127; first time sunk by air attack 132; flat-top 130

drive 50

Dryad (HMS) 148

early maneuver warfare exercises 113

Egypt 106, 168, 195

Eilat (Israeli destroyer) 168

Elliott, Gen. George 54

Ellis, Maj. Earl (Pete) 54, 55, 56, 57, 58, 60, 66, 74, 88, 111, 112, 198

Engel, Gregory 169, 170, 171

engines of change 3, 52–3, 59–61, 65, 71–2, 79, 87–8, 105, 108, 117, 121–2, 145–6, 155–6, 165, 179–80, 182–4, 193

EPLARS (enhanced position location reporting system) 180

Europe 68, 108

evacuation operations 106

Evans, David 99

exercises 37, 40, 46, 92, 109; early maneuver warfare 113; experimental tank force 50; field 90; fleet 129, 137, 147; free-play 90; game board 123; mobile warfare 44, 45; peacetime 94; tank supporting other arms 44; unrestricted 44

exotic systems 11

expeditionary forces 53, 54, 62, 102, 108; East and West Coast 60; smaller, argument for 72; *see also* MAGTF; MEF; MEU/SOC

experiments 42, 121; approval of wide array of 37; championing 39; continuous 176; fleet 30; limited 50; mechanized forces 92, 93; new tactics and techniques 90; operational 138; secretive 40; tank 36, 41, 44, 46, 50

exploitation 40, 44, 50, 176; rapid 41

Far East 51, 59, 121

Fedayeen resistance 101

Feland, Brig.-Gen. Logan 55

Fenton, Mun 183

firepower 81, 88, 89, 96; armored forces 33; decreased, consequences of 72; destroying enemy bombers 135; maneuver and 48; offensive 158; overwhelming 49; superior 87

First Air Fleet (Japan) 138–43

first strike 158

fixed-wing aviators/nonaviators 70

Fleet Problem IX 127

FLEX (Fleet Landing Exercises) 65

flexibility 73, 88, 92, 102; armored forces 33; strategic 171

flood relief 106

flying boats 73

FMF (US Fleet Marine Force) 51, 52, 57, 74, 83, 86, 104; and amphibious assault warfare 58–65; control and composition reconsidered 81; counterterrorism capabilities of selected units 105; creation of 112, 113; dispersion for 72; integration of aviation technology into 79; intellectual process that would drastically alter doctrine of 88; Pacific 93; promotions 77; restructured 81; Special Operations Capabilities in 106; stressing its base defense capability 114

FMFM-1 *Warfighting* 86, 90, 98, 99

follow-up units 48

forcible insertion/entry 108, 109

Formosa 59

fortifications: impregnable 37; invulnerability of large fortresses 49; lethal turrets 26

Fort Pickett 90, 113

Foster, John 168, 169, 170

France 23, 27, 47, 54, 56, 57, 61; air
mission invited to Japan 133; attrition
warfare 98; colonial territories 136;
culture 49; incumbent technology
failure 37; interwar warfare
developments 7; Marine aviation unit
organized for 80; military texts (1930s)
88; model of methodical battle 85;
reliance on old frameworks 37; senior
military leaders, interwar period 26; *see
also* Maginot Line
Franks, Gen. Tommy 101
Freedom Banner (exercise) 109
free-play training 100
frigates (nuclear-powered guided
missile) 159
Fritsch, Gen. Freiherr Werner von 32, 39,
41, 42, 86
From the Sea (US Navy white paper) 164,
165, 172, 173, 175, 187
frontal assault 87
Fuchida, Lt.-Cmdr. Mitsuo 143
fuel 4
Fukudome, Vice-Adml. Shigeru 138
Fullam, Capt. William F. 54
Fuller, Gen. Ben H. 59, 60, 112, 113
Fuller, Vice-Adml. Sir Cyril T. M. 128
Fuller, Col. J. F. C. 32–3, 36, 37–8, 50, 51
Fuller, Capt. William 53
full production 10
Furious (HMS) 133

gadgets 11
Gallipoli (1915) 51, 56
garrison duty 60
gas turbines 167
Gatling guns 27
Geiger, Lt.-Gen. Roy 72
Genda, Cmdr. Minoru 137–8, 143
Germany 7, 48, 85, 114; advances
derailed 26; amphibious tactics
employed to defeat 67; armored
warfare 19, 20, 32, 35, 36, 37, 38–47;
culture 50; defeat of 51; forces
reconnoitered and bombed by seaplanes
133; General Staff College 90; invasion
of Poland 58; maneuver warfare 87,
98; military history 88; tactics and
organization 93; *see also* Blitzkrieg;
Hitler; Guderian; Luftwaffe; Panzer
Force; Wehrmacht
gimmickry 2
Gingrich, Newt 90
Ginther, James 81

Glenn, Lt. W. Lewis Jr 161
Godley, Gen, Sir Alexander 33
Goldman, Emily 14
Goldwater-Nichols legislation (1986) 194
GPS (Global Positioning System) 171
Gray, Maj.-Gen. Alfred 86–91, 94–101,
105, 107, 110, 111, 113, 114–15, 193,
197, 199, 200
Great Depression (1930s) 136
Greenwood, Col. John 92
Gregson, Lt.-Gen. Wallace (Chip) 102
Grenada 89, 96, 100
ground forces 38, 81; superior 80
Guadalcanal 72, 80
Guam 59
Guantanamo Bay 52
Guderian, Heinz 23, 27, 32, 36–7, 38, 39,
41–5, 46, 47, 49
Gudmundsson, Capt. Bruce 99
guided missile ships 159, 177
Gulf War (1991) 11, 100–1, 164, 173, 194
gunnery 13, 117, 123; artillery 114;
battleship 122, 128; dramatically
improved 1; elevating gears 16, 117;
giant 121; heavier, with longer ranges
28; new way to aim and fire 116;
stationary 118; steady stream firing
124; support 60; US shortcomings
in 118
gunships 76, 82, 113, 114

Haig, Field-Marshal Douglas 33
Haiti 54, 106
Hammes, Col. T. X. 99
Harding, Warren G. 123
hardware 12, 14, 15; special-purpose 177
Harpoon missile 146, 168
Harrier VSTOL aircraft 158
Hart, Gary 87, 88, 89, 91, 93, 94, 98, 111
Hartz Mountains 40
Hawaii 128, 136, 138, 158
Hayes, Capt. Bradd 6, 13–14, 164, 172,
173
Hayward, Adml. Thomas 145, 146, 148,
149, 150, 167
heavy vehicle transport 66
helicopters 52–8, 70–8, 81, 82, 83, 108,
112, 147, 158; disguising the arming of
114; use as both gunship and infantry
transport 113
Helldivers (film) 130
Henderson, Rebecca 15–16, 22, 34
Heywood, Cmdnt Charles 54
Higgins, Andrew Jackson 66

high-performance ships 159
high technology 1, 11, 26
Hirschberg, Gen. Karl von 45
Hitler, Adolf 5, 34, 38, 39, 42, 43, 45, 48, 49
HMX-1 (experimental helicopter squadron) 73, 74
Hoffman, Lt.-Col. Jon 51, 67
Hogaboom, Maj.-Gen. Robert E. 74, 82, 112, 113
Holcomb, Gen. Thomas 63, 64, 66, 114
Holland, Capt. Jerry 148
Holloway, Adml. James L. 145, 152–3, 167, 170
homeland defense 178, 184
Hong Kong 117
horizontal alliances 11
Hosho (Japan's first aircraft carrier) 133
hostage recovery/rescue 104, 105, 107, 108
Huchting, Adml. George 177
hull-mounted sonar 147
Hundley, Richard 21
Hunter Killer Group 147

ICBMs (intercontinental ballistic missiles): land-based 169; sea-based 169, 171
ideological struggle 8–9
ignition 4
Inchon 95
incremental improvement 5
incumbent technology failure 20, 35–7
India 50
Indian Ocean 108
infantry 8, 81; advocacy of tank replacing 50; armored development 33; armored vehicles no longer tied to pace of 44; aviator willing to fight for causes that helped 83; close air support to support penetration 80; colonial 51, 53; dependence upon speed 48; helicopters a tool to help 74; integration of aviation into assault divisions 79; mobile 82, 110; motorized units 44; naval 51, 152; pushing through enemy defensive positions 43–4; superior, conventionally-trained 3; support weapon for 41, 42, 43, 44, 45, 49; tactical manual rewritten 34; tactics 95; vulnerability to enemy ground fire 76
infantry-artillery paradigm 36
infiltration 39, 48, 50

informal structures 30
information technology 176
initiative 40, 50, 87, 145, 176; importance of seizing and maintaining 96; prized 86
innovation: championing 8; civilian 6, 8, 20; component 22, 24; contrasting hypotheses about 12; defensive 52, 158–9; doctrine-driven 5, 8–9, 11, 12–14, 24, 164; external causes 2–3, 47–9; incremental 16, 17, 21, 23, 24, 25, 26; insights into 24; internal causes 2–3, 8–14; linkage 22, 23, 24; major 1, 13, 14, 15, 24; minor 14, 15; modular 16, 17, 21, 25, 26; offensive 52, 53; peacetime 8, 9, 10–11; radical 16, 17, 21, 22, 25, 26; small groups 30–1, 60; social 15; tactical 13, 47, 51; three-point-shot 2; top-down 107; traditional engine models 4–8; vertical assault 79; wartime 67–8; *see also* architectural innovation; disruptive innovation; sustaining innovation; technological innovation
Inoue, Vice-Adml. Shigeyoshi 134, 142–3
insertion mission 108
Inspector of Target Practice (US Navy) 118
insubordination 118
integrated air defense 6
intellectual breakthroughs 57, 58, 61, 65, 73, 81, 112, 113
intellectual process 53–6, 59–61, 65–6, 72–5, 80–2, 95, 105–6, 108–6, 117–18, 122–5, 146–8, 152–3, 156–8, 196; Aegis 166; civilian 97–8; surface land attack 172–3; Tomahawk 168–9
intelligence 9, 47, 55
intensive predeployment training 104
Internet 179
interoperability 177, 178
interventionist forces 64
inventions 3, 9, 10
Iran 87, 105
Iran–Iraq war (1980–88) 108
Iraq 173; terrorist threats 164; *see also* Operation Iraqi Freedom
Ireland 50
isolationism 59, 64, 113
Israel 87, 168
Iwo Jima 52, 67

Jackson, Capt. (later Maj.) Tim 90–1, 111, 113, 115

Japan 54, 56, 58, 59, 102; air fleet concept 140–2; amphibious tactics employed to defeat 67; battleships 18-inch gun 21; carrier warfare 3, 17, 18, 126, 132–44; defensive concentrations 67; landing assault on Chinese positions 66; outbreak of hostilities with 52; serious potential enemy from naval viewpoint 53; *see also* Pearl Harbor
Johns Hopkins University 179, 180, 181, 182
Johnson, Adml. Jay 173
Johnson, Louis 71
Jones, Maj. (later Gen.) James 86–7, 98, 111
Jordan, William 134
junior officers: career paths 114–15; developing 143; promotion 99, 121; protection of 50, 99, 110, 121
Junior Officers' Tactical Symposium 90
Jutland 29

Kaga (Japanese battleship) 134, 135
kamikaze 143
Kansas 57
Kapos, Erv 156
Kasumigaura Aviation Corps 143
Katzenbach, Edward L. 23
Kazan 40
Kelly, Gen. Paul X. 94, 105, 106, 107
Kelso, Adml. Frank 172, 176, 199
Kennedy, John F. 3
Kidd. Vice-Adml. Isaac 161
Kier, Elizabeth 3, 7, 49, 121
killing zone 27
King, Adml. Ernest 131
Kissinger, Henry 160
knowledge: acquiring and applying 37; assimilation of 30; component 22, 30, 36; doctrinal 22; embedded 35; expert 93; *see also* architectural knowledge
Korea 71, 74, 77, 88, 108, 158; set piece battles 86
Kriegsakademie 90
Krulak, Gen. Charles C. 51, 86–7, 90, 91, 97, 100, 102, 111, 113, 114, 193
Krulak, Gen. Victor 53–4, 59, 60, 61, 62, 65, 66
Kuckuk, Lt.-Col. Paul (Lester) 101
Kuroshima, Capt. Kameto 138
Kurth, Ronald 12
Kuwait 98, 194

Laird, Melvin 155
land campaigns 53, 61
Landersman, Capt. Stu 147–9, 150
landing craft 65–7, 70
landings: amphibious 75; submarine 73; three-echelon concept 68; unopposed 51, 53
Langley (USS) 122, 124, 125, 128, 129
Lasswell, Col. James 99
Lautenschlager, Karl 13
LCVP (landing craft vehicle, personnel) 66
Leader, Capt. Chuck 95
Leahy, Adml. William D. 64, 114
Leavenworth 57, 63
Lebanon 94, 105
Lee, Gen. Harry 96
Lejeune, Gen. John 52, 54, 55, 56, 57, 58, 60, 62, 112
Lexington (USS) 122, 125, 126, 129
Liberia 106
Libya 164, 170
Liddell Hart, Lt. Basil 33, 35, 38, 44, 50, 57
Lind, William S. 85–94, 96–101, 110, 111, 114, 193, 200, 201
Lindsay, Brig.-Gen. George 33
linear logistics system 47
linear warfare 28, 43
liner conversion 66
linkages 30, 34, 35, 40–1; component 36; new, problem of 64; old 46; rigidity of 36
Linn, Maj. Tom 99
Linsert, Maj. Ernest 66
Lipetsk 40
Little Big Horn, battle of (1876) 27
Locke, Capt. (later Cmdr) Walter 168, 169, 170, 171–2, 198
logistics 8, 110, 114; executing 33; failure 46–7; multipurpose combat ship 153; problems of Blitzkrieg 43; uncontested support 157
Long Beach (USS) 167
'low cost bets' 10
Low Countries 5
low-end ships 159
LSTs (landing ships, tank) 66, 67
Luce, Adml. Stephen 122
Luftwaffe 44
Lupfer, Timothy 48
Lutz, Gen. Oswald 35, 45–6

MacArthur, Gen. Douglas 59, 60, 67

McCutcheon, Lt.-Col. Keith 73–4, 75–6, 77, 81, 82–3, 112–13, 114
machine guns 27
Maginot Line 1, 26, 28, 37
MAGTF (Marine Air-Ground Task Force) warfare 70–1, 79–84, 87, 98, 104, 106, 108, 111, 112–13; capabilities extent within 105; disruptive innovations 114
Mahan, Adml. Alfred Thayer 122
mandated islands 59
Maneuver Rules (board game) 123
maneuvers: all officers expected to display 40; annual 36; armored 34, 46; chart 123; critical aspects of 37; disruptive 39; emphasized 38; firepower and 48; fleet 135, 137; innovative concepts 34; new doctrine that emphasized 36; potential impact underestimated 36; single or double envelopment 47; stunningly swift 98; tactical system of 40; tank 33, 34, 35, 41
maneuver warfare 40, 85–103, 114; efforts to develop and push the adoption of 112; negative effect on developing 111; one of the most vocal opponents of 110; potential operating methods 113; tenets of 176
Maneuver Warfare Board 89
Maneuver Warfare Correspondence courses 111
Maneuver Warfare Handbook 87, 89, 90, 91, 110, 111, 193
Manhattan Project (US 1942) 10
Manila 51
Maradim Island 106
Marine Corps Advanced Research Group 81
Marine Corps Association 55
Marine Corps Equipment Board 65–6, 81
Marine Corps Gazette, The 55, 92, 93, 99
Marine Corps reserve 63
Marine Corps University 95, 98
Marine Corps Warfighting Center 86
maritime action groups 164–75
market share 26
Marquesa Islands 106
Marshall, Andrew 14
Marshall Islands 59, 71
mass aerial strike 136
mavericks 5, 6, 38, 39, 49, 116, 193; amphibious warfare 111; can cause sustaining architectural innovations 21; can cause sustaining innovations 28; civilian 114; civilian intervention to

assist 47; intellectual 57, 98; no room for 98
Mayaguez (SS) 106
MCC (Modular Command Center) 183
MEB (7th Marine Expeditionary Brigade) 108, 109
mechanization 33, 34, 44, 92–3, 108; heavy reliance on 96; operational and tactical development 49
Mediterranean 155, 158, 160
MEF (Marine Expeditionary Force) operations 100, 101, 102
merchantmen 158; conversion 66
Metcalf, Vice-Adml. Joseph 171
methodical warfare 26, 28
MEU/SOC (Marine Expeditionary Unit [Special Operations Capable] Force) 104–7, 110, 198
Mexico 54
Meyer, Capt. Wayne 166–7, 168, 198
Micronesia 53, 55
Middle East 107
Midway 52, 58
Milestone Decision 184
military demand 9
Military Reform Movement 87, 88, 89, 91, 92, 93, 94, 98
Miller, Col. Ellis Bell 62–3
Millett, Allan 62, 64
Milligan, Brig.-Gen. Robert F. 89
Miragoane, Lake 106
misdirection 31, 97
missiles 179; air-to-surface 166; guided 10, 13, 159; high-speed 166; long-range 146, 168; saturation attacks 166; sea-launched 168; Soviet 167; subsurface-to-surface 169; surface-to-air 168; surface-to-surface 168, 169; tactical 146; *see also* ballistic missiles; cruise missiles
Mississippi (USS) 123, 128
Mitchell, Lt.-Col. (later Gen.) William 6, 82, 122, 124, 125, 128
Mitsubishi carrier fighter/medium bomber 134–5
mobile warfare 36, 37, 43, 48, 96; air bases 122; armored 50; Blitzkrieg differed from 47; combined-arms 48–9; disguised 40; disruptive 41; exercises 44; light forces 68; product champion of 39; sound and intellectual foundation 42; tank as key part of 49

mobility 41; air 70, 71–8, 111, 112, 114; armored forces 33; enhanced 108; helicopter 73, 112; lack of 1, 26; several ways to achieve 72–3; strategic 108, 109; superior 40; tactical 81
modeling 30
Modern Military Strategy for the United States, A (1978 white paper on defense) 111
Moffett, Adml. William A. 122, 123–4, 125, 126, 127–30, 183
Moore, Capt. Scott 89, 90, 99, 115
Moorer, Adml. Thomas 157
Morison, Elting 159–60
Morrow Board hearings 128
motorization 34, 39; engineers 46; ground forces and air forces 40; infantry 43, 44; motorcycle platoons 45; support 48; transport importance for combat troops 46
MPFs (Military Prepositioning Forces) 104, 107–10, 198
MTF (Mechanized Training Force) 93
multi-carrier task force doctrine 127
multiple narrow landing points 67
Mundy, Gen. Carl 86, 97, 106, 110, 111, 172, 176
Murray, Williamson 7, 23, 27, 34, 35, 36, 38, 40, 42, 47, 49, 121, 123
Myatt, Lt.-Col. James 89, 99, 105, 107

Nagumo, Vice-Adml. Chuichi 139, 143
Naples 161
national interests 121
National Security Act (US 1947) 79
NATO (North Atlantic Trreaty Organization) 152, 157, 158
NAVAIR (US Naval air community) 183
naval auxiliaries 158
naval aviation 12, 24, 80, 168; breakthroughs 166; ceaselessly and enthusiastically promoted 143; critical role 127; development of 123, 134; languished 133; major coup for 128; Moffett's task of selling 129; pivotal champion of 132; product champions of 121, 122, 123, 124; progress of 124; *raison d'être* 134; strategy of 128; strike force 13; tactics of 128; Yamamoto's plan for building 141
naval bases 51; defending 54, 58; marooned and indefensible 59
naval construction limitation 59
Naval Proceedings 64

naval warfare 130, 132, 153; centerpiece of 168; new theory of victory in 171; redefined 122; three-dimensional 146
Naval Warfighting (manual) 176
NAVSEA (US Naval sea community) 183
Netherlands 136
network-centric warfare 176, 177, 178–9, 183; surface and air, linking of USW to 184
Neustadt, Richard 119, 120
new entrants 37
New Guinea 67
Newport's war gaming 123
New Providence Island 106
new technologies 2, 11, 15, 34, 81, 171; embraced 37; employing in old ways of fighting 26; experimenting with combinations of 86; failure to adopt and exploit in a new and unsuspected way 26; help to perform existing missions better 116; incumbents of 37; lack of 42; modern militaries investing aggressively in 20; most are sustaining 26; Navy investing aggressively in 3; new doctrine to suit 139; normally assimilated to old doctrine 24; offensive potential of 39; propulsion 12; recognizing, linked in a novel way 20; senior military leaders understanding of 37; sound and intellectual foundation for use of 42; used in a disruptive way 35
Newton's first law of motion 118
New York Commodities Exchange 91
Nicaragua 60
night flights 143
Nitze, Paul 168, 195
Nixon, Richard 170
noncombatant evacuation operations 106
Normandy 71, 72
North Africa 67, 68
North Carolina 89
NTPS (Near Term Prepositioning Ships) 108
nuclear-powered ships 153, 159, 166, 167
nuclear weapons 6, 169; delivery capability 11; propulsion systems into US submarine force 13; tactical 80

obsolescence 125, 129, 136, 155, 156, 166
Ockenden, Stephen E. 164
O'Donnell, Gen. Andrew W. 93

offensive warfare/operations 40, 48, 52, 55, 56, 60, 65; advanced 113; amphibious 57; beach 58

officers: all expected to display initiative, exploitation, and maneuver 40; infantry 42; influential 55; intellectual 171; mid-grade 11, 107, 113, 171; opposed to transformation 41; overage 64; promoting and protecting 56; tactical action 150; *see also* junior officers; mavericks; senior military leaders; younger officers

oil crisis (second) 108

Okinawa 72, 102

O'Neil, Admiral Charles 118

Onishi, Rear-Adml. Takijiro 138, 143

ONR (US Office of Naval Research) 183

OODA (observe, orient, decide, act) cycle theory 88

Operation Plan 712D *see* War Plan Orange

Operations: *Crossroads* (1946) 71; *Desert Storm* (1991) 85, 89, 96, 100–1, 164, 165, 194; *Eastern Exit* (1991) 106; *Enduring Freedom* (2001) 101, 102; *Iraqi Freedom* (2003) 85, 87, 96, 100–1; *Mousetrap* (1951) 74; *Overlord* (1944) 71; *Sharp Edge* (1990) 106

OPEVAL (Operational Evaluation) 184

OPNAV (US Office of the Chief of Naval Operations) 160, 161, 183, 185

opposed assault 61

organizational culture 2, 4, 7, 49–50, 121, 195

organizational theory 4, 5, 6, 19

O'Rourke, Ronald 177, 178

Osaka 141

Osprey (vertical take-off/landing plane) 22

Owens, Adml. William A. 164, 172

Ozawa, Rear-Adml. Jisaburo 140–2

Pacific 52, 54, 66, 80, 93, 96, 117, 130, 142, 149, 155, 176, 183; basic carrier doctrine (PAC-10) 127; Central 59, 67–8; conflict in 158; Japan's emergence as dominant military power 132; movement of naval forces from 158; multicarrier operations 146; set piece battles 86; South/Southwest 67; surface community 150; US emergence as a military power 132, 136; US strategic position 122; Western 53, 121, 136, 158, 160; *see also* CincPac

Palau (USS) 74

Palmer, Rear-Adml. Leigh 12

Panama/Panama Canal 96, 127, 129

Panzer Force 27, 46; championing of 39, 43; contribution to creation of 34, 35; development of 32, 40, 41–3; disruptive maneuver warfare doctrine 39; dummy tanks replaced and converted into units 42; Reconnaissance Battalion 45; success in forging of 49

parachute operations 73

Pate, Gen. Randolph 74, 112, 113

Patton, Gen. George S. 96

Pearl Harbor 121, 132, 136–7, 138, 141, 143

Peattie, Mark 142

penetration 41, 48; blocking 81; combined-arms, relying on surprise 48; deep 33; infantry, close air support to support 80

Pentagon 156, 160, 185

Pentium 179

pentomic division 24

performance 1, 4, 21, 31; as a measure of effectiveness 24–7; improved 1, 25, 26, 70; increased 13; most favorable 33; spotty 27; testing 10; trajectory overshoot 28–9

Persian Gulf 87, 106, 108, 109; *see also* Gulf War

Petrie, Capt. 'Rusty' 172

Peznola, Maj. Michael N. 94

Philippines 51, 53, 59, 106, 121, 136

Pile, Lt.-Col. Frederick 34

Pinatubo, Mount 106

Plan 1919 (combined-arms scheme) 32–3

Poland 27, 47, 58

Polaris 6

political process 56–8, 64–5, 66–7, 76–7, 82–3, 98–102, 106–7, 109–10, 119, 130, 150, 160–1, 181–2, 183–4, 189–90, 197–8; political process surface land attack 173; setting the stage for 31

popular films 130

Port Darwin 132

Posen, Barry 3, 4–5, 6, 9, 10, 23, 38, 39, 41, 42, 47, 48–9, 111

power projection 155, 157, 159

Pratt, Adml. William 125

PRC-117 (multiband radio) 180

Priboloff Islands 106

Price, Col. Charles 60

Prince of Wales (HMS) 132

Princeton Marine Corps study (1951) 51
Pringle, Capt. 'Dirk' 161
product champions 4, 5, 9, 23, 31, 52, 63,
    70, 75, 97, 104, 118–19, 160, 165, 169;
    Aegis 197; air mobility 72, 76, 112;
    amphibious warfare 55, 56, 58, 98;
    armored warfare 36; attempt to capture
    political power 130; Blitzkrieg 49;
    broad implications in securing resources
    32; civilian 94; CVBG 152; CWC
    150; disruptive innovation 116, 172;
    knowledge that may actually handicap
    37; MAGTF warfare 98, 112–13;
    maneuver warfare 85, 86, 89, 91, 97,
    99, 114; mobile warfare 36, 39; must
    switch to a new mode of learning 30;
    naval aviation 121, 122, 123, 124; naval
    combined arms warfare 145; protecting
    47, 49; support of 49; surface land
    attack warfare 172; sustaining
    innovations 107; Tomahawk 168, 169,
    197; vertical envelopment 74
Professional Military Education
    System 99
professional soldiers 38
Project 60 (defensive sea control warfare)
    155–63, 166, 169
Project Hindsight 9
promotion 99, 143; fast track for 97; lack
    of 161; moral equivalent of 62; new
    pathways for young officers 70, 98;
    policies 9; refused 43; selection
    system 64
proximity fuses 13
Prussian army 92
Puerto Rico 53
Puller, Lt.-Gen. Lewis B. 96

Quallah Battoo, capture of (1832) 106
Quantico 55, 61, 62, 64, 71, 73, 81, 93,
    94, 109, 173; Lind banned from 91,
    100, 111, 114, 193; small innovation
    groups 112; top position at 99;
    Van Ripper appointed to head
    schools 95

radar 13; *see also* Aegis
Radford, Lt.-Cmdr. Arthur 127
'radical' technologies 13
radios 41; introduced 33; low cost 180
RAF (Royal Air Force) 6, 7, 134
raids: air, first carrier in history 3;
    clandestine 104; deep penetration 33;
    dive-bombing 127; limited objective

106; maritime-based 107; ramp-bow-
    type boats 66
Ranger (USS) 126
Raytheon St. Petersburg 181, 182
RCA 166
Reagan, Ronald 105, 107, 108
'real infantry' Marines 80
rearmament 42
reconnaissance 44; air arm 134; armor-
    plated cars 45; aviation 133; clandestine
    104; fleet 134; long-range 139
Record, Jeffrey 94
reef-crossing vehicle 66
Reeves, Adml. Joseph 122, 123, 124–5,
    126, 127–9
refueling and repairing facilities 53
Reichswehr 38, 48
reinforcements: enemy 80, 81; rapid 158
Republican administration/party (US)
    59, 123
Repulse (HMS) 132
rescue 74, 105, 107, 108
reserve forces 68
resistance 48, 101
revolutionary approach 38
Rhine 66
Rickover, Adml. Hyman 6, 152, 166, 167,
    169, 171
risk-taking disruptive groups 30–1
rivalry 118; international 117, 122;
    interservice 19, 52, 53, 59, 71, 72, 79,
    122, 125, 128, 133–5, 165, 186–7, 190,
    193–4; intraservice 3, 19, 53, 80, 110,
    135–43, 164, 191, 193, 194–5
Robert E. Lee 95
rogue states 173
roll-on/roll-off ships 108
Rommel, Field-Marshal Erwin 96
Roosevelt, Franklin D. 63
Roosevelt, Lt.-Col. James 63
Roosevelt, Theodore 53, 117, 118
Rosen, Stephen 3, 5–6, 8–10, 11, 13, 14,
    15, 19, 20, 24, 26, 36, 56, 57, 58, 61,
    107, 121, 199–200
Ross, Andy 14
Royal Navy 3, 66–7, 128; Admiralty 118;
    carrier aviation 28, 122, 123, 126, 129,
    132; cruiser construction limits
    circumvented 134; experiments with
    aviation in reconnaissance roles 133;
    gunfire accuracy 117; Maritime
    Tactical School 148
Rumsfeld, Donald 101, 152–3
Rundstedt, Gen. Gerd von 46

Russell, Maj. Bill 55
Russell, Maj. (later Gen.) John 52, 54, 56, 57, 58, 59, 60, 61, 62, 63, 64, 65, 73, 75, 80, 112, 113–14
Russia 47

Saddam Hussein 194
SALT I agreement (1972) 168–9
Sampson, Adml. William T. 52
San Diego 149, 150
Sands, Jeffrey 156, 160
Santiago, battle of (1898) 51, 117
Santo Domingo 54
Saratoga (USS) 122, 125, 126, 127, 129
SAW (School of Advanced Warfighting) 99, 101
Scharnhorst reforms 92
Schlesinger, James 160
Schmitt, Capt. John 90, 91, 99, 111, 113
Schneider, Greg 182
Schneiderman, Bernie 146–7, 150
schools: Army 62, 63, 64–5, 96; flying 40; Marine Corps 57, 61, 62, 63, 65, 92, 96, 100, 112, 193; primary officer development 94; tank 40; *see also* Amphibious Warfare School; SAW; TBS; US Naval War College
Schreiber, Lt.-Col. K. D. 90, 110, 113
Schultz, George 160
Schultz, Rear-Adml. Paul 182–3
scientific invention/scientists 9, 10
Scott, Adml. Percy 117, 118
screening force 127
Sea-Control Ships 158, 159
SEAL (Sea, Air, Land) teams 105, 106
sealift: heavy reliance on 157; prepositioned 109; strategic 108, 110
seaplanes 128, 133
Sea Strike (war plan) 146
Sea Wolf program 28
secrecy 57
security 4, 6, 106; changes in 5, 52, 53, 123; international 132, 135; national 172; naval stations 54; random 158; shifts in 87; strategic 59, 121, 165; structural changes in 9; threats to 5, 7
security guards 51
Seeckt, Gen. Hans von 32, 36, 38, 39–41, 42, 43, 46, 47–8, 49, 86, 156
Selection Bill 64
self-discipline 86
Sempill, Sir 134
senior military leaders 4, 6, 11, 20, 21, 23, 27, 28, 34, 42, 79; adoption of

helicopter assault 72; approval of wide array of experiments 37; architectural knowledge requires explicit management by 30; biased toward stagnation 2–3; directly attacked for not transitioning to a new way of war 37–8; generally view all innovations as sustaining 31; identification of disruptive innovation 35; innovative approaches to warfare 5; key to peacetime innovation 8; 'mainstream' 5, 9; managment of innovation 8, 13, 15; most influential 12; opposed to use of tanks 49; pitted against each other 8–9; prefer sustaining innovations 30; preparation for war in interwar period 26; top-down influence of 11; trajectories of performance improvement 70; understanding of new technology 37
Seventh Amphibious Force 67
Shanghai 66
Sheehan, Kevin 24
shells 13
Shenandoah (airship) 128
Shepherd, Maj.-Gen. Lemuel 72, 73, 75, 77, 81, 82, 112, 114
Sherman, Lt.-Cmdr. (later Adml.) Forrest 127
Shinsu-maru 66–7
shipbuilding 12, 64, 66, 142, 167, 170
ship-to-shore movement 55, 56, 65, 66, 70, 112; contested 68; future 74; greater degree of surprise and speed 72; new tentative doctrine for 73
SIAP (Single Integrated Air Picture) 178
signal troops 46
Sims, Lt. (later Adml.) William 23, 63, 117–18, 119, 122–3, 125, 198
simulations 30, 124, 125, 137; strategic 123; tactical 123
SLCMs (submarine-launched cruise missiles) 169, 170
SLOCs (sea lines of communication) 109
small innovation groups 30–1, 64, 72, 74, 81, 82, 89, 90–1, 105, 109, 152–3, 156, 159, 165, 187–8, 191, 196; carrier warfare 124; Colonel Gary Anderson 113; drawn from US Navy and Marine Corps staffs 172–3; importance of 112; propositions 34–5, 39–40, 125–7, 180–1; SOC 107; vertical envelopment 73
small wars 51, 53, 54, 56

Smith, Capt. Edward 165, 173
Smith, Maj.-Gen. Holland ('Howlin'
    Mad') 55–6, 58, 60, 67–8
Smith, Lt.-Col. Ray 89, 100
Smith, Vice-Adml. Leighton 'Snuffy' 172
social change 99; mild 13
software 12, 179; sensor-netting 177
Solipsys Corp. 179, 180, 181, 183
Somalia 106
SOPs (standard operating procedures) 3,
    23, 24; helicopters operating within
    MAGTF 82–3
SOSUS (underwater cables on the ocean
    floor) 147–8
Southern Operations (Japan) 136
Soviet Union 24, 28, 49; anti-ship missile
    159; confronting the threat 194; fall of
    165; follow-on strikes 170; invasion of
    Afghanistan 108; military technology
    developments 9; naval aviation
    breakthroughs 166; naval threat 150,
    152, 155, 157, 158; nuclear missile
    arsenal 165; risk from missile attack
    167; SALT I agreement with US (1972)
    168–9; secret arrangements with 40;
    targeting problem for planners 171;
    western, proposed attack on 146
Spanish-American War (1898) 52, 53,
    117, 121
specialized task force operational
    concept 153
special operations forces *see* MEU/SOC
speed 36, 41, 50, 72, 73, 74; helicopter
    lift and 70; infantry dependence upon
    48; technological improvements in 44
Spinny, Franklin Charles 98
squadrons 108; fighter 75, 129;
    helicopter 73, 74; multipurpose
    vessels 104
SSNs (attack submarine nuclear) 148,
    159, 195; missiles launched from 170
stability 3; organizations strive for 24;
    scarce 23; strategic 170
Stackpole, Maj.-Gen. H. C. 172
stagnation 2–3; explained 6; root cause of
    4; tactical Tomahawk 171
stalemate 49
state behavior 4–5
static positions 49
steam engine invention 10
strategic deterrent 157
strategic locations 104
strategy 47, 75, 93, 158, 194; air in its
    relation to Fleet 124; airmobile 76;

civilian, for promoting disruptive
    innovation 201; decisive encirclement
    43; 'delay and defer' 184; direct or
    nondisguised engagement 142;
    disguising 61–4, 75–6; innovation 51;
    intellectual 136–9; managing
    uncertainty 9, 10; maneuver warfare
    90; means for accomplishing 159; naval
    52; naval aviation 128; offensive 137;
    political 139–40; wait-and-react 136
strike force 13, 24, 80, 127; capable of
    rapidly exploiting breakthrough 44;
    landing 51; primary 129
strike warfare 126, 147, 152, 172
Studt, Col. John 88
Stulpnagel, Gen. Otto von 44
Styx cruise missile 168, 195
submarines 8, 16, 72, 73, 156; attack 28,
    153, 155, 171; ballistic missile weapon
    systems 6; cruise missile attacks 173;
    diesel 94; direct-support 147; hunting
    149; missiles launched from 168;
    oversized 28; search for and
    prosecution of 147; Soviet 147, 155,
    158; strategic ACMs on 169; tactical
    fighter and attack aircraft to find 148
Suez Canal Crisis (1956) 106
Sumatra 106
Sun Tzu 97
supply lines: long 47; potential threat to
    53; relatively short 47
support forces 68
surface land attack warfare 164–75
surface ship escorts 127
Suribachi, Mount 52
surprise 41, 72; combined-arms
    penetration relying on 48; tactical 68;
    tank attack 32
surveillance 104
sustaining innovation 1, 2, 13, 14, 25,
    28–9, 37, 96, 104–20, 135, 171, 190–1;
    ability to foster a culture of 19;
    acquisition system supporting 184–5;
    advanced base force 52–8; amphibious
    warfare 41–3, 52; armored warfare 41;
    carrier battle group concept 152–4;
    CEC 177–8; champions of 46; civilian
    intervention causing 49, 107, 110;
    disruptive innovation disguised as
    31–2, 64; external causes 195–6; good
    management of 4; helicopter warfare
    52–8; institutionalizing 46; internal
    causes 198; managing 1, 26; maneuver
    warfare 85; maverick officers can cause

28; potential disruptive effects of 165;
promoting 45; senior military leaders
prefer 30
Swartz, Peter 153
system architecture 16, 21, 22; unique
15, 17

tactical collaborative network 176–91
Tactical Training Group Pacific 149
tactics 39, 47, 83, 86, 89, 91, 93, 166; air
123, 124; amphibious 67, 72, 108;
armored 34, 41; battle group 148, 149;
CWC 150; decentralized 41; defensive
56, 137; fundamental changes in 116;
gunnery 118; helicopter employment
73; infantry 95; infiltration 48, 50;
maneuver warfare 89, 90; naval
aviation 128; Naval War College
128; tank 44, 50; vertical 74; *see
also* TTP
Taft, Robert Jr 87, 91, 111
tanks 34, 39; attack and pursuit 41;
banned in German army 36; conscious
adaptation of 41; coupling with cavalry
43; dummy 40, 42, 44, 45; employment
as a reconnaissance vehicle 44;
experiments with 33, 41, 50;
incorporating with mobile warfare
doctrine 43; independent corps 50;
infantry support 42, 43, 44, 49;
invention/introduction of 24, 27, 28, 32,
36; key part of mobile warfare 49; large
45; loss to friendly artillery fire 36;
maneuverability of 44; maneuvers 35;
might extend infantry's capacity to
exploit tactical situations 46; need to
continue to develop expertise 36;
possibilities of use in mobile warfare
37; superior in protection and
armament 37
Tarawa 66
target ships 71
Task Force Sea Angel 106
TBS (The Basic School) 90, 95, 96, 100
TCN (Tactical Component Network)
178–86
technological change 13; modest 35
technological innovation 9–14;
characterized 21; classic case of 27;
disruptive 184; how senior military
leaders manage the impact of 8; impact
on the organization 15; important 12;
landing craft development 65–7;
sustaining 26

telescopic sights 16, 117
Tentative Landing Operations Manual 61,
62, 65, 80, 86
Terrible (HMS) 118
terrorism 164; catastrophic, defense against
186; counteraction program 107
theater of operations 54
think-tanks 193
Thomas, Lt.-Gen. Gerald C. 71
threats 4, 5; external security 7;
significant, countering 106
throttles of change 53–6, 65–6, 72–5,
80–3, 88–91, 105–7, 108–10, 117–18,
122–31, 146–50, 152–3, 156–61,
166–73, 180–2, 183, 196–8
Tomahawk cruise missile 146, 153, 165,
168, 197, 198; first time used against an
enemy 173; land attack 164, 170;
strategic 169, 170, 171, 172; tactical
169, 170, 171, 172
Tonkin, Gulf of 157
torpedo warfare 133, 138
Towers, Capt. Jack 127
Towers, Lt.-Cmdr John 12
Trainer, Mark 184
Trainor, Lt.-Gen. Bernard 86, 88, 92, 94,
96, 97, 101
treaty enforcement 106
trench warfare 27, 38
Trident II 6
Trimble, William 124, 126
Trincomalee 132
troop transports 158
Truk 143
Trüppenfuhrung, Die (manual) 41
Tsingtao 133
TTP (Tactics, Techniques, and
Procedures) 95
TU-22M 'Backfire' bomber 166
Turner, Adml. Richmond K. 67–8
Turner, Rear-Adml. Stansfield 156, 157,
161, 201

uncertainty 9, 10
underground forces 68
undersea warfare 160
United Kingdom *see* Britain
United States *see following headings
prefixed* 'US'
US Air Force 9, 79, 83, 114; Global
Positioning System 171; guided
missiles 10; insistence on control of all
aviation assets 81; interservice rivalry
between Navy and 165

US Army 9, 55, 56, 71, 90, 92, 93;
  aviation 128; belief that Marine Corps'
  primary mission was fighting alongside
  113; Infantry School 96; maneuver
  warfare adopted 94; manuals, doctrine,
  and training techniques 61; shift to
  maneuver warfare 88; Tables of
  Organization 58; urged to adopt
  maneuver warfare 87
US Army Air Force 71
US Army War College 57, 62
US Bureau of Aeronautics 123; bill
  creating 124
US Bureau of Aviation 124, 128, 129
US Congress 52, 64, 80, 88, 168, 169,
  183, 185; courted for appropriations
  129; creation of new unified command
  105; decision for special operations in
  low-intensity conflict 106; FMF seen as
  provocative and interventionist 113;
  House Naval Affairs Committee 53,
  124, 127; influence on military matters
  111; lobbying 160; Meyer's influence
  in 167; Reform Caucus 88, 111;
  Research Service 177; Rickover's
  influence in 167, 169, 171; SOC
  legislation 107
US Department of Defense 92, 93, 98,
  171; initiative to link all air radar
  sensors 178; Project Hindsight 9
US Department of the Navy 53, 62, 108,
  123, 165
US Marine Corps 5, 8, 22, 51–115, 152;
  small innovation group with Navy
  172–3; *see also* FMF
US National Security Council 160
US Naval War College 13, 30, 55–6, 59,
  93, 122; carrier simulations 125;
  cooperation between Bureau of
  Aeronautics and 123; games 124, 125;
  Strategic Studies Group 150; Tactics
  Department 124, 128, 150
US Navy 5, 67, 72; aviation strike force
  13, 24, 80; Bureau of Construction and
  Repair 66; Bureau of Navigation 117,
  118; Bureau of Ordnance 116–17, 118,
  166; carrier operations 121–31, 137;
  continuous aim gunfire 16, 116–20;
  coordination between Marine Corps and
  68, 109; cruiser construction limits
  circumvented 134; disruptive
  innovation 121–31, 145–51, 155–91;
  experiments with aviation in
  reconnaissance roles 133; General
  Board 12, 53, 54, 58, 126, 129; guided
  missiles 10; identified by Japanese
  Navy as its sole enemy 137; innovation
  advocates 11; Lind's genuine hostility
  to 94; nuclear doctrine 6; obsession of
  building larger battleships with bigger
  guns 28; Office of Aeronautics 12;
  Operation Crossroads (1946) 71;
  overwhelmingly superior strength 136;
  political techniques to successfully
  introduce aviation into 12; SEALs 105,
  106; Sea Wolf submarine program 28;
  submarines 6, 13; sustaining innovation
  116–20, 152–4; three distinct branches 8
US Senate 53, 91, 111; naval affairs
  committee 124

Van Riper Lt.-Gen. Paul 91, 95, 98, 99
Vandegrift, Gen. Alexander 72, 73, 112
Versailles, Treaty of (1919) 36, 40, 42, 44
vertical alliances 11
vertical envelopment 71–8, 112
victory: new theory of 9, 11, 136, 171;
  stunning 36; tactical 32
Vietnam 74, 82, 83, 86, 87, 88, 90, 155,
  156, 157, 158, 161; counterinsurgency
  doctrine 24; focus on strike warfare
  145; sea control 159
Virginia *see* Fort Pickett; Quantico

Wake 52, 58, 59
Wallace, Ensign John 183
Wallace, Gen. William Scott 101
Walt, Gen. Lweis W. 96
Warfighting (manual) 111, 113; *see also*
  Naval Warfighting
Warfighting Skills Program (maneuver
  warfare correspondence course)
  91, 113
war games 30, 36, 37, 121, 123, 124, 125,
  135, 137
warheads 169, 170
War Office (UK) 33, 34
War on Terrorism 102
War Plan Orange 55, 57, 59
Warsaw Pact 87
Washington Conference (1921–22) 59
Washington Post 53, 182
Washington Treaty for the Limitation of
  Naval Armaments (1922) 121–2,
  126, 134
Watanabe, Cmdr. Yasuji 138
Watkins, Adml. James D. 171
Watt, James 10